MINISTRY OF EDUCATION SCIENCE MUSEUM

SCIENCE SINCE 1500

A SHORT HISTORY O
MATHEMATICS, PHYSICS, CH
BIOLOGY

By H. T. PLEDGE, B.A.

Reprinted with minor corrections - - - 1940
Reprinted - - - - - - - 1946
Reprinted - - - - - - - 1947

PREFACE

In view of the increasing number of manuals on the history of special scientific subjects published or projected by the Science Museum, it seemed desirable that a background study should be made. The present book, however, tries to fill a niche not only in the Museum series but in the rather small group of studies of the history of modern science as a whole which have been published anywhere, or at any time. The presence of this niche is soon explained. There now exists a considerable production of virtually professional work in the subject, in the midst of which Sarton's vast "Introduction" towers like a monument. But Sarton is still at work on the 14th century. For the modern period his present scale of treatment is clearly impossible. What is happening, instead, is that workers are clearing up obscure individual points; and by doing so are gradually, but in the end considerably, altering our view even of the well-known landmarks. There is thus a need, every decade or so, of short co-ordinating surveys. The present one has been carried down in elementary fashion to the last year or so to avoid an arbitrary end-point; but the writer is keenly aware that his account of recent years can be little but a reflection of his personal interests.

Such surveys must of necessity grow more technical as time goes on. Thus, an attempt has been made here—for the first time, it is believed, in a general book—to discuss 19th-century mathematics, so obviously essential in understanding 20th-century physics. The result is a book aimed primarily at the "specialist in other subjects," the scholarship candidate, the university student, the research worker.

Certain classes of fact suffer especially under abbreviated treatment. It is possible to give a brief exposition of principles: it is not possible to give any proportionate impression of the vast masses of facts, indi-dividually unimportant, which make up expert knowledge in any region. It is only possible to indicate from time to time where the reader should be especially on his guard.

Brevity has been secured by suppressing biography and the human sciences, sociology, anthropology, psychology, economics. To balance this, the story itself has been thrown against a background of the human and economic factors in successive periods rather than against the more intellectual and philosophical backgrounds which have mainly interested writers of previous general works, such as Sir W. Dampier. At the same time, the reader must not expect—what he will find in the works of Prof. L. Hogben—a treatment in which social needs take primary place. Primary place is given to the technical tradition by which alone those needs have been supplied.

The plates have been chosen to emphasise not so much major figures and aspects, as neglected ones. No originality is claimed for their choice. The charts and maps illustrate some geographical, personal, and instrumental factors. Many dates of births and deaths have been relegated to the index.

ACKNOWLEDGEMENTS

Apart from special acknowledgements made at various points in the book, thanks are gratefully offered for the criticism and other help given by the following, who, however, are not to be held responsible for errors and ineptitudes which remain: Mr. C. H. KEBBY, of the Imperial College of Science and Technology, and Dr. A. PRAG, as to mathematics; Prof. W. H. McCREA, of Belfast University, as to mathematical physics; Dr. H. J. EMELEUS, of the Imperial College, as to chemical apparatus; Dr. S. E. JANSON, of the Science Library, as to recent chemistry; Miss F. L. STEPHENS, Mr. G. TANDY, Mr. A. C. TOWNSEND, of the Natural History Museum, as to points in biology; Mr. W. ADAMS, of the London School of Economics, as to background sections; Dr. F. A. B. WARD and Mr. F. G. SKINNER, of the Science Museum, for Graphs I and II, and to M. H. MINEUR for Graph III; and to friends and relatives for help with the proofs. Miss D. BUTCHER assisted in the compilation of the index. Finally, Dr. J. NEEDHAM, of Cambridge, read through the whole work in proof and made many suggestions.

ANALYTICAL TABLE OF CONTENTS

Although it is hoped that the artificiality of subject divisions is made apparent in this book, the reader looking up any main subject in the Index will find a run of chapters indicated in which he can follow that subject with a minimum of interruption.

An asterisk in this Table of Contents indicates a short separate background study inserted between chapters.

LIST OF ILLUSTRATIONS

SCHEME OF GRAPHS

The accuracy of instruments at any period is at once one of the greatest limiting factors in scientific discovery and an expression of the advances already achieved.

SCHEME OF CHARTS

THE CONNECTION OF MASTER AND PUPIL

EXPLANATION

The charts are designed to illustrate how large a number of noted scientists fall into a continuous network of master and pupil; though it is not asserted that all the links indicate profound influence. The linkage is especially close in the experimental and biological sciences: mathematicians seem to learn more from books, less from persons.

In the third chart, drastic simplification has been necessary to avoid overwhelming complexity. Indeed one great teacher appears as if he had only one pupil. It is hoped to publish a fuller chart elsewhere.

To complete the connections, a few scientists not mentioned in the text have been introduced; but none who are not of real note. Dropping this latter condition we might, with extensive research, have succeeded in connecting the three charts into one and also in bringing an even larger proportion of the chief men into the one network. So far as we have pursued the subject here, men without noted masters or pupils are still numerous and eminent; several great teachers have no noted masters. A disproportionate number of the greatest teachers have been of German origin, even when (Albinus, Guinter, F. Sylvius) they taught outside Germany.

SCHEME OF MAPS

EXPLANATION

Maps II–VI show birth-places which cluster remarkably in particular periods and regions. It is this clustering which makes the maps of birth-places worth while: the dominance of certain research-centres is better known. The maps, however, are not intended as proofs of the existence of clustering: this must be taken on trust, since to save space the relatively blank regions bordering the clusters have only been shown in some of the cases.

No full analysis of the reasons for the particular clusters is possible in this book. The writer hopes to attempt it elsewhere, together with a more elaborate ranking of names. But it will be seen that none of the clusters is wholly national. Neither wealth nor population account for all of them. For instance, 19th-century Ireland was neither wealthy nor populous. Present (early 1938) frontiers are shown, as contemporary ones often varied even within the period of one map.

Only a small proportion (about 20–25 per cent.) of the scientists in this book occur in the clusters. Whole countries, and particularly France, receive much less than justice. A few names not in the text appear in the maps.

CHAPTER I

SCIENCE AND PRE-SCIENCE

THE traditions of civilisation, which had grown in Egypt and Mesopotamia, converged on the Mediterranean Sea, reaching a climax in the Near East as regards industrial arts, in Greece as regards cultural quality, in the Roman Empire as regards extent. After Rome there was a long ebb before the revival which we call the Middle Ages. The Greeks had an instinct for pure mathematics but a distaste for its applications; their curiosity ran to the observational sides of biology but it had no decisive effect in the experimental science of physics.

After the decline of Alexandria, Greek science lingered in Southern Italy and in Byzantium, and revived and spread, with the fiery religion of Islam, east to Bagdad and west to Spain. Many words with the tang of the alchemist's or astrologer's den recall the four centuries when Arabic was the language of scholarship: alembic, elixir, zenith, zodiac, algebra, almanac. When, in the 9th and 10th centuries, Arabic science and culture were at their height on the south of the Mediterranean, Europe, on the north, was still in relative darkness. Mohammedanism and Judaism, like Christianity later, selected from Aristotle those features which best suited them, and produced a scholastic theology: al-Ghazzali (1058–1111) for the Muslims, Maimonides (1135–1204) for the Jews. At about the same time the contrary tendency, sceptical philosophy, reached its height in Averroes (b. 1126). Then, in the 12th century, there came a Mohammedan decline and a corresponding rise in Christian Europe. Our "First Renaissance of Learning" culminated in the 13th century in the scholasticism of St. Thomas Aquinas (1225?–74). Science was not yet a primary activity, but still, there was a revival. It came from several sources: from Italy through the medical school of Salerno and the law school of Bologna, from Spain through the translators from Arabic, who were often Jews.

But it was not only in learning that Europe revived by the 12th century. Trade grew, and towns, and with them far more technology and hygiene than is commonly realised. Water power spread. The improved harnessing of horses must (exaggerated claims apart) have released human labour and opened up new possibilities of bulk transport of goods. But perhaps the most fundamental technical developments or redevelopments were the magnetic compass and the clock. These devices, which now exercise such complete sway over our lives, enabled the men of cloudy northern regions to find their way in space and, by legitimate extension of meaning, in time. Lewis Mumford and others have linked them with the change of the eternity of religious man into the time of secular man, of space as order into space as opportunity for expansion and independence. For the 14th and 15th centuries saw the increasing break-up of the theoretical religious unity of the Middle Ages, while new continents began their challenge to navigating man. With this new opportunity, the original scientific tradition from Islam, now largely in Jewish hands, started into new life in Portugal in the 15th century.

During the same century another fundamental technical advance was achieved. Printing multiplied the means both of propaganda and of enlightenment. A second renaissance (often called the Renaissance simply) was meanwhile coming into prominence in Italy, with stimulus from the third relict of Greece, Byzantium. The 12th-century move-

MAP I

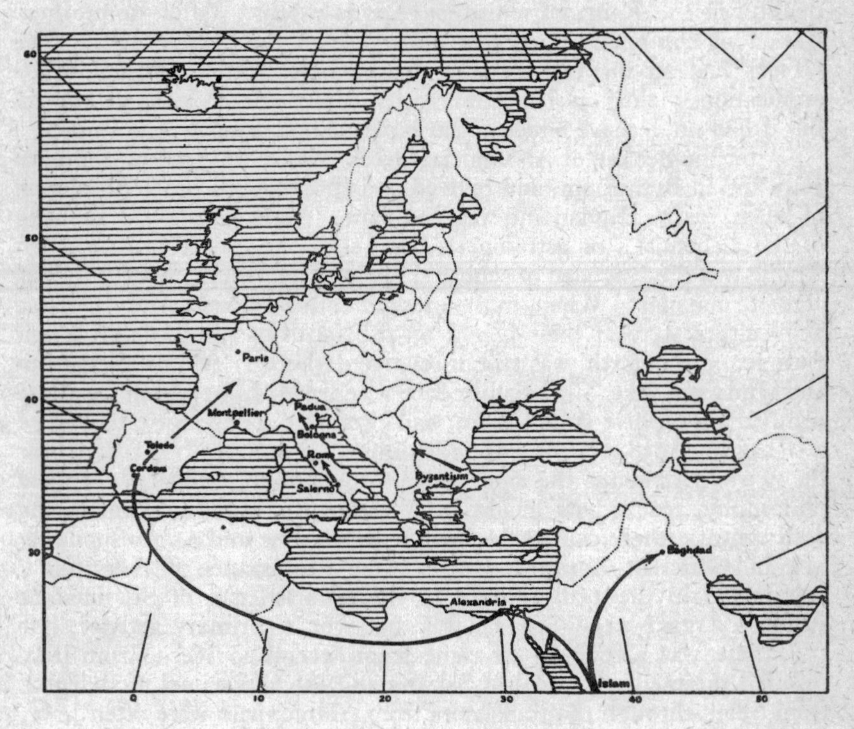

TRACKS OF SCIENCE FROM THE END OF CLASSICAL TIMES TO 1500.

The Explanation on p. 10 should be consulted.

ment had been religious ; the newer one, preferring Plato to Aristotle and Averroes alike, scorned the old pedantries and advertised its own recovery of the elegance of classical times. Because science, no less than law or theology, must risk pedantry if accuracy is in question, the implied casting off of authority was for it by no means pure gain. The second Renaissance was not itself a scientific (rather an artistic) age : if its thrusting spirit sped science over early obstacles, it did so in a manner which left many problems in mathematics and physics in a state demanding profound reconsideration four centuries later.

We see that twice within European history, before the Mediterranean climax and before our present one, we can trace man's emergence from earlier stages of civilisation. The second case was modified by remnants of the first, but the analogy between the two remains. Thus, at a decisive stage of each came maritime exploration, the ancients, like the moderns, feeling down the African coast; and, with this, ideas of the sphericity of the earth and of the sun as the centre of revolution. Beginning our story in the middle has this cautionary advantage: it reminds us that, however far back we were to push our beginning, we should be starting not from a blank but from an existing corpus of achievement, a separation of verified knowledge from fancy and magic. In Greece we should find the organised medicine of Hippocrates in conscious opposition to the quackery of the Aesculapians. In the Roman Empire we should find not only alchemy but the burning of alchemical books.

Starting in the middle reminds us, too, of the continuity of history. Some would deny the second Renaissance all validity as a division. Such judgements seem to depend on what feature interests us: it is easy to detect continuity in the growth of knowledge *after* an advance has been made. The brevity of the present book will rarely leave room for discussion of precursors discovered after any main enlightenment of the scientific world. It will force us to erect divisions—to see something characteristic of a new age in Renaissance adventurers like Leonardo da Vinci, ready and able to turn their hands to anything, from painting to military engineering, in request at the brilliant, murderous courts of the Italian merchant princes. We suspect a new individualism, in contrast to the old corporate life of manor, guild, or monastery.

There is much in common between the mediaeval monkish scholar and the present-day research worker: books and manual work, quiet concentration on a small field, corporate action. Scholarship has grown right down from the Dark Ages whenever war has allowed. The Hundred Years War interfered with it before 1500, the Thirty Years War after that date. And yet there is such a thing as the pre-scientific spirit, and, though this was by no means universal in 1300 nor absent in 1900, we must shortly sketch it before we end this panoramic introduction by a brief note on individual sciences about 1500.

Pre-science

We say that mediaevals copied books rather than nature, a process perhaps natural when printing had not replaced the copying of manuscripts. But even to-day we consult tables to avoid redetermining a melting-point. It is, in part, a question of available energy and stimulus: if there are resources and motives to go fully into a question and to find out more facts about it, this step will eventually be taken. But it is not irrational to look first at work already done, and to rearrange one's own thoughts rather than, experimentally, to rearrange nature. Steps like separating yellowness from all yellow objects are of this kind: they only appear foolish when they have been completely accomplished.* We shall see at the end of this book that some of the proudest feats of modern science would be open to suspicion if suspicion this were.

Finding out more facts may easily bear the paradoxical appearance of trying to simplify a problem by first making it more complicated; and until this has once or twice been done with striking success, it is not an obviously sensible process. It is, in fact, arguable that the greatest single fact about the universe elicited by science is the frequent success of this proceeding, from mathematics with its imaginaries all down the scale.

Science crystallised from several different sources and in many different places. The crystals had both to separate themselves from the matrix and to grow together into a compact mass. Science receives many more influences from the general mental atmosphere of its time than scientists always realise; but the *confusion* of the two is disastrous to it. So long as science cannot give autonomy to its criteria not only of what is true but of what is interesting, it is fatally hampered. Anatomy was weak so long as it was part of the theology of macrocosm and microcosm. Mechanics and, in a different way, embryology were hampered so long as they were mixed with the philosophy named after Aristotle. On the other hand, there were debts; this philosophy, in fact, was one mother-liquor of science. Another was Renaissance art itself. Artists like Leonardo da Vinci (1452–1519) were, as we shall see, among the most enterprising workers in anatomy, optics, and mechanics.

This lack of specialism went with the absence of a compact self-conscious scientific group to take up and confirm the discoveries of genius. No one did for Leonardo what Halley did for Newton—made him write up his results in mechanics and saw them through the press. Leonardo's notebooks remained unpublished. In the interval between the two men that nervous integration of the scientific world which is one of the main themes of this book had taken a long stride forward.

From Roger Bacon (1214–94) until well into our period, discoveries were made and lost and made again. Leonardo's own break with Aristotle in mechanics had been anticipated by Buridan (*c.* 1297–*c.* 1358), and had to be re-made, this time finally, by Galileo. Like Roger Bacon, Leonardo urged the primacy of mathematics; but, also like Bacon, he did not in his own person end the isolation of that greatest legacy of Greece from the vast fields of application which awaited it.

Again, in spite of the mariner's compass, the magnetic knowledge of Peter Peregrinus in the 13th century was only recalled by Gilbert in the 16th. The isochronous property of the pendulum, said to have been known to certain of the Arabs, was lost and found out again later; so that although it was used for astronomical time measure in the 15th century,† it had to be rediscovered by Galileo at the end of the 16th, while clocks did not gain from it for another half-century still.

* From this point of view, however, mediaeval "realism" had gone too far, making yellowness not merely a convenience of thought but something more real than individual cases of it.

† Not, however, in clocks.

In fact, a most important case of this isolation of things capable of fruitful union was that of craft knowledge and theorists' knowledge. The Greeks have always been cited as cases of this. Recent workers in the sociology of science have stressed that experimental science arose from the theorists' taking account of the crafts. On the other hand, the crafts have often failed to learn from the theorists almost down to our own day. The Renaissance got rid of an extreme instance of the former sort of isolation, the maps of the world which the monks used to make with the aid of travellers' tales. These bore no relation to the very practical Mediterranean pilots' charts of which, by the 14th century, many makers existed. Paracelsus, also, strove to draw on the science implied in the mining in the Tyrol, as Agricola did in that at Joachimstal.

Two other products of isolation and of other early difficulties may be noted, mysticism in inappropriate places and a cult of secrecy. Mediaeval men craved order, in science as well as in life ; and when they were halted in finding true laws, they had recourse to symbolism. Things with the same number, for instance, were grouped together, after the fashion of the number-mysticism of the Greeks (see Chap. IV).

In Greek chemistry there had been four "elements," fire, air, water, earth, themselves compounded in pairs of hot, dry, cold, and wet. To them, according to Empedocles, corresponded the four elements responsible, by their balance or otherwise, for health and for the variety of human constitutions. These were the blood (supposed to originate in the liver) and a sanguine temper, black bile (from the spleen) and a melancholic temper, phlegm (lungs) and a phlegmatic temper, and yellow bile (gall bladder) and a choleric temper. Before the dodecahedron was discovered there were also four regular polyhedra : the tetrahedron, octahedron, icosahedron, and cube.* These, then, must correspond to the four chemical elements. But then the fifth (and last) regular polyhedron was discovered. It was a genuine achievement, but what of the scheme ? An ingenious idea was needed : the fifth figure represented the entire universe ! Undoubtedly, if we follow Eddington, integer properties have a deep significance ; but this significance is perhaps rather the last, than the first, thing we find out.

Such ideas may seem to be mere muddle ; and yet abstract clarity, the logic of the *enfant terrible*, is one of the virtues of the pre-scientific stage. In scholasticism there was often a strong clarity ; and Leonardo da Vinci leapt at once beyond perpetual motion and the like to Newton's first and third laws and to Archimedes' clear conception of fluid pressure. Only, without mathematics, he failed to develop them.

This pre-scientific symbolism held in solution some features of symbolism in the modern sense. To this we come in Chap. IV, where

* In the Middle Ages other fours were added, the evangelists and the natural virtues. Seven—four plus the number of the Trinity—is the richest : the musical notes, the planets of the astrologers, the alchemists' later elements, the days of creation, the deadly sins, the gifts of the Holy Ghost, and the supernatural virtues. As regards the planets and the elements, the sun stood for gold, the moon for silver, Jupiter often for tin, Saturn for lead, Mars for iron, Venus for copper or tin, and Mercury for its name-metal or for others. Holmyard notes (*Nature*, 13 March, 1927) that the alchemists relied very largely on pictorial representation.

it appears as one main element in scientific method. It is plainly related to the realism which we have mentioned. The sheer insolubility of certain problems endowed them with a kind of mystical, hypnotic, fascination (the *mysterium fascinans* of mystical theology), and this sometimes grew into the unhealthy secrecy of the alchemist's or the witch's den, a gloating avarice of knowledge, to be distinguished from the secrecy of fear, of gestation, or of commercial gain.

Some Individual Sciences

With these general facts in mind, let us briefly consider the state of the individual sciences about 1450–1500. Our notes will be in inverse proportion to the extent of their subjects, for the great explorers of the 15th century, the astronomers, mathematicians, and biologists of the 16th, have the succeeding chapters to themselves.

Physics, as we shall see later, had not yet defined its fundamental units of thought. Its individual branches formed, therefore, a scattered collection. Sound was much occupied with a semi-mystical arithmetic of music. Magnetism, in spite of the stimulus of navigation, was in abeyance (see Chap. III). Electricity was confined to a few mysterious observations on amber and the like. Heat was, in the main, part and parcel of alchemy. We have noted that, like biology, optics and mechanics interested the great artists. To us, their optics is little more than geometry—than perspective. But then both perspective and anatomical accuracy were part of the new phase of artistic naturalism which was setting in, with men like Botticelli (1444–1510), Dürer (1471–1528), and Leonardo da Vinci. Such phases are recurrent in the history of art and are commonly associated in discussions of the subject with that very consultation of nature which is one side of the scientific spirit.* The geometrical side of perspective was studied by a series of artists, Brunelleschi (1379–1446), Leonardo, Alberti, and Dürer; and this study, as we shall see in later chapters, had important effects in mathematics and physics.

The geometry of refraction already had a long, if intermittent, history: from Ptolemy's experiments, through Ibn-al-Haitham (965–1020) on the refraction of lenses and of the atmosphere, to Roger Bacon's project of a telescope (*c.* 1250). By 1300 convex spectacles were spreading, especially from the Venetian glass-works, but concave ones were not common until the middle of the 16th century, and were even then highly valued objects.

There was, however, some physical, and physiological, optics as well. Leonardo studied the eye, though he thought the liquid centre portion the main image former. But the old view that vision is a matter of tentacles coming out of, as well as of light entering, the eye had still not lost its hold.

For their movement-studies these artists called in the science of mechanics. But they did little to advance it. Leonardo's work, being unpublished, produced little effect. In spite of the classical texts of

* We cannot discuss the point here, but it is not quite a simple one. A perspective drawing gives at first sight nature as she seems, not as she is—like Ptolemy, not Copernicus (see Chap. III).

Vitruvius, Hero, and Ctesibus, established theory ended with statics. In dynamics perpetual motion was still sought, " gravity " and " levity " still made things seek their " natural " places, and the naïve point of view held sway that force is needed to keep up a uniform motion. It was otherwise with practice as opposed to, and largely isolated from, theory. We have mentioned the clock ; many early clock mechanisms are remarkably well-balanced in their action, though without the pendulum their accuracy could only be low. With the growth of trade, however, punctuality began to be demanded in other matters than religion. The Renaissance metal workers achieved a virtuoso's skill with mechanical toys ; and when jointed-metal work was going out for armour, the great surgeon Ambroise Paré found it (*c.* 1560) a new niche in artificial limbs. Thus craftsmanship was not unready for the demands of mechanical theory in the 17th century.

In passing to chemistry we find, too, not a body of exact ideas, but an alchemical literature and an oral tradition of craft knowledge. There were balances and weights of considerable accuracy.* The chemical crafts already made and used in some quantity the fundamental reagents, such as sulphuric and nitric acids. Distillation had been practised from the 11th or 12th century. The distillation of spirits became a considerable trade in the 16th century. The German mining and metallurgy of the 16th century were already old. Italian glass dated from at least the 12th century, and by the end of the 16th had reached an artistic climax. Gunpowder was already an industry, though gun barrels were still very crude. Drugs † were a chief article of long-distance trade, and were responsible for a great diffusion of chemical technique. Toxicology, though little understood, was much practised. Like the chemical weapons of to-day, it contributed its quota to the tradition of chemical skill. In all this much future chemical knowledge was implicit, but we shall not attempt any catalogue of the substances and reactions which were already familiar ; for the word " familiar " could not have its present sense. It is because we describe them in terms of certain fundamental notions such as compound, mixture, atom, that we are " familiar " with substances at the present day. These notions were not formulated until long after our period opens.

The proximate reason for this seems to have been a perfectly specific one, the lack of clear conceptions of the nature of gases and of experimental technique for handling them. But this was itself only an aspect of a deeper deficiency of ideas : the concept of the *physical*, of bodily materiality as we understand it, was lacking. Ideas were still, for the speculative mind, more real than material things.

We sometimes speak as if this was the " lowest," and perhaps by implication the most primitive, of our ideas—this of a material something existing independently of the observer and underlying, essentially unchanged, the changes and varieties of matter, such as fusion, solution, chemical reaction. But in fact this idea, so far as science goes, has been only slowly and recently formed, and the tracing of its growth will be

* See Graph I.
† Not synthetic drugs.

one of our main preoccupations. It only began to grow definite in the 17th century; it did not mature until the early 19th; and already, in quantum and relativity theory, it sometimes seems outmoded.

Without it, in those early days, men were helpless: helpless in physiology, as we shall see in Chap. II; helpless in physics proper. Without it, optics and astronomy all through the 16th century were largely exercises in geometry. Heat was not a branch of physics but a chemical element; while the alchemists made what seem to us gross confusions of material substances and abstractions or qualities such as colour. Thus, the ability to make a substance *look* like gold seemed to them at least an approach to actual transmutation. At the opposite extreme, they held that, whatever the appearances, each firing or distillation refines away more of a thing's external "dross" or "accidents" and brings us nearer its quintessence or "soul," which is the elixir of life or the philosopher's stone.

The Greeks, having analysed the world into the fiery, the liquid, the earthy, and so on, did not use these as mere convenient abstractions,* but rather treated individual substances as particular cases of these underlying realities. The alchemists identified the realities with particular material substances: sulphur, mercury, and salt respectively. Nor was this mere verbiage. They gave such symbolic correspondences practical effect. That between the planets and the chemical elements constituted a connection of alchemy with astrology: and so, when one dealt with a metal, its planet had to be in the proper position. The conjunction of fire and water among the elements was correlated by Renaissance geologists with the fact that volcanoes are always near the sea. The conjunction of fire and sulphur was connected by Paracelsus with the preparation of the very "fiery" substance, ether, with the aid of sulphuric acid. It was long before men found anything better than such apparently not unpromising starts; and, as has been implied above, they only did so after decisive advances had been made in physics.

The service—and it was a very real one—of all the centuries of alchemy was that of trying all the possible permutations and combinations of all the substances known, mineral, vegetable, "spiritual," "aerial," of clearing away those which did not interact, and of slowly fixing attention on the less accidental features of those which did.

We note in conclusion two or three individual men who, soon after 1500, were trying to free chemistry from mysticism and to bring it into touch with those craft-sources from which it was to draw such wealth. Paracelsus (*c.* 1493–1541) was perhaps a typical Renaissance braggart. It has still to be proved that he knew of hydrogen as such; while it was another, Valerius Cordus (1515–44), who gave the first non-mystical account of the sulphuric preparation of ether. But Paracelsus was also a lifelong voyager in search of facts, with a hot tongue in his head for the pomposities of contemporary scholarship and medicine. Agricola

* A proceeding which would not have erred, but which, it should be noted' would not by itself have brought science into being. For science needs abstractions *narrow enough* to issue in predictions disprovable (if false) by experiment, not wide enough to be certainly usable in some sense or other.

(1490–1555) the metallurgist was a much quieter figure, a collector of information and specimens, a genuine practising expert in his subject, an exemplar of the scientific virtue of sobriety.

Paracelsus, who was more a doctor than a chemist, was one of the first to insist that the body must be viewed as primarily chemical. Titley has suggested that by reacting from the mixed prescriptions of his time, he prepared Boyle's stress on the idea of a pure substance, so vital for chemistry. He favoured mercury and other mineral drugs at the expense of the vegetable decoctions of the time. In fact, like Agricola, he was a keen mineralogist, and had glimmerings of the ideas which later developed into geology ; such as that of the succession of the rocks. But a few solitary voices were not enough. In the 16th century a new burst of religious fanaticism flooded Western Europe. After these men, that isolation of which we have spoken swallowed up chemistry the science ; even while the Rosicrucians, and princes like Augustus of Saxony and the Emperor Rudolph II, followed chemistry the mysterious, the fascinating, the secret.

CHAPTER II

BIOLOGY BEFORE THE MICROSCOPE

As science now gets into its stride in several subjects at once, it is necessary to state the plan of narration adopted. Each chapter treats a major advance, its precursors and its effects. The chapters, therefore, overlap chronologically. This is the price which has to be paid for a reasonable continuity of exposition in any one subject. Fortunately, the number of subjects at peak at any one time rarely exceeds two or three.

If this were an encyclopedic work of reference, it might be best to treat all the sciences of each period in one chapter, thus securing complete cross-sections and giving not only the ideas in flower at any given time, but also those not destined to germinate for centuries. For a work of the present size, this would involve a prohibitive amount of duplication and would interfere with the central object of the book, which is, to keep the interrelations of the sciences foremost. Interrelations with the technology and general conditions of each period must also be understood if the history of science itself is not to seem at many points arbitrary. Hence short background studies have been inserted at intervals.

Natural History

We have already noted how at the Renaïssance the descriptive sciences such as botany and anatomy were bound up with art. But each had also its independent workers. There had always been doctors ; while it was for drugs that mediaeval herb gardens had been laid out. We shall consider these beginnings of natural history before describing the great achievements of the 16th century in the more serious, medical study of anatomy.

Natural history is an observational, not an experimental science, and as such had made some headway in ancient Greece. Aristotle was a keen zoologist, fish, then of extreme economic importance, being among his chief interests ; while his disciple Theophrastus (380–287 B.C.) did corresponding work in botany. An intelligent man in an enlightened age, Aristotle saw that living creatures form a series of increasing complexity. His scale, however, was only roughly the same as ours. It ran from lower plants, via higher plants, through molluscs, arthropods, reptiles, birds, and (collaterally) fishes, mammals, men. It is improbable that his view of this scale came very near that of a modern evolutionist.

But it was not this particular classical tradition which the Middle Ages and the Renaissance followed in natural history. The mediaeval bestiaries had derived rather from the travellers' tales of the Roman,

Pliny the Elder (A.D. 23–79), and their illustrations were not copies from nature, but lively imaginative elaborations of the descriptions given in these classical texts.*

One of the main ideas of all this literature was the anthropocentric one of which pre-Copernican astronomy is another case: the assumption that plants and animals exist not for and in themselves, but for their use to human beings. Teleology crops up everywhere in pre-scientific ages, the idea that we have disposed of a plant or an organ when we have expressed our keen sense of its suitability to its purpose. Unfortunately, it is just cases of unsuitability—of *ill*-health—which interest doctors. In fact, this recital of harmonies, like that of number-simplicities, has the fatal feature of all pre-scientific thought: it closes rather than opens the subject. We shall never understand the nature of science as a living movement, and its contrast to philosophy, unless we realise that all through its history, but especially to-day, it has been quite as important to the scientist to keep on finding problems as to keep on finding solutions. Control, not contemplation, must be the object even of the pure scientist, if he wishes to find out further truths. Its means are mainly the finer and finer *analysis* of a plant or organ into simpler elements (preferably those of physics or chemistry), not its perception as a whole. *Divide et impera* in knowledge as in politics. Our present subject, the collection and classification of specimens, is but an aspect of this.

It is easy to preserve dead plants by merely drying them, but to preserve dead animals (except their skins) needs special methods, not available at this early stage. The collection of *live* animals remained possible, and by the end of the 15th century this had become a choice form of ostentation for princes with venturesome mariners and curious minds. Lisbon, the capital city of the earliest explorers, had one of the earliest zoological gardens. By the 16th century, private individuals were following the example—the rich merchants of Germany and later of the Netherlands. At the same period, Italian universities like Padua and Pisa laid out botanical gardens for scientific study, and the practice spread to Germany and the Low Countries.

We have already mentioned Agricola, in whom so many scientific traditions of early Germany find illustration. Like the Frenchman, Belon (1517–64), and the Swiss, Gesner (1516–65),† he formed biological collections. These two, and Aldrovandi, were zoologists. Early botanists included the Germans Brunfels and Fuchs and the Lowlanders L'Eclus and L'Obel. These men, and especially Gesner, got into a habit of great significance in the history of science, that of corresponding with scientists in other countries and of exchanging specimens. In the Middle Ages the official organisation of the Church had secured intellectual communication in the most esteemed subjects. Science had no such means, but by unofficial channels these naturalists

* Rufinus, 13th century, has been mentioned as a partial exception to the practice of copying not nature but books.

† Gesner was one of the earliest to express that love of the mountain scenery of his native land which in the 18th century led De Saussure to found alpine geology.

began to build up some impression of the prodigious variety of natural forms.

They absorbed all that the classics could give, and prepared for their successors' emancipation from a not unnatural error—that of supposing that the Mediterranean flora and fauna described by Theophrastus were necessarily those of Germany in the 16th century. Gradually observation grew autonomous, and ceased to be made largely with a view to collation with classical texts.

In this way, a vast bulk of scattered material gradually became available. The labour of its collection well suited the German temper. Brunfels (1530), Fuchs (1542), and others wrote huge books, often with beautiful woodcuts in the new naturalistic manner; taking advantage of the recent advances in wood-engraving. Illustration has a certain importance for our purpose, and we revert to it later.

After collection some sort of classification and nomenclature were necessary; and these were an obstacle destined to slow up botanical development for a long century to come. Aristotle, Theophrastus, and Galen, in their various spheres, had been halted by the same difficulty. It was a twofold one. In the first place, the desire to have a pigeon-hole for *every* species or mineral is the root motive of classification, and sooner or later this requires the classifier to abandon all cherished traditional divisions (such as that by uses) and all pleasantly superficial ones (such as those by the largest or brightest features). But it is a hard step to admit this after a lifework based on them. In fact, it was only in the slightly formal and cold-blooded atmosphere of the 18th century that the necessary effort of abstraction was finally made. Even then the effort was so great that it ossified the minds which made it; and that *provisional* quality which from the modern point of view is the essence of classification was not achieved. Through all this early period that quality would not even have been admitted as a merit. Men were still apt to feel that some uniquely "natural" scheme existed, if only they could find it.

An important early worker was Cesalpino (1519–1603), who classified plants by their flowers and fruit (1583), and who made the first serious attempt to include the whole plant kingdom as then known. Few of his divisions have survived, but at the hands of his more northerly contemporaries in Holland and Germany, some of our modern ones (such as umbelliferae, leguminosae, gramineae) became quite definite. Before this (1542) Fuchs had published a glossary of which many of the terms are still in use. Later, Jung (1587–1657), though he published nothing, made great advances as to technical terms and approached a binomial nomenclature for species. This was of capital importance. For, to come to the second point, until we have decided on what features to concentrate, we shall lack technical terms; and anyone who has tried to compare, say, two flowers, in detail, without these, will know that their absence practically halts the process at the start. These troubles were at their worst in botany; for whereas function * (derivable, at least for the higher animals, by analogy with

* Of the parts in the whole creature, not, of course, of the creature in the whole universe.

ourselves) provides at once a basis for the division of an animal body into convenient parts, function is by no means so obvious in plants.

We shall try, in this book, to keep the knowledge of plant, animal, and human biology in parallel, but the profound historical and practical contrasts must not be forgotten. The modern unified biological point of view must not be imposed on these early workers.

The early German school did not itself reach any decisive discoveries. These were destined to come only when, in the next century, lenses had increased the fineness of men's vision of detail. Meanwhile Germany came to be torn by religious war, and Holland rose to commercial dominance. The private museums of its rich drug and spice shippers carried on the tradition of natural history which culminated in the microscopists of Newton's time, to be treated in Chap. VII.

ANATOMY

In this book, artists have been mentioned before doctors in connection with anatomy. And this was not unfitting. Unhindered by tradition, the artists at once achieved a luminous accuracy which left medical orthodoxy far behind. The Renaissance was a hey-day of art, a world in which impersonal goods like drains and machinery had not turned men's minds from emotional things. It was not a hey-day of medicine. Corruption and inertia lay heavy on the orthodox faculties. Quacks, sow-gelders, even executioners and torturers (as having some anatomical experience !) were consulted. There were heroes, as always, but even they did not look to research, only to charity.

The first effect of the Renaissance was not wholly a good one. As in botany, there came immensely more accurate and inspiring Greek texts of the classics. It was not only that these formed excellent sticks with which to beat the brusque Renaissance innovators who insisted on looking at facts, not books. It was that the collation of these texts and their translation into Latin occupied the gifted men of a whole generation, and turned them from research.* These were the " medical humanists," such as the Englishmen Linacre and Caius, translators of Galen ; Leoniceno and the great Rabelais (1490–1553), translator of the Aphorisms of Hippocrates ; Guinther or Winter (1487–1574) the teacher of Vesalius and Servetus. Broadcast by the printing press, which from about 1450 had begun to drive the great manuscript trade out of business, these translations suppressed until far into the 17th century the fact that biology is only secondarily a matter of books.

But with all this said against the doctors, it remains true, of course, that the artists alone could never have made the science of anatomy. The medical interest is primary. It was only when the minds of the artist and of the scientist united in the strong personality of Vesalius that permanent effect was produced, a quasi-sexual union of contrasting cultural traditions which often precedes major advances.

* Even in abstract science Copernicus was, not without justice, accused by Kepler of interpreting Ptolemy, not Nature. This is possibly what young abstract scientists largely do to this day : solve problems arising in their elders' theories rather than in " reality."

To understand the medical anatomy and physiology of 1500 it is necessary to go briefly behind our period and to glance at the subject from its earliest days. The name of Empedocles (*c.* 490–430 B.C.) is used as a label for several theories, those inevitable, if ultimately sterile, theories of " spirits " or " pneuma " or " breath of life " which we shall see dogging biology all through its history. Empedocles also spoke of that ebb and flow of blood along any one vessel which, right up to Harvey's time, was preferred to the concept of uni-directional circulation.

In the great figure of Hippocrates and the writings of his medical school at Cos (*c.* 460–360 B.C.) teleology was discarded and a truly positive, experimental spirit appears for the first time in recorded history. These wonderful writings have been an inspiration to doctors ever since. And yet the immediate harvest was poor, owing to the intense difficulty of the subject. The Hippocratics had few instruments. Their botany concerned not the structure but the healing properties of plants. Their anatomy showed little effective research. Their greatest practical principle was that of the *vis medicatrix naturae*, that nature is her own best healer and that the wise physician only tries to remove adventitious obstacles from her path. To this we find medicine constantly reverting, and always with great benefit.

With Pythagoras and Plato there came more theorising, notably another idea traceable through all subsequent history, that of a deep analogy, or one-to-one correspondence, of the heavens (macrocosm) and the human body (microcosm). This haunted the Middle Ages, distorting both the things compared.

Aristotle, Plato's pupil, represents another reaction to the scientific outlook. He has been Europe's great mental disciplinarian. He will, however, appear in these pages in such apparently opposite rôles—philosopher and biological scientist—that a word of explanation is necessary. He is, perhaps, the water-shed of these two studies, philosophy and science. After him, it was appreciably harder not to feel that the two studies were incompatible. Observation, indeed, accords well with philosophy—spectatorships both. But experiment, and also mathematics, do not. They are *activities*. Experimental chemistry had to live down Aristotle's elements. Mathematics, too, as we shall see, was hindered by the logical scruples of the Greek philosophic mind.

Even at the scientific height of the ancient world, the Museum of Alexandria (*c.* 300 B.C. onwards for three or four centuries), and *a fortiori* through the long twilight which followed, the fundamental fact about anatomy and physiology was that they could scarcely be founded on close inspection of the interior workings of the human body, since dissection was rarely allowed by law. It may seem strange that in those times of hazard, brutality, and war such little matters should have been difficult. Perhaps it is only when anarchy dies down that civil and religious sentiment can afford to relax. But there was another difficulty. In the hot countries of early science, and in the absence of preservatives, the steps of dissection are dogged by decay. Thus even the great Alexandrians Herophilus and Erasistratus, reckoned respec-

PLATE II

LINKS OF SCIENCE AND ART II. RABELAIS. Literature and medicine.

PLATE III

Valerius Cordus. A neglected early scientist. *Reproduced, by kind permission, from an engraving in the Surgeon-General's Office, U.S.A.*

tively the founders of anatomy and physiology, made mistakes which suggest scant experience of human bodies.

Unlike Aristotle, however, they firmly seated the mind in the brain, not the heart. They also divided nerves into sensory and motor, and had a clear view of muscular shortening. After them came a long decline. The practical Roman genius culminated in the great practitioner Galen (A.D. *c.* 130–*c.* 200). In him, the ancient world was no longer confining itself to observation and theory, but was experimenting persistently; only, not with the decisive success which could really set physics, and so physiology, on their way. Thus, by section of the ocular nerves and the spinal cord, Galen roughly localised several nervous functions.* But Galen was an inveterate teleologist, and it has been suggested that it was for this reason that his work so easily and completely became the medical bible of later ages. Codified, with smatterings of Hippocrates, by the great Arab Avicenna (979–1037), translated into Latin for European use by Gerard of Cremona (*c.* 1180), Galen was not really displaced from orthodoxy until long after our period begins.

The Salernitan medicoes of the 10th century had made dissections of pigs, but it was perhaps one of the rather few links of science with *law* which finally brought in the dissection of human beings: postmortems at the great law school of Bologna early in the 14th century. Thereafter on rare and special occasions there were instructional dissections before students; but to illustrate Galen, not to improve him. If Galen said that the uterus had seven cells, seven cells it had. Apart from such exceptions as Mondino (*c.* 1316), the professors did not demean themselves to dissect in person, but discoursed while a menial operated.

But with the Renaissance, times changed. Vigorous men like Michelangelo and Leonardo did not hesitate to dissect bodies whenever they could. Definite anatomical achievement was not long in following. About 1535 Andreas Vesalius (1514–64) appeared on the medical scene. A Fleming with German blood in his veins, son of the Court apothecary of the Holy Roman Emperor, he came of a long line of doctors, and was already dissecting as a boy. He worked at Paris under J. Sylvius (a famous Galenist) and under the great humanist Guinther. He went on to Venice (1535) and finally to Padua, the university of the Venetian region—comparatively tolerant, perhaps for commercial reasons. He became most learned in Galen, but current anatomy did not impress him, and he was soon giving vent to revolutionary views. From 1537–43 he lectured at Padua, soon to large audiences. In the latter year he published his "De Fabrica Corporis Humani." In view of the book's full-blooded scorn of tradition, it was no wonder that it was by no means greeted by the orthodox as the masterpiece that it was, but that on the contrary it secured a century's contempt for its author's memory. The effect of its reception was to

* His subjects are said to have been chiefly barbary apes. Throughout the Middle Ages, which would have shuddered at the suggestion of an ape-like ancestry, doctors were unconsciously treating their patients as apes!

disgust Vesalius with research. In fact, with it, his scientific work was done.

Dissecting in person, inventing or borrowing new instruments at need, inventing much of the technique (for instance, the mounting of skeletons) which is still used, Vesalius revolutionised anatomy. The feature of his book was its wonderful illustrations. In the Middle Ages there had arisen a tradition of manuscript anatomical sheets. These were what we should call mere diagrams, the body being shown in a squatting posture with indications of the correspondence of the parts of this " microcosm " with those of the " macrocosm " (the universe). Bones, muscles, arteries, veins, nerves, and sometimes reproductive organs, were put on separate diagrams. They were grossly inaccurate, but the point here is rather that they were diagrams, not naturalistic plates. These latter were the innovation of Vesalius in anatomy and of the " German fathers " in botany. It was an essential advance, but it must be noted that it was only a preparatory one. At the present time, both kinds of illustration are used, diagrams having returned for many purposes. But until the details had been accurately depicted, it was not safe to caricature them for the sake of emphasis.

It was emphasis, abstraction, caricature which gave the mediaeval wood-cut its liveliness compared to some of the naturalistic drawings. In those drawn for Vesalius there is no lack of life ; but this is itself due to a touch of the abstract—to the vivid sense of the body as a living, functioning whole which is impressed on them.* This Renaissance perception was not far from the modern ideas of organism. Nowadays we have a second idea, that of the evolutionary growth of the race, to balance the isolation of the individual which " organism " is apt to imply. The Renaissance was without this idea, although it was beginning to be alive to two related ones : the growth of the individual —embryology—and the comparison of man with other animals—comparative anatomy. In Coiter (1534–76), and others of the Vesalian tradition, the comparative instinct was alive, and Ruini of Bologna produced (1598) a fine equine anatomy.

Before we pass to Vesalius's successors in the Paduan line which led up to Harvey we must note two great men on the medical side. Of one, Paracelsus, we have already spoken. He was an early appreciator of the medical value of cleanliness, and was one of the first to note that miners (for instance) have characteristic (" occupational ") diseases. In his use of the vernacular he resembled our second figure, deservedly more beloved of doctors, the great French army surgeon Ambroise Paré (1510–90).†

Paré had the thrusting touch of the Renaissance more mildly than Paracelsus, but he was likewise a great innovator and was likewise

* We cannot mention many further individuals in the long line of anatomical illustrators which Vesalius initiated. Almost as great and almost as early was Eustachius (1520–74). But much of his work was never published, so that the decisive factor was that of Vesalius.

† The greatest scientific value of war has not lain in research into weapons like explosives, but into remedies for their effects. Rulers indifferent to the fate of their civilian subjects were keenly alive to that of their soldiers, and we shall notice many decisive biological advances made by military surgeons.

snubbed by the orthodox faculties. For example, he made advances in surgical instruments and in the treatment of labour. In him the Hippocratic doctrine of the *vis medicatrix naturae* was reasserted. An accidental shortage in camp of those ghastly mediaeval means of sterilisation,* boiling oil and the cautery, gave him instances of wounds healing by themselves which (through his insistence) doctors have never since quite forgotten. He was, in fact, the first to stress the vital question of sterilisation, which, three centuries later, was the keynote of such great events.

Physiology

We now revert to Padua and to the decisive advances in physiology which followed those in anatomy. But surgery should still be in our minds ; for the fact that surgery is one of the primary branches of medicine illustrates the fundamental fact in the history of physiology, its dependence on the growth of mechanics, physics, and chemistry. A large proportion of experiments in physiology depend on the simplest of mechanical things, a knife. Another large class depend on some of the simplest chemical compounds : calomel or Epsom salts are instances. The particular case of this dependence which we must now examine centred round the action of the heart and the circulation of the blood. To understand this we must briefly describe the errors of the ancients about it. It should not seem strange that it was on this most vital of the bodily systems that men were so long at fault. The more vital an organ, the harder it is to watch in action.

Herophilus and Erasistratus had distinguished veins and arteries, and Herophilus had noticed the pulsations of the former, which he regarded as an intrinsic property, not as due to the heart. Erasistratus taught that every organ has three sets of vessels, veins, arteries, and nerves. Now whenever (through the accident of a wound) an artery can be seen in the dead, it contains little blood, while the veins are full. From a post-Harveian point of view the reason is clear and the observation trivial. Not so before Harvey. The fact was correlated with the " pneuma " and the " ebb and flow " of Empedocles. Galen's doctrine ran roughly as follows. Blood, formed and charged with " natural spirits " in the liver, ebbs and flows under the heart's action between the liver and the right ventricle of the heart. But a little penetrates the (valveless) septum into the left ventricle, there to meet air from the lungs (in breathing we draw the basic principle of life, "pneuma," from the general or World Spirit) and become " vital spirits," which are distributed by the arteries—little blood and much spirit. Some goes to the brain, where it becomes " animal spirits," distributed by the nerves (then thought hollow to permit this), and so to muscles and elsewhere. It was especially in these " spirits " that the lack of clear

* We think of the Middle Ages as notably wanting in cleanliness, but the idea of infection, with the practice of checking epidemics by quarantine, is one of the few capital traditions in medicine which they—and we—derive from Scripture and not from Galen and Hippocrates. With the cautery before us, we see that the truth was as much that the Middle Ages were terribly strict, as that they were terribly lax, in sterilisation. Penance and the ascetic life are examples of this on the mental plane.

physical and chemical ideas was most fatal. Spirits appeared in the most diverse sciences. We shall find Kepler using them in the unpromising region of astronomy at the end of our present 16th century. They were rampant in chemistry until the 18th century. Verbal ambiguity illustrates the trouble. " Spirits " (vinous) were known to affect " spirit " (mind) ; and the " spirits " in the nerves and arteries were vaguely thought of, like the breath (of life) itself, as what we should now call vapours. The evaporation of liquids must have been a puzzle, and probably delayed the establishment of the fundamental ideas of materiality and of the conservation of mass.

Again, the common metaphor for the action of the blood on the tissues was that of irrigation, already familiar to men from agricultural

CHART I

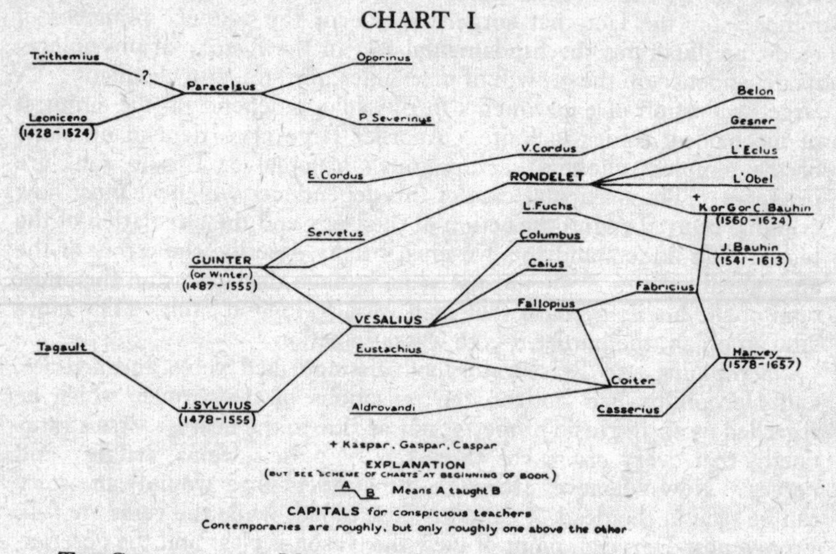

THE CONNECTION OF MASTER AND PUPIL. *Mainly Sixteenth Century.*

practice. The nearly-empty arteries were thought to transport no more blood than could be absorbed by the tissues (as a field absorbs rainwater) and be got rid of by perspiration and excretion. Physics, with its stress on quantitative accuracy, would have sought to test the point.

Leonardo had felt a difficulty about the action of the heart, and so had Vesalius himself. Attention, in fact, was gradually concentrating on the point. The unhappy Spanish religious fanatic Servetus (1509–53), a pupil of Guinther, stated (1553) that it is in the lungs that the air mixes with the blood from the heart to form arterial blood.* This is an awkward point if it is not from the lungs but from the aorta that the arteries spring. Then Fabricius of Acquapendente, pupil of Fallopius,†

* Anticipated by Ibn-an-Nafīs in the 13th century.

† Fallopius was himself a pupil of Vesalius (see Chart I). Fabricius, Harvey's revered teacher, was the last of the great Paduan line. He was Harvey's precursor in embryology also ; while his exposition of the muscles and bones as levers was in line with the new interrelation of biology and physics.

learned from the Venetian Sarpi of the existence of the valves in the veins (1603). Viewing them in the light of the contemporary revival in mechanics, he showed by experiment that they hinder flow except *to* the heart. As, however, he was still thinking of Galenic ebb and flow, he supposed that they were there merely to prevent accumulation in the extremities. Between about 1570 and 1590 Cesalpino threw out (in controversy with Galenists) the view that the blood rather *circulated* than oscillated. The suggestion was unproven and uninfluential, but Harvey acknowledged Cesalpino as a precursor.

William Harvey (1578–1657) had made several continental journeys in search of teaching before he hit on Padua, where he studied 1597–1602. For him, the heart and blood question had become crucial. He knew all its literature. By 1616 he was giving his great discovery in lectures. But not until 1628 did he publish it. At first disputed as well as acclaimed, it was generally accepted before his death.

" De Motu Cordis " is a classic of science. Negatively, it was unencumbered by metaphysics, a virtue then exceedingly rare. Positively, it stands for the entry at once of the quantitative and of the comparative methods into biology. As to the latter, having looked at the heart's action in dozens of species, Harvey had the crucial advantage over earlier dissectors of finding that in cold-blooded animals, this action is slow enough to be visibly analysable on mechanical lines. After the heart has contracted, the aorta expands. There is an obvious resemblance to a pump. Now pumps were in the air at that time. The dominant commercial power, the Lowlands, one of the dominant regions of England, East Anglia, depended on hydraulic engineering for their very lives. Not long after Harvey's work, the Dutch engineer Vermuyden drained the Bedford Level. Mines (for instance in Germany and in Cornwall), growing deeper and deeper, were growing more and more dependent on pumps. As we shall see in Chapter V, the physics of the middle of the century was consummated by the construction of the *air* pump. It was as a pump (" bellows ") that Harvey saw the heart and its set of one-way valves (" clacks "). He showed that the one-way action of these valves implied that the blood, entering from the great vein (the *vena cava*) into the right *auricle*, is pushed by the latter's contraction into the right *ventricle* and thence by another valve via the pulmonary artery into the lungs, then back (after aeration) into the *left* auricle and so into the left ventricle and by another valve into the aorta. Though the one-way action of the valves was of course known, it had never been adequately analysed before. With the mysterious penetration of the valveless septum abolished, the way was clear for the perception that it is the whole of its content of blood, not a small percentage, which the heart regularly throws into the arteries. So large an amount could never be accounted for by vague " irrigation." It must return to the heart, via the veins, for repetition of this complete *circulation*.

The proof was pursued into every detail but one : the stage in the circulation when veins and arteries lose themselves by subdivision into the tiny " capillary " vessels. And here again there was a dependence on physical advance—a lens was needed to show these vessels and their action. This had to wait for the second half of the 17th century.

This great discovery had many consequences. One was the investigation of the rise of sap in plants ; provoked by the effort to find an analogue of the circulation of the blood. But the most important consequence was the impulse to substitute mechanical and chemical for " spiritual " explanations of vital action.* " Spirits," indeed, remained until 1800 in the great mystery of the nerves. But the 17th century saw them abolished in most regions of internal anatomy. In the torrential rush of blood there was simply no room for them in the veins or arteries. Having once put them aside, Hofmann, Wirsung, de Graaf, and others worked out (from about 1640–65) the action of the pancreas and of several other internal mechanisms. We shall see in a later chapter how such successes as Harvey's and these, and the muscular levers of Fabricius, led to premature crudities of mechanical and chemical explanation (" iatromechanics " and " iatrochemistry ").

This chapter is headed " Biology before the microscope." There was another point, besides the last link in circulation, at which biology, with Fabricius, Aldrovandi, and Harvey, reached the limit of what was possible without the microscope. This was the question of the development of the individual from egg and womb.† Alcmaeon of Croton (*c.* 500 B.C.) was among the first of embryologists. A Hippocratic writer fixed on what has ever since been the classic subject of experiment, hens' eggs opened at daily intervals. Both Aristotle and Theophrastus were obsessed with the problem of reproduction, and Aristotle formulated the characteristic theory that the female furnishes matter, the male only form. Epicurus, on the other hand, followed by Galen, held both contributions to be material. Previously, maternity had often been denied, woman being likened to a field sown by man. The formation of the individual from nothing, or nearly nothing, looks like an event without a cause ; so that the subject has ever since been an ideal one for metaphysical speculation. One form of this is the riddle : which came first, the hen or the egg ?

The ova and spermatozoa of human beings could not be recognised without a lens, and the seeds of many small animals and plants could not even be seen. But in the deer-parks of Hampton Court and Windsor, Harvey studied (often in the company of Charles I) the earliest mammalian embryos ever watched by man. He did not view development as always the unfolding of the already existent—as the preformationists would suppose ; but in the absence of the microscope the proof was not convincing, and here we must leave the subject until Chapter VII.

* Spiritual in one sense, but in a very degraded one, the word spirit being understood half materially.

† J. Needham notes, however, that lenses were used in embryology as early as 1618. He also notes that another great help here, artificial incubation of birds' eggs, was not successfully practised in Europe until the 19th century. It was known in ancient Egypt.

CHAPTER III

ASTRONOMY BEFORE THE TELESCOPE

PHENOMENA so definite and simple that mathematical abstractions can at once be read into them hardly occur in untutored nature. Experiment as well as observation is necessary to exhibit them. To this rule there is one exception which strikes all early men, the simpler movements of the heavenly bodies, the sun, the moon and the planets. Even at the Renaissance stage, astronomy is the first exact science to make decisive advances. In Babylon about 4000 B.C. the sky seems to have been under expert observation. In China, not long after, the equinoxes and the solstices—that is, the solar year—were determined. At the same period the Egyptians were making careful observations of the sky, and orienting the pyramids with surprising accuracy.

The heavens have always had the fascination of the mysterious, the supernatural ; and this is at first their uppermost aspect. Astrology—itself no loose or random body of doctrine, but a strict discipline—is only one instance of this. Even when men consciously contrasted it with true astronomy, that astronomy still served partly ritual purposes. It was for fixing services and feasts, for ecclesiastical calendars, that time measurement was used. In secular matters, primitive races have, and need, little idea of time ; and for some way even through the Middle Ages an hour was for the majority a constant fraction not of a day but of daylight.

It was only when long-distance voyages began to be undertaken that we reach a first-rate secular use of astronomy. Astrologers had been carried on ship-board in the Middle Ages, but in the 15th century the new race of great explorers began to consult navigational astronomers instead. Such astronomy was highly technical, but the very much earlier popularity of such books as the " Sphaera Mundi " of Sacrobosco (d. 1256) reminds us that there had always been another kind of astronomy, that the general principles of the subject have always interested all educated men. In fact, from the very beginning of its history right down to the present day, astronomy, like some other sciences, has remained two subjects, one a matter of central ideas, the other a difficult mathematical and observational technique known only to professionals.

We have said that astronomy gives us examples of mathematically simple phenomena. But this simplicity will not survive much further investigation. Each fresh observational or instrumental advance challenges the ingenuity of the theorist. Consequently, from the earliest times, man has found himself led on from puzzle to puzzle, with little prospect of an end. The continuity and complexity of the process forces us to make, as in the last chapter, a brief survey of the state of our science before the modern period opens.

At an early date the Egyptians and Babylonians seem to have viewed the universe as a rectangular box ; and Menon,* observing the practice of Indian astrologers to this day, has ingeniously connected certain early units of angular measures with the practice of viewing orbits, or the universe, as square. But we need not go back earlier than times when geocentric circular orbits were the unit of thought. We say geocentric, but, as noted in the first chapter, it was really on his own self that early man naïvely centred the universe.

Greece began to receive Babylonian astronomy about the 7th century B.C. The Babylonians themselves continued to make advances, and just possibly knew of the precession of the equinoxes by about 350 B.C. Gradually an orthodox world-picture developed. The heavenly bodies and their motions were contrasted, as incorruptible and immutable, with the corrupt and ever-varying things of earth. Yet it was the earth which was the centre of the system of crystal spheres which held planets, sun, moon and stars alike. The fixed stars had the outermost sphere to themselves.† Each planet, and also the sun and moon, was embedded in a separate sphere ; and for its orbit only the " perfect " figure, the circle, was good enough.

When observation began to disagree with this scheme astronomers, and especially the great Hipparchus (*c.* 130 B.C.), mended it by supposing that the planets revolve in circles not actually about the earth but about points themselves revolving circularly about the earth. This device (" epicycles ") became the dominant one in the orthodox astronomy of the ancient world, which Claudius Ptolemy (fl. A.D. 127–51) codified in his great work the *συνταξις*, or Almagest, as it came to be called. Ptolemy also published a table of chords for astronomical computation.

Hipparchus is the summit of ancient astronomy, drawing together Babylonian observation and Greek geometry (Euclid, fl. 300 B.C.), discovering the precession of the equinoxes,‡ initiating trigonometry, perfecting or inventing instruments at need. He began the practice, later so significant for mathematics, of specifying the positions of places on the earth's surface by two " coordinates "—longitude and latitude. In all this, as in anatomy and physiology, Alexandria was the centre.

Revolutionary ideas had, however, been suggested before the time of Hipparchus. Aristarchus of Samos (fl. 280–64 B.C.) had put forward the first heliocentric theory. By his time there had been exploration down the African coast and the earth was recognised, at least in professional circles, as round, not flat.§ Aristarchus pointed out that this, taken together with lunar eclipses, suggested a sun vastly larger than the earth, and therefore more likely to be fixed. At the same time he

* Menon, C. P. S., 1932, " Early astronomy and cosmogony," London.

† The *space* between it and the next sphere, not merely the surface of the outer sphere. See McColley, G., *Annals of Sci.*, Vol. 2, pp. 354–6.

‡ Usual attribution. But see above.

§ Holland Rose has argued that Phoenicians may have circumnavigated Africa *c.* 600 B.C. Eratosthenes, librarian of the Museum of Alexandria (*c.* 273–192 B.C.) made an estimate of its circumference less than five per cent. out. A later, erroneous, much smaller, one by Poseidonius, was destined to put heart into Columbus.

criticised another assumption which had always been made—the assumption that the " fixed " stars are not necessarily very far off. He asserted that only their immense distance could explain their apparent immobility. The orthodox view was that as (in pre-telescopic days) stars were of about the same conspicuousness (size, brightness) as planets, there was no reason for assigning them vastly greater distances. As these questions arise even more forcibly within our period proper, we shall defer their consideration.

Aristarchus did not develop his theory far enough to demonstrate its incomparable practical simplicity—to show that, in fact, the orthodox epicycles were only the earth's orbital motion transferred separately to each of the planets. It was consequently not adopted by the computational astronomers, and it was the work of Ptolemy which, all through the Middle Ages, enjoyed an authority like that of Galen in medicine.

But, as with Galen, much of the classical technique was lost. Even the Arabs and Jews have little new to their credit. Their new computing methods, however (see next chapter), did have effect : extensive table-making became possible. And, astronomy having a practical use, it was tables, and also navigational treatises, which dominated it from 1000 to 1500. The Arabs had drawn up a set of trigonometrical tables as early as 850. They recognised the superior convenience of sines over chords for this purpose. It was the 12th century before the Christians began to copy them, and the 13th before they began to work on their own. It was long before their accuracy equalled what the Arabs achieved in their final phase about 1400–50.

Trigonometrical calculation is not easy, and, as the medical humanists sought help in the newly-recovered antiquity, so did the computers. The 15th century Germans, Purbach, or Peurbach (1423–61), and his pupil J. Müller (Regiomontanus, 1436–76), came to Italy, accordingly, to collate the new texts of Ptolemy. But at this point, where our period really begins, the subject splits into two parts which we shall treat separately : first, the Arab-Jewish tradition in Spain and Portugal, with its practical issue in navigational astronomy and the great explorations ; second, the theoretical astronomy of Germany, with Copernicus and his successors as its great luminaries.

The Explorations

Roger Bacon (1214–94) had supported the theory of the sphericity of the earth, a belief which had receded with the end of the ancient world, but which again came to be generally accepted in educated circles by the 15th century. On this basis, men naturally dreamed dreams ; and in particular the Portuguese prince, Henry the Navigator, dreamed of missionary and commercial enterprise, of a short sea route to India unhampered by the Moslem power.

From the 12th century ships and their rig had been improving. So had the means of laying and keeping a course. There had come, first, magnets floating on water, then (13th century), box compasses. We have already seen how the magnetic knowledge of Peregrinus fell into oblivion, but there was sufficient practical knowledge of the variation of declination for Columbus to find the first place where that quantity becomes zero.* The Arabs, too, had done something to provide navigators with astronomical instruments more portable than those of the ancients. In their day a mariner would have a cross-staff, or later an astrolabe or quadrant, to give the maximum height of the sun. No satisfactory determination of longitude was possible, but by means of tables of solar declination, or of pole-star altitude corrections, the latitude could be obtained.

Henry, from about 1420, and after his death John II (in 1482), assembled all the mathematicians and navigators they could find, to improve this technique and to construct charts. Nearly all were Jews. Better tables of the sun's declination were needed, so Henry set up an observatory near Cape St. Vincent. His mariners discovered the Azores in 1419, and by 1456 had pressed so far south along the African coast that Cadamosto had to examine the southern constellations for a substitute for the pole-star. It was at one time thought that the Portuguese in their great voyages used the 1474 German tables of Regiomontanus; but these were not based on the same value of the obliquity of the ecliptic as those of the Portuguese, who used the work (1473–78) of the Jewish professor of astronomy, Zacuto of Salamanca.

While the Portuguese hoped to reach India round the south of Africa, the Genoese Columbus was pressing the Spanish crown to support an attempt on the "direct" Atlantic route. Here the prevailing winds seem adverse, and incapacity to determine longitude is serious; but at length in 1492 Columbus reached the Bahamas, only to be treated as a failure for not having reached India. This the Portuguese Vasco da Gama did in 1497, via the Cape of Good Hope, as originally planned. We cannot pursue the later voyages, such as Cabot's from Bristol (e.g. 1497), and Magellan's circumnavigation of the globe (1519–22). They may all be regarded both as the dawn of a new age and as the last flick of Arab technology and of the quixotic traditions of the mediaeval chivalric spirit.

The profound social and economic effects of the explorations cannot here be discussed. But they illustrate a point which we shall notice again and again in the relations of science to economics and industry: how hard it is to say which is cause and which effect. The *theory* of a spherical earth made a mercantile project possible; this again paid for the further scientific research which was necessary. The theory seems to have started the ball rolling; but the theorists were not usually disinterested men, any more than the mercantile venturers like Prince Henry were free from romantic, quite uncommercial, theories.

We must pursue the scientific side, and in particular geography and the theory and practice of mapping. These long voyages forced to the front the question of the effect on mapping of the sphericity of the earth. How, for example, to set out on a plane map the course which would be shortest on a sphere; in general, how to represent a sphere on a plane. This raises profound questions of which we shall see

* The Chinese, probably in the 11th century, had known that the compass does not point true north.

mathematical repercussions centuries later. Its immediate effect was the invention of projections.

The work of Nunez (1502–78) a Portuguese of Jewish origin, paved the way for this. Nunez was a great early scientist, who invented a cumbrous precursor of the vernier. By 1537 he had realised that the loxodrome—the course of a ship cutting meridians at a constant angle—is not a circle but a spiral, and that this explained some errors found in the charts made up to then. The well-known cylindrical projection of Mercator (1512–94) followed in 1568. This secures that angles equal on the sphere shall be equal on the map. Meanwhile (1530) the Dutchman Regnier Gemma Frisius (1508–55) suggested the present method of finding longitude by means of difference of times. Thus the vital question was reduced to that of constructing clocks both accurate and able to stand the motion of a ship ; though, as such clocks were not available for two centuries later, efforts on other lines continued to be made.

As to maps, we have mentioned in the introduction that practicable coastal charts existed in the Middle Ages. From about 1450, ocean charts began to be made. The Portuguese official cartographers were a secret body, for these early trade routes had the most vital commercial importance ; but gradually their knowledge spread to Spain, England, and elsewhere. Early cartographers were extremely slow in responding to new information, being often half a century late.

Geography, in the sense of the accumulation of detailed knowledge of countries, is a true branch of descriptive science. But, it is one of the many in which individual details are too numerous for a book like this, and we can give it no continuous history. Up to the middle of the 18th century long-distance exploration was mainly coastal. Later it was mainly concerned with the continental interiors. But intensive local mapping began in Europe in the 16th century. In England, for instance, the redistribution of monastic lands led to a great deal of surveying. The Digges, father and son, were only two of those interested in surveying in the second half of the 16th century. From 1550–1650, the Netherlands, owing to their increasing trade importance, were the centres of cartography, the modern technique of trigonometrical triangulation being definitely established by the Dutchman, W. Snell, in 1617. During the two centuries following, maps of ever increasing accuracy were made of most great Western European countries, the great triangulation of France being begun in 1683. It took more than a century to complete.

Astronomy

We now turn to astronomy proper. If the great explorations were the main advances in human technical activity in the 15th century, what we now approach was one of the two main advances of the 16th.

Astronomy was " in the air " at the end of the 15th century. The Julian calendar, for example, was much discussed ; for, in the fifteen hundred years since its adoption, Easter had grown notably out of place. Reform was necessary. We cannot discuss this except to note that the chance to impose the improved calendar on a united Christendom was lost through ecclesiastical schism and delay. In fact England only adopted the reforms in 1752. Among the scientists concerned were Copernicus and Clavius. The chief seat of the astronomical revival was the Germanic region, tutored by Italy. It was at Nürnberg (1472) that Regiomontanus, with Walther (1430–1504), erected his observatory, equipping it for the first time with a weight-driven clock.

On theological grounds Oresme and Buridan had advocated a moving earth, and Nicholas of Cusa (1401–64) had, like Leonardo, reached many modern views on motion in general, if indeed he had not reached beyond them. It is convenient to treat a few of these general points here because abstract clarity about them has often, in history, been found easier before, than after, specific cases of them (such as the planets) have been investigated in full. Cusa asserted that as no two bodies are strictly alike, strict knowledge is impossible ; he denied the strict truth of the then fundamental dogma of the circularity of the heavenly motions ; he denied that the earth is the centre of the universe, or is exempt from the universal reign of motion. He had probably a clearer hold than Copernicus on that modern view of space which by itself almost makes the relativity of motion an inevitable concept. This idea of space is too subtle to be profitably examined in full at this stage : new views of it must be allowed to emerge in their natural places from time to time. But we may recall what was said about time and space early in Chapter I.

Pointing out that continuity of motion, though thinkable, cannot be shown in practice, Cusa adumbrated an idea which, much clearer in the post-Copernican Bruno (1548–1600), fell thereafter into abeyance until apparently revived in the very latest theory of Eddington (see end of Chap. XXII). This idea was that the very possibility of measurement as a logically significant proceeding implies discreteness or atomicity. In the eyes of these men, and of Platonists like the astronomer Novara (1454–1504), this discreteness, like the Copernican system with its (partial) relativity, depended on replacing the absoluteness of motion by that of number.* This again seems to be close to what was effected by Hilbert in the axiomatics of geometry (see Chap. XIII).

The whole replacement of dogma by criticism in European thought, the realisation that our own point of view is not the only possible one, is often connected with the events of this period—the explorations and the Copernican theory. This may be justifiable if the length of time needed for these early theoretical clarities to reach external application and internal exactitude is duly allowed for. For instance, we spoke in the first chapter of the early absence of our idea of objective, physical materiality. Naïve notions here were threatened by the new relativity concept in its various ramifications, and the two must have hardened out in relation to one another. The first adjustment lasted for several

* That is, of partially physical, by purely abstract, entities. For all this, see (e.g.) Heath, L. R., " The Concept of Time," Chicago, 1936. Novara was an influence on Copernicus, who studied in Italy off and on from 1497 to 1506. So was Nicholas of Cusa.

centuries, objectivity and relativity being so conceived that relativistic invariance * became an important part of the concept of the objective. It has only been very lately that physicists have been forced to try to find a new harmonisation of the two, because only lately has the question of their application to *the means of observation themselves* become inescapable.

For a long period yet the line of advance was to lie in the gradual emergence of the somewhat naïve physical-objectivity idea, and in its use to correct the even more naïvely Pythagorean applications of the idea of the supremacy of number, or at any rate of mathematics. For, all through the 16th century there was, quite apart from the Copernican system, a growing preoccupation with the question, what *kind* of law connecting the different heavenly bodies was desirable. Copernicans and Ptolemaics alike were to be divided on the point. The crude Pythagorean sort is illustrated, at the beginning of the period, by Leonardo's examining the sky for simple polygons among the fixed stars, as if they were a mere pattern on that crystal sphere in which they were stuck like bosses : the *physical* idea of forces between the planets is absent. Mathematical simplicity and beauty was the chief advantage which Copernicus himself urged for his return to Aristarchus.†

We shall see Kepler taking the same view even at the end of the century ; but in him a desire for another, a physical kind of law is also struggling into expression. Meanwhile the merely numerical attitude was losing its tinge of number-mysticism or realism, and growing into the practical, professional, distinction of formula (or numerical rule of computation), and ultimate "law of nature." This distinction, as we shall see, served the astronomers a useful turn ; for it enabled them to accept the Copernican system for calculation without asserting the heretical doctrine that it represented the constitution of the universe. Moreover, the distinction can still be drawn to this day. Fundamental research like Einstein's is apt to leave the practical mathematics of calculating astronomical tables quite unaffected.

The manuscript of Copernicus was complete about 1530, and about then he privately circulated a short sketch, omitting all proof, among his friends. This received papal sanction and even encouragement, for the grim days of the Counter-Reformation had not yet dawned. Copernicus was already a man of reputation. He had an unrivalled grasp of Ptolemaic doctrine. He was consulted, as we have seen, about the calendar. His main book came out at his death in 1543. To it Osiander added an anonymous preface designed to shield it from

* Or more generally, the property of appearing the same to different observers.

† It must be remembered that the Ptolemaic system was not rigid, but was constantly being modified to meet new facts. Fresh epicycles had been added until it had lost all formal simplicity and beauty, and roused little enthusiasm. In this it resembled many decaying mediaeval things. The very treatises on it were traditional in form, and gave defences of the stationary earth purely because ever since Ptolemy such arguments had formed part of a proper astronomical treatise.

persecution by claiming its central idea as a mere mathematical fiction for practical use. It was, in fact, at once put to such use by Reinhold, to supersede the existing planetary tables,* and the book escaped official condemnation until long after, in Galileo's time. But, on the other hand, in no country did it make many Copernicans. The needed reform was too great to be carried through singly even by this great man.

It was by no means the planetary theory as we know it which Copernicus put forward. No physical theory of forces at work was propounded. The " perfect," circular motions of the planets were not questioned, though the finite spherical boundary to the fixed stars was apparently abolished.† It was not round the sun at all, but round the centre of the earth's orbit, that the planets circled. Thus the earth retained an exceptional position.‡ Nor did Copernicus—so Dreyer has pointed out—realise the fact which gave his system its greatest advantage, the fact that the planes of all the planetary orbits pass through the centre of the sun. Kepler was left to discover this. In addition, on the new system, a planet had not only its yearly revolution and daily rotation but a third, a conical, motion which Copernicus confusedly supposed to be necessary (as in a conical pendulum) to keep a circling planet's axis pointing to the pole-star. And to account for the " first inequalities," epicycles and eccentric circles were still necessary; thirty-four circles, in all, to account for the observed motions of his eight bodies, earth, moon, sun, and five planets.

Nor did Copernicus support his theory with crucial observations. His chief contention, as we have said, was that it was simpler than that of Ptolemy; and yet its real simplicity was still to be demonstrated. Owing, also, to the fewness of his observations and his too-great reliance on Ptolemy, he largely failed to find out the best sizes to assign to the orbits. But he did refute objections. Ptolemy's—that the earth's rotation would make it fly apart—he easily turned against those who maintained the colossal daily sweep of the fixed stars. Another ancient objection persisted for a century. If the earth is in motion, why does it not leave the air behind and cause bodies to fall far from the vertical? Copernicus answered, of course, that the air and the bodies are moving too; but he did not himself rise superior to the Aristotelian dynamics which made this answer seem unsatisfactory. Until Galileo's time no clear answer could be given, and even then grave difficulties were put on one side. The objection seemed fatal to Tycho Brahe (1546–1601), the next of the great triumvirate of planetary theory, who accordingly always denied that the earth could be in motion, and professed himself anti-Copernican. Tycho also felt the difficulty that he could detect no annual parallactic motion of the fixed stars.

Tycho Brahe's service to astronomy was an opposite of that of Copernicus: he was an observer of immense patience and skill. In

* The new (" Prutenic ") tables, 1551, were, however, little improvement on the old.

† See McColley, *loc. cit.*

‡ Relativity was of course imperfect so long as *any* body, even the sun, was regarded as fixed.

1569 Pierre de la Ramée had called for an astronomy without theories, based *ab initio* on observations. Brahe, then in his early twenties, answered him that theories were necessary to give direction to observations. But theories were not the mainspring of Tycho's life.* He was a Danish noble, and by the favour of his king was able to erect a costly observatory and to carry astronomical instruments to the greatest accuracy of which they were capable without lenses. With them he pursued the heavenly bodies in detail right across the heavens, a crucial innovation on the former practice of observing only special points in their orbits. The latter practice was a natural one on the assumption that the orbits were circles or compounded of circles. By his remorseless attention to detail, Tycho discovered that this vital assumption of the old orthodoxy did not always hold. It was certainly not true, he found, of the comet of 1577. The perfect motion had been impugned.

Quarrelling in his last years with the Danish king, Tycho transferred himself and his instruments to the protection of that great patron of science and pseudo-science, the Holy Roman Emperor Rudolph II. Also under that patronage was a very young man, already a Copernican, whose collaboration with Tycho was destined to become a classic of scientific method. Kepler (1571–1630) took over the vast corpus of Tycho's observations, and supplied their needed complement of theory to furnish the second great landmark in modern astronomy, the empirical laws of the planetary orbits.† It was his great service to compare the observations with theory after theory of the shape, size, and periodic time of the orbits, until he found the ones which fitted :—

1. (1609) The planetary orbits are ellipses with the sun in one focus.
2. (1609) The line joining a planet to the sun sweeps out equal areas in equal times.
3. (1619) The squares of the periods of the planets are as the cubes of their mean distances from the sun.

And yet in Kepler's work we find a forcible reminder of how close to mere number-mysticism were even the greatest men of that age. Casting about for the pattern on which the Creator might have worked, Kepler conceived the notion, of which he remained enamoured all his life, that the planetary spheres were successively (proceeding outwards) the circumscribed and inscribed spheres of the five regular polyhedra : the octahedron, icosahedron, dodecahedron, tetrahedron, and cube. Mediaeval ideas already mentioned will be recognised in this theory. It was largely in its service that Kepler put himself to his heroic labours. And when the agreement was only moderate, he developed a further doctrine of the same kind, the " music of the spheres," by which he related the orbits to that other Pythagorean mystery, the numbers of the five notes of music.

* Ultimately (in 1587) he published one of his own. Like much professional work of its epoch, it used the convenience of the Copernican idea while denying it " reality." It long rivalled that of Copernicus.

† He was employed by the Emperor in the compilation of the " Rudolfine " tables (1627) intended to supersede the Prutenic tables.

Yet, while he held to this pre-physical outlook by preference, he also, as we have said, felt the need of physical ideas. He felt that because the planets did not move as they should—uniformly in circles—forces were needed to account for their motion. It was the same contention as Newton's later, but for Newton the norm had become the straight line. We shall pursue this contrast in later sections. Here we only observe that had circles proved mathematically as convenient as straight lines we could not now quarrel with Kepler on this point; * but that *then*, the matter was far from being seen as a mere balance of convenience. Further, if the circles had not been brusquely cleared away, the other baffling perplexities of the problem, described in our later chapters, would almost certainly never have been cleared up.

What, then, were the forces which Kepler imagined; for he had no proofs? In the first place, as mentioned in the last chapter, he imagined "moving spirits" in the planets. Secondly, he imagined a force emanating from the sun which he traced, in part, to magnetism. Magnetism, which was coming to the front at that time, had interested him before he read the "De Magnete" (1600) of Gilbert (1540–1603). Even in Gilbert's mind the subject was associated with Copernican theory, and not only with navigation. Gilbert revived the results obtained in the 13th century by Peter Peregrinus of Picardy. Peregrinus had spoken of what we should call "poles" of strongest attraction which aligned and attracted needles; and he had distinguished these poles by the crucial property that like ones repel, and unlike ones attract, each other. To explain the directedness of solitary lodestones he had stated that the *heavens* have poles; and it was by asserting instead that it is the *earth* which is the magnet in the case, that Gilbert conceived himself to be supporting Copernicus.† It is perhaps needless to add that even in Gilbert's book the whole subject was mixed up with mere exercises of the imagination.

We have now surveyed the progress of astronomy (that is, practically of exact science itself) so far as this was possible, first, without clear dynamical and physical concepts; second, without the telescope's spectacular discoveries; third, perhaps, without the increased facility in computation obtained about 1620 by the rise of logarithms. The empirical laws of planetary motion had been discovered. It remained for Galileo to bring the subject to the attention of the world at large and make general and unmistakable the break with the old orthodoxy.

* Indeed, for computational purposes, circles are in a sense used to this day, in that trigonometrical functions are used for the successive approximations of the exact theory.

† G. Mercator before him, in 1546, had stated that the earth has a magnetic pole, but this was in a letter unpublished until 1869. Dip seems to have been known in the 16th century.

CHAPTER IV

MATHEMATICS BEFORE THE CALCULUS

MATHEMATICS, like other science, came to Europe chiefly through the Arabs, who had had it from Babylonian, Indian, and Greek sources. The Greeks themselves had derived its humbler aspects—geometry for surveying and architecture, arithmetic for the counting house—chiefly from Babylonia and Egypt. The central idea of "pure" mathematics, that of making the subject an abstract one, with general theorems proved from axioms, seems to have been chiefly their own; though fresh discoveries about the earlier civilisations constantly tend to whittle down such assertions. It was pure mathematics which was the main achievement of Greek science. We must briefly describe it, since otherwise the European development cannot be understood.

Early thinkers do not seem to have begun with geometry only, or with any single branch of mathematics, but with figures and numbers together.* The most conspicuous features of, say, a cube (perhaps of stone in the builder's yard) are not more geometrical than numerical. A cube has 8 angles, 6 faces, 12 edges. Now, further, $\frac{1}{8}=\frac{1}{2}(\frac{1}{6}+\frac{1}{12})$, which is the formula for what the Greeks called the harmonic mean, from its connection with music. Thus, arithmetic, geometry, and music are linked, a procedure typical of early science. When men, long before Pythagoras, found that certain sets of three integers like 3, 4, and 5, having the relation $3^2+4^2=5^2$, gave right-angled triangles, what we should call number-theory and geometry advanced together. When they drew a diagram , we should say that geometry and algebra advanced together.† The rich confusion or matrix was only gradually split up, chiefly by an enormous development of geometry. The lines on which this was done will serve to introduce the rough

* The Chinese and Japanese seem to have followed the same line, reaching very early the idea of a *spatial array of numbers*, such as the magic square. This gave them Pascal's triangle as early as 1300, and the idea of a determinant as early as Leibniz had it in Europe (1693). A simple illustration of their nicety of symbolic taste is their writing $a+b+c$ in the form of a triangle, $c\ {}^{a}\ b$, since in $a+b+c$ the symmetry of the letters is lost, b being differently placed from the other two. Magic squares, like many other problems in the theory of numbers, so easy to state, so hard to solve, have always exerted a strong fascination, and drawn men towards number-mysticism.

† This geometric algebra, a geometric mode of representation of $a^2+2ab+b^2=(a+b)^2$, is as typical of them as Descartes' algebraic geometry (see Chap. VI) is of us.

division of the post-Greek development adopted in this chapter. First came the " quadrivium " :

(1) *Arithmetic.* This meant theory of numbers, not practical calculation. The latter was a separate subject, and was called *Logistic.* It was hardly thought worthy of an educated man.

(2) *Geometry.* By this was meant practically *plane* geometry as opposed to

(3) *Sphaeric*, which was the geometry of, or rather on, the sphere. This was very largely the mathematics of astronomy.

(4) *Music*, by which was meant chiefly the Pythagorean numerical aspects of the subject, already instanced. Its presence in mathematics is of course only an aspect of the comparative Greek lack of physics, and consequently of physiology of the senses.

In addition, there was *Geodesy*, which was what we should call measurement and surveying ; and *Catoptric*, which was the theory of mirrors. The equality of angles of incidence and reflection being the main piece of knowledge here, the subject was practically a branch of geometry ; as was *Scenographic*, i.e. scene-painting, which would now have been associated with perspective. We shall not add anything, on these two subjects, to what has been said in the introduction.

The quadrivium survived for instructional purposes into the Middle Ages, but its " geometry " was often degraded into geodesy (or even geography) together with certain propositions from Euclid without their proofs. Its " arithmetic " and " sphaeric " became largely the theory of the sun-dial, the calendar, or the " houses " of astrology. " Music " became a series of rules by which church music was more and more strangled as the Renaissance approached. It was not until the 16th century, when both the theory of numbers and music proper freed themselves from this connection, that either could begin to progress once more. Nevertheless, we shall see in Chapter VI further gains of mathematics, and also of physics and physiology, from the connection with music. The indirect connection of the subjects, in the musical passion of many mathematicians and in certain similarities of the types of mind, is perennial. " Logistic," that is, computation, on the one hand, and pure mathematics on the other, will form the main division which we adopt for convenience in this chapter in tracing mathematics up to 1600–50.

For our purpose it is instructive to note briefly how one of the strongest characteristics of the Greeks, love of rigour, prevented them from developing on the lines which we feel to be the most profitable. For merely to say that they failed to discover symbols—had some algebraic knowledge but only in geometric shape—is to miss much of the point. What seems to have happened is that " Pythagoras' " theorem introduced his followers to lengths (such as what we write $\sqrt{2}$), which they could not express in the positive integers to which they had such a devotion. As among moderns, *problems* (such as this) were a main spur to the development of new points of view ; and the re-definition of ratio by Eudoxus (*c.* 408–355 B.C.) to enable reasoning to be extended to these " irrational " numbers was only a symptom of the new aim to which this problem led them. For they began to leave aside measurement and computation (occupations, after all, more suited to slaves than to cultivated men !) and to concentrate on the relations of the lines, points, and so on, in the abstract. This procedure—only consummated in the 19th century, if then—gave them strength in geometry, but it not only made algebraic symbolism seem unnecessary until their flower was long past but it inclined them to the view that *nothing valid* lay outside the closed circle of propositions which Euclid systematised. Much of what passed for mathematics in Europe until very lately would have struck them as profoundly untrustworthy.

There was here a point of view widespread among Greek thinkers quite outside mathematics, the view that science is not essentially progressive, but is part of something finished and complete : philosophy, or the education of a cultivated gentleman. In tune with this, the Greeks demanded of mathematics not only logic but a beauty of form which Europe did not regain until the 18th century. This, and that mystical-realist Pythagorean tinge which only their most advanced thinking quite left behind, ruled that they should often be interested rather in all the " harmonies " or properties of the regular figures or of certain especially " perfect " numbers, than in the most general figures or numbers which possessed any one of these properties.

Lacking our symbols they lacked system : much that had appeared detached to Babylonians appeared to the Greeks as part of the integrated whole of their geometry, but much that now appears part of the integrated whole of analysis appeared to them disconnected. Thus, their points of view often have, even for us, a refreshing variety about them, both in geometry and in the theory of numbers, reminding us how easy it is for words like " generality " to take on a narrow, conventional meaning. The Pythagorean definition, for instance, of an even number is, one which can be divided by one operation into parts greatest in size (halves) and least in number (two) ; while an odd number cannot be so divided. Again, an odd number cannot be divided into parts similar, nor an even one into parts dissimilar, in respect of oddness or evenness (two itself excepted).

The Greek Geminus observed that there are only three lines, the straight line, the circle, and the cylindrical helix, and only two surfaces, the plane and the sphere, any part of which fits any other part. And Plato defined a straight line as one such that when it is looked at endways the middle of it covers the ends. This definition is unsatisfactory as it stands because of its implication of the properties of light, but it is not necessarily incapable of interesting development. Finally, though it would be wrong to attribute to the Greeks modern projective and topological points of view, they developed the theory of conics in precisely the shape most suited to the former ; while much of their geometric-algebra, whether algebra proper (continuous quantity) or number-theory (discrete quantity) was essentially topological.* Thus a cube continues to have 8 angles, 6 faces, 12 edges, however much the sides are bent or stretched.

* In the more elementary, combinatory sense.

We have spoken of the re-definition of ratio to make it possible to reason about irrationals ; but the idea of infinity took also, even at an early stage of Greek mathematics, a much more awkward form than that, namely continuity. Zeno's puzzles were designed as a serious criticism of the Pythagorean view of this question. One was the " disproof " of the possibility of continuity of motion by considering Achilles in pursuit of a tortoise. Achilles can never get ahead of the tortoise, since by the time that he reaches where the tortoise was when he (Achilles) started, the tortoise will always have moved on. Nor was Zeno without effect. Not only did he put an end to the tendency to regard lines as made up of points, but he was a considerable factor in the general caution of the Greeks as to this dangerous but immensely profitable field. It was only at the height of their development that a very few men like the great Archimedes took a further step (*c.* 250 B.C.). We defer the discussion of this step to Chapter VI, where we consider the corresponding European development. In both regions, practical applications suggested that illegitimate reference to motion or flow which had been condemned by Zeno, and therefore prompted the dismissal of Zeno as an idle sophist. But symbolism or its absence had importance here too. In Europe, the temptation to let results run away with rigour was much greater, because in the interval symbolism, with its infinite suggestions for formal development, had come into being.

We have said that at the tail end of their history the Greeks did introduce symbolism. The great Diophantos (*c.* A.D. 250), working in the theory of numbers, used a letter to represent an unknown ; but it was typical of his limitations that he did not extend the practice to two or more unknowns. His innumerable results remain individual, and lack the systematisation brought by symbolism. They were due to his amazing ingenuity rather than to this new method. In fact, he developed so few of the features of a really suggestive symbolism that it will be best to leave this subject to the next section, when the European development is considered. And yet no great advance was made on his work in number-theory itself until Bachet's French translation of him in 1621 gave Fermat the lead.

In the not wholly unrelated subject of permutations and combinations, Greeks seem to have been less interested. Early in the Christian era, the matter is mentioned in the Hebrew Cabala. By 1321 it was known that (as we should write) ${}_nC_r=\frac{n\,!}{\overline{n-r}\,!\,r\,!}$, while the Arabs, before that, knew " Pascal's " triangle for ${}_nC_r$ in its aspect as the coefficient of $x^r y^{n-r}$ in $(x+y)^n$, n integral. That is, they had the array

$$\begin{array}{ccccccc} & & 1 & & 1 & & \\ & 1 & & 2 & & 1 & \\ 1 & & 3 & & 3 & & 1 \end{array} \quad \text{etc.}$$

Another connected subject, that of probability, had been discussed for simple cases in the 15th century.* The problem of the division of

* Unlike the moderns, the ancients do not seem to have seen a new branch of mathematics in such practices as insurance.

the stakes after an unfinished game was reached in Pacioli's "Suma" (1494), but it was to remain unsolved until the time of Pascal and Fermat, although with Cardan and Galileo the subject had already acquired its connection with the "applied science" of gambling. But we turn to the development of algebra.

Pure mathematics, especially algebra

In Babylon, rules were known for solving linear and quadratic equations; these rules took a geometrical form among the Greeks, a much more symbolical one among the Hindus. But in the absence of symbols the matter did not easily present itself to these peoples as it does to us. In the additional absence of a clear conception of negative and fractional quantity, two linear equations are not always easily identified as of the same class of problems. The difficulty is even greater with quadratics. The chief early method for linear equations, "false position," is as different from ours as two methods for so simple an operation well can be. Two random suggestions were tried, and the root was then distant from each an amount proportional to the divergence of the values thus obtained, from the required value.

It was the Arab trade with Italy which brought Europe its first hints of algebra. Leonardo of Pisa (about 1200) is an early name, but we shall not pause upon the Middle Ages except to remark that they realised the next algebraic problem to be solved. This was the cubic equation, which the Arab mathematicians believed to be, in general, insoluble. Even to define the problem was in itself an achievement; for it is far from easy without symbols to see that *degree* is the crucial thing about equations, and that it can be treated as a number just like any other. It was such symbols—the practice of writing letters or other signs for numbers and operations—which were the decisive contribution of the 16th century. To this period we now turn.

We are accustomed to alchemist—and astrologer—scientists, to academic and commercial scientists, to amateur, even to noble, scientists. But actually villainous scientists are novel. Yet we meet them in these Italian Renaissance mathematicians. As with poets and musicians (those other seekers of beauty in ideal realms), the first step was painful and violent. The great French poet Villon was King of Vagabonds in Paris. Our own Marlowe was stabbed in a tavern brawl. The mathematician, Cardan (1501–76) was a scandal even in the 16th century. He had one son guilty of wife-murder, he personally cropped the ears of another, he cast the horoscope of Christ, and in general divided his time between intensive study and intensive debauchery. And yet he thought out means to teach the deaf and dumb to read and write.

Another adventurer of vile temper and morals was Ferraro (1522–65), who started as Cardan's servant, and was (probably) poisoned by his own sister. A third, Fontana ("Tartaglia," the stammerer, 1500–57), who earned his living by teaching, and by writing arithmetics, was among the first scientific military engineers. Military engineering was then the dominant branch of the subject except in mining districts.

In 1505 Scipio Ferro of Bologna communicated to a pupil a solution of one form of cubic equation. Later, but specially 1535–41, Tartaglia gave solutions of all the various forms proposed. His solution seems to have been stolen by Cardan, who added (1545) a discussion of negative and " imaginary " roots. This latter had no immediate effect, for, though imaginaries had been discussed by the ancients, the decisive advances in understanding them were still far in the future (see Chap. XII).

Diophantos had known the rule of negative signs in multiplication, but he had never conceived negative *quantity*, only expressions like $(a-b)$ so long as $a>b$. Cardan's practice, and that of his age, was very similar. All missed the decisive simplification which could be brought into algebraic formulae by using independent negative quantity. Thus Cardan gave separate solutions for the equations which we should write $x^3+px=q$ and $x^3=px+q$, because his symbolism did not enable him to view them as the same.

After the cubic, the biquadratic. Here Cardan and Tartaglia failed, but Ferraro succeeded. He solved first a particular case proposed, and then the general case.

The quick succession of these two triumphs made it all the more surprising that for a quarter of a millennium no corresponding insight was gained into the quintic. It was long before mathematicians reconciled themselves to defeat on this point, and recognised that they must immensely widen and deepen their knowledge of quantity in general before they could understand the quintic. In fact, the widening and deepening proved, as is often the case in mathematics, so thrilling that it quite obscured its original provocation.

Symbolism was advancing apace by the time of the great French mathematician Viète (1540–1603), who was also magistrate, lawyer, and privy counsellor. Viète was one of the first to recognise the desirability of having a different set of letters for unknowns and for constants in literal expressions. Let us note down some of the more definite stages in the growth of symbolism, constructing a chronological series of ways of putting what we should write $1+3x+3x^2+x^3$. First of all $1+3N+3Q+C$, where no attempt is made to represent the *relationship* of N, Q, and C. Then, $1+3x+3x$ quadratus $+x$ cubus, Viète's improvement. The *numerical* character of the relationship is still obscured. Next, 1⓪+3①+3②+③, Stevinus, 1585. This fully expresses the numerical fact, but would be awkward if there was more than one unknown. It suggested to Stevinus—and such suggestiveness is the essential service of symbolism—a very significant question. This was, why only integral indices? He suggested fractional ones.* Another, $1+3x+3xx+xxx$ (Harriot, who lived 1560–1621), was not as suggestive as that of Stevinus. Finally, $1+3x+3x^2+x^3$, Descartes, 1637.† This is like ours, but was used only for integral values of the index.

* Occasionally used earlier by Oresme. Stevinus' notation was connected with his notation for decimal fractions (see next section).

† To be strictly accurate, he wrote x^2 as xx. This he did because x^2 was no economy, two signs being needed in each case. He thus sacrificed one object of symbolism, the suggestiveness of uniformity, to another, succinctness. Posterity has not followed him.

By 1659, the English mathematician, Wallis, was suggesting (in words, not symbols) meanings for negative and fractional indices ($a^{-n}=\frac{1}{a^n}$, $a^{\frac{1}{n}}=n^{th}$ root of a). In 1668–9, Newton took the step of considering, generally, *any* index. Newton's binomial expansion, with its coefficient $\frac{n!}{\overline{n-r}!\,r!}$ (our notation) could be given an interpretation when n was not an integer or positive. Here Newton greatly profited from symbolism. On the other hand we shall see in Chapter VI that his notation for the calculus was not as suggestive as that of Leibniz. We shall also see what an influence this fact had on English mathematics.

The story of mathematics at this time is made complex by the lack of swift and regular scientific communication. We have seen that Harriot's notation was not as good as Stevinus', though it was later. It was the same with negative quantities: Girard (1595–1632) had our modern idea of $+$ and $-$ as directions along a line, but his work had little influence. And yet in the first half of the 17th century, the business of standardising what we should now call ordinary school arithmetic and algebra went on rapidly, notable names in this connection being Harriot, Oughtred (1574–1660), Bürgi, Descartes, and Napier.

Before leaving symbols, we note that in the 17th century they made it possible for number theory to shake off mysticism and enter on a period of spasmodic advance. The subject ought perhaps to be reserved for the 17th century chapter, but its isolated nature makes its placement largely a matter of choice. Its hosts of individual results make its early stages unsuited for treatment in a book like this, and our few notes on it are therefore placed here.

Number-theory benefited much less than "continuous" theories from symbolism. It is an exceedingly difficult subject, and before the great work of Gauss, about 1800 (see Chap. XII), it had hardly even classified its problems. It might even be called, up to that date, the empirical part of mathematics; for numerical experiment (always its charm for amateurs) remained one of its chief resources, and unprovable theorems one of its fascinations. Its chief early worker, Fermat (1601–65) gave, often without proof, a number of brilliant new results. Among these were:

$a^{m-1}-1$ is divisible by m, if m is prime and a is prime to it.

$2^{2^n}+1$ is always prime (since disproved, e.g. for $n=5$)

and the "great" theorem, or rather conjecture:

$$x^n+y^n=z^n \text{ in integers, } n>2.$$

This last has never been proved, and indeed the proof or disproof of Fermat's results has been one main preoccupation of workers in the theory of numbers ever since his time.

We must, however, return to earlier periods, and trace the history of computation—of methods designed to deal with individual numbers, not number in general; two things which, as the Greeks were forced to conclude, are very different. It might seem that the former, as the simpler, should have come first. As, however, our section on it ends with logarithms, the most advanced idea yet considered, we have placed it last.

COMPUTATION

We think of calculating devices as a recent innovation on paper calculation with ciphers; and so, of course, as regards complex automatic machines, they are. But in the simple form of numbered rods and abaci, or of sand tables essentially similar to abaci in method, they long preceded cipher calculation in Europe. The position was analogous to that with symbols for quantity in general: ciphers were used for record, not as the device, or algorithm, by whose own laws of combination calculation was carried out. This is very evident in some of the early scales of notation, and particularly in the Roman numerals with which Europe came to be burdened. The use in these of the subtractive, as well as the additive, principle (IV as well as VI), together with the use of several radixes (V as well as X, C, L), made them highly non-algorithmic, that is, not admitting of simple manipulative rules.

Counting on the hands, it is natural that 5, 10, and 20 should always have predominated among radixes.* The Babylonians used, for different purposes, both 10 and 12, the number 60 thus acquiring a special significance for them. Seven, eleven, and other radixes have also been used.

The use of calculating devices often has this disadvantage: we in effect rub out each line of working in going on to the next. Hence, perhaps, the development by the Arabs soon after 800 (perhaps earlier) of checks such as "casting out the nines," the importance assigned to which now puzzles beginners in arithmetic.

No doubt the ever-increasing trade of the Italian and other mediaeval cities made Europe look out for a more convenient notation for numbers than the cumbrous Roman with which it emerged from the Dark Ages. It was the same impulse which led men to develop cursive script for ordinary use. At all events, after about the year 1000 they slowly began to accept the Arabic-Indian † numerals. These had by then been completed by a sign for zero, a vital addition where their use for calculation was concerned. These had only one radix, ten. This simplicity made it possible for the writers of commercial arithmetics in Florence (*c.* 14th century onwards) to develop rules for cipher calculation. At the same time, perhaps, the coming of a cheaper writing material (paper) made it less necessary to reserve writing for record only.‡

These changes were long only local. Indeed, the abacus is still

* Rules were long current for actually calculating on the fingers.

† Or perhaps Greek (see *Isis*, Vol. 19, 1933, p. 182).

‡ It is not entirely fanciful to point out that the coming of paper calculation brought geometry into algebra in a new sense long before Descartes brought algebra into geometry. Obliterating our last step, as on calculating rods, strings out our operations in one dimension only. Paper gives us two. We have only to reflect on simultaneous equations to realise how many symmetries, and so developments, these two dimensions enable us to see. Sense of form, one pole of the mathematical mind, hitherto confined to geometry, can enter analysis.

used in the East, and was not generally abandoned in England until late in the 17th century. Arabic numerals were long regarded, by dignified official and religious bodies, as monstrous innovations.

The Florentines brought in multiplication tables, though at first only up to 5×5 ; while in the 16th century further commercial centres, Holland, Frankfurt, Lyons, began to take up cipher arithmetic. Division, however, was still apt to be a skilled job. Theoretical interest, meanwhile, moved on to harder regions—the notation, and especially the decimal notation, for fractions. Sexagesimal fractions $\left(\frac{a}{60}+\frac{b}{60^2}+\frac{c}{60^3} \ldots\right)$, from ancient Babylon were then still usual.* The first European to begin to see decimals as a means of extending whole-number rules to fractions, was C. Rudolff of Augsburg, 1530. Nor did he get far. Even Stevinus, to whom the definite theory is due (1585) did not use a very good notation.† It was largely Briggs (1617), Kepler, and the other logarithm-table-makers, who standardised the methods. For example, the practice of writing in the denominators only disappeared slowly, and the full manipulative neatness of the method was long in being grasped. The *point* did not become universal until the 18th century.

We now turn to logarithms, the introduction of which so vastly facilitated computation at a time when the progress of exact science was peculiarly bound up with this. So subtle an idea could not have occurred to minds not filled with the best theory as well as with the best computing practice of the time, but it was the practice rather than the theory of logarithms which advanced rapidly during the first half of the 17th century.

No doubt the wish to reduce multiplication and division to the much simpler and less risky processes of addition and subtraction must have occurred hundreds of times to the laborious table-computers of the 15th and 16th centuries. A partial means existed in what we should now call a 1 : 1 correspondence ‡ of the arithmetic and geometric series known to the ancients, for instance :

$$\begin{array}{l} 1 \cdot 2 \cdot 3 \cdot 4 \cdot 5 \cdots \\ 3 \cdot 9 \cdot 27 \cdot 81 \cdot 243 \cdots \end{array}$$

* The trigonometrical table-makers avoiding them where possible by choosing very large radii.

† He suggested decimal weights and measures.

‡ This idea of 1 : 1 correspondence has great importance in mathematical history (see Chap. XIII). One unadvertised practical use of it, in which mathematicians have always been involved, is codes and ciphers. Viète solved Spanish codes for the French king, Wallis solved Royalist codes for the Roundheads, Thomas Young (*c.* 1800) deciphered the Egyptian hieroglyphics. Sir Alfred Ewing headed a naval decoding department in the War of 1914–18. Cardan made a Braille-like " code " for the deaf and dumb, for whom Wallis also worked. This side-line of mathematicians is not surprising, for symbolism itself is a code, and deciphering is an exercise of the reason. Moreover, codes subserve that secrecy of which we have spoken in the introduction. Leonardo's note-books were in cipher ; Dürer is said to have concealed a statement of the canon of proportion in his picture " Melancolia." There was similar secrecy right down to Newton's time.

For addition in the first corresponds to multiplication in the second. As early as 1544 Stifel sensed this; and on its basis the Swiss Bürgi (1552–1632), court watch-maker to the Holy Roman Emperor Rudolph II, actually made (1611, published 1620) a table of antilogarithms of integers. Obviously, however, it is to discrete rather than continuous quantity that such an idea applies; * and though any table necessarily refers to a discrete series of values, it was his introduction of continuity, which marks Napier (1550–1617) as the real originator of logarithms. Starting from the above correspondence, he reached the idea of the correspondence of the simultaneous motions of two points, one proceeding uniformly, the other slowing down so that its velocity is always proportional to the distance x which it has still to go before reaching a certain point on its path. If y is the distance meanwhile travelled by the other point, $y = \log x$.†

He had communicated a summary of his idea to Tycho Brahe in 1594, for it was to astronomers that he hoped that it would be useful (he published it in 1614). So much, indeed, was this connection uppermost in the minds of computers of that time that Bürgi and Napier calculated the logarithms not of numbers but of the trigonometrical ratios already tabulated. We cannot give Napier's methods (published 1619) but they were immensely laborious. It was Briggs (1556–1630), a strong advocate of decimals for all purposes, who brought in the base ten. Briggs was the first to make an extensive table of the logarithms of numbers (1–20,000 and 90,000–100,000). The Dutchman Vlacq filled in the gap 20–90,000. The influence of Kepler, who used Bürgi's tables and who had himself been convinced by Briggs, did much to spread the use of logarithms. These early tables were not without errors, but the pervasiveness and finality of much of the work of these early table-computers is a striking thought.

Before leaving computation, we may note that the 17th century saw developments in instrumental methods also. Napier's "bones" (1617) were not logarithmic scales, but merely the last refinements of the old calculating rods. In 1620 Gunter made a single logarithmic scale which, with a pair of compasses, could be used to multiply. In 1622 Oughtred brought out the much more convenient slide-rule, with two logarithmic scales sliding over one another. The practical connection was with surveyor's practice. But true automatic machines avoid the element of error involved in logarithmic scales, and are essentially "rods" of the older kind, provided with automatic means of moving up the next line of digits when the last has run its gamut. There had been suggestions before, but Pascal's (1642) was the first actual construction. In the early 1670's Leibniz, and also Morland, followed.

* *We* at once see logarithms as indices (e.g. in the above geometric series). But this is precisely what these early workers, lacking the full indicial notation, failed to do. Euler (1748) was the first to regard logarithms as indices. Their interpretation as the area under an hyperbola came earlier, in 1668 (Mercator).

† This procedure is, of course, open to many of Zeno's criticisms. That is why its usefulness in computing was so important: it became necessary to reopen the question (see Chap. VI).

PLATE IV

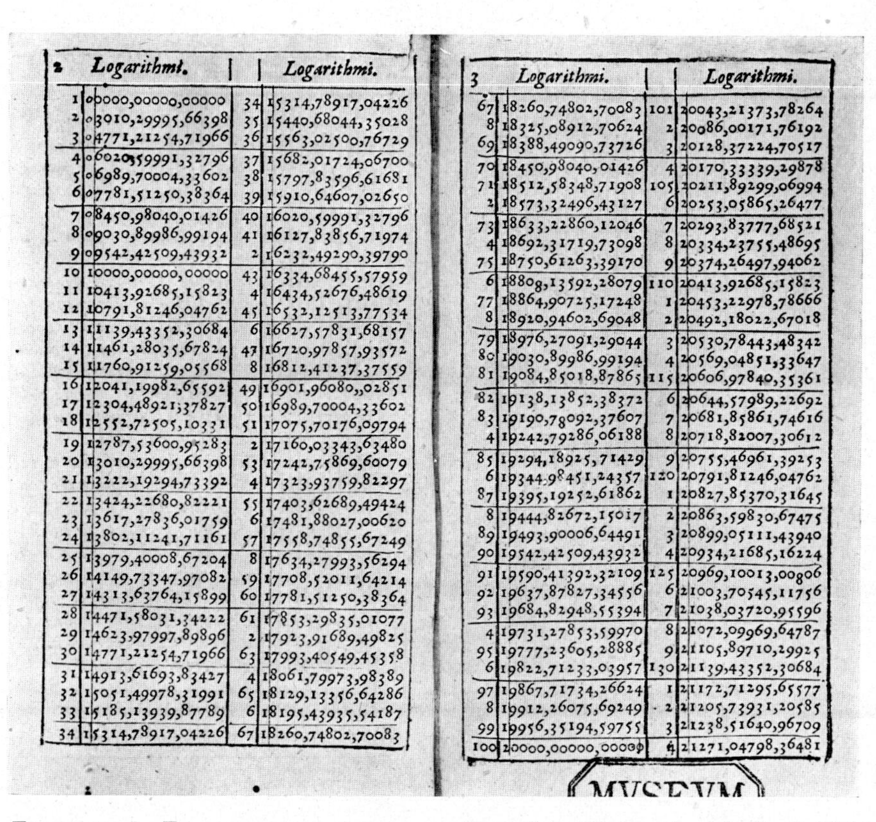

2	Logarithmi.		Logarithmi.
1	00000,00000,00000	34	15314,78917,04226
2	03010,29995,66398	35	15440,68044,35028
3	04771,21254,71966	36	15563,02500,76729
4	06020,59991,32796	37	15682,01724,06700
5	06989,70004,33602	38	15797,83596,61681
6	07781,51250,38364	39	15910,64607,02650
7	08450,98040,01426	40	16020,59991,32796
8	09030,89986,99194	41	16127,83856,71974
9	09542,42509,43932	2	16232,49290,39790
10	10000,00000,00000	43	16334,68455,57959
11	10413,92685,15823	4	16434,52676,48619
12	10791,81246,04762	45	16532,12513,77534
13	11139,43352,30684	6	16627,57831,68157
14	11461,28035,67824	47	16720,97857,93572
15	11760,91259,05568	8	16812,41237,37559
16	12041,19982,65592	49	16901,96080,,02851
17	12304,48921;37827	50	16989,70004,33602
18	12552,72505,10331	51	17075,70176,09794
19	12787,53600,95283	2	17160,03343,63480
20	13010,29995,66398	53	17242,75869,60079
21	13222,19294,73392	4	17323,93759,82297
22	13424,22680,82221	55	17403,62689,49424
23	13617,27836,01759	6	17481,88027,00620
24	13802,11241,71161	57	17558,74855,67249
25	13979,40008,67204	8	17634,27993,56294
26	14149,73347,97082	59	17708,52011,64214
27	14313,63764,15899	60	17781,51250,38364
28	14471,58031,34222	61	17853,29835,01077
29	14623,97997,89896	2	17923,91689,49825
30	14771,21254,71966	63	17993,40549,45358
31	14913,61693,83427	4	18061,79973,98389
32	15051,49978,31991	65	18129,13356,64286
33	15185,13939,87789	6	18195,43935,54187
34	15314,78917,04226	67	18260,74802,70083

3	Logarithmi.		Logarithmi.
67	18260,74802,70083	101	20043,21373,78264
8	18325,08912,70624	2	20086,00171,76192
69	18388,49090,73726	3	20128,37224,70517
70	18450,98040,01426	4	20170,33339,29878
71	18512,58348,71908	105	20211,89299,06994
2	18573,32496,43127	6	20253,05865,26477
73	18633,22860,12046	7	20293,83777,68521
4	18692,31719,73098	8	20334,23755,48695
75	18750,61263,39170	9	20374,26497,94062
6	18808,13592,28079	110	20413,92685,15823
77	18864,90725,17248	1	20453,22978,78666
8	18920,94602,69048	2	20492,18022,67018
79	18976,27091,29044	3	20530,78443,48342
80	19030,89986,99194	4	20569,04851,33647
81	19084,85018,87865	115	20606,97840,35361
82	19138,13852,38372	6	20644,57989,22692
83	19190,78092,37607	7	20681,85861,74616
4	19242,79286,06188	8	20718,82007,30612
85	19294,18925,71429	9	20755,46961,39253
6	19344,98451,24357	120	20791,81246,04762
87	19395,19252,61862	1	20827,85370,31645
8	19444,82672,15017	2	20863,59830,67475
89	19493,90006,64491	3	20899,05111,43940
90	19542,42509,43932	4	20934,21685,16224
91	19590,41392,32109	125	20969,10013,00806
92	19637,87827,34556	6	21003,70545,11756
93	19684,82948,55394	7	21038,03720,95596
4	19731,27853,59970	8	21072,09969,64787
95	19777,23605,28885	9	21105,89710,29925
6	19822,71233,03957	130	21139,43352,30684
97	19867,71734,26624	1	21172,71295,65577
8	19912,26075,69249	2	21205,73931,20585
99	19956,35194,59755	3	21238,51640,96709
100	20000,00000,00000	4	21271,04798,36481

FACSIMILE I. EARLY LOGARITHM TABLES. *Pages from the rare tract in the British Museum,* BRIGGS, H., *Logarithmorum Chilias Prima,* 1617. Errors occur in last figures of logarithms of 37,118,131. *See Logarithmetica Britannica, Pt. IX.*

PLATE V

NAPIER OF MERCHISTON. Mathematician. *From a rare print by Delaram, in the British Museum.*

THE SEVENTEENTH CENTURY

The 17th century was a time of great activity in many cultural realms. A man born in English court circles towards the end of Elizabeth's reign might possibly have known not only Harvey, but Shakespeare and later Milton. If he had travelled the courts of Europe, as young nobles then did, or if he had been a religious refugee, he might have met, in Italy, Galileo, in the Netherlands, the Jewish philosopher Spinoza, the Frenchman Descartes (refugees both), and some of the greatest painters in human history. Proceeding late in life to France, he would have found a country which, in culture and population—in all but commerce—was the centre of Europe. There he might have met the German Leibniz and the Dutchman Huygens, like him engaged upon their travels. Returning, a very old man, he would have heard that his own country had not been idle. It had produced Newton.*

The scientific prominence of the Netherlands and of Scandinavia would not have surprised him. The Baltic trade was important, for timber, hemp, tar, and other marine stores of that century of growing overseas trade. The Baltic was also a corn-exporting region. After the fall of Spain at the end of the 16th century, the Netherlands were the trade centre of Europe for two thirds of a century. England's turn was not yet, but between the England and the Holland of the time there was a certain geographical unity. The observer, running his eye on a map up the English Channel into the North Sea, would have noted, standing opposite the Lowlands on the other side of the narrow sea, the then wealthy region of East Anglia, marked by the fenlander Newton and the fenland university of Cambridge. Other sources of wealth before the time of coal and iron were the West Country sheep runs and the Cornish tin mines. The former gave us Glisson, Wren, Willis, and Sydenham, the latter, Lower and Mayow, and also Newcomen and Savery.

The idea of a nation would not have been as dominant with him as with us. When the glory of the Renaissance ended, Europe was devastated by religious dissension which everywhere cut right across national boundaries. Even centralised France was a tangle of local jurisdictions and conflicting loyalties. Order was terribly needed, and it was to be the special business of the 17th century to supply it. Much was subordinate to war. Even mathematics contributed cryptography, range-finding, and the geometry of fortification. One of the earliest aims in the new science of statistics † was to find the proportion of a population able to bear arms.

But the pace was too hot to last, and, among other forces of order, science raised its head in a new and corporate sense. The first learned societies were literary, not scientific (descended from the Florentine academies of 1433 and 1442). There was good reason for this. The

* J. Needham has pointed out to the writer that John Aubrey almost fills the bill.

† New as a science. Not new as an occasional, and often faulty, practice in state-craft.

MAP II

BIRTHPLACES OF SCIENTISTS

Please consult Explanation on p. 10 before studying the Map

(*a*) *Holland and East Anglia, chiefly in the 17th century.*

Number on Map	Name	Birthplace *
1	Barrow	London
2	Boerhaave	Leyden
3	Caius	Norwich
4	Digges (father and son)	Barham
5	Frisius, G.	Dokkum
6	Gilbert	Colchester
7	Girard	Holland
8	Graaf, de	Schoonhaven
9	Groot, de	Holland
10	Gunter	Hertfordshire
11	Hales	Bekesbourne
12	Halley	London
13	Harvey	Folkestone
14	Helmont, Van	Brussels
15	Huygens	The Hague
16	Landen	Peterborough
17	L'Eclus	Holland
18	Leeuwenhoek	Delft
19	Linacre	Canterbury
20	Lippershey	Middleburg
21	L'Obel	Lille
22	Mercator	Holland
23	Newton	Woolsthorpe
24	Oughtred	Windsor
25	Ramée, la	Cuth
26	Ray	Braintree
27	Snell	Leyden
28	Stevinus	Bruges
29	Swammerdam	Amsterdam
30	Sylvius, F.	Leyden (work-place)
31	Sylvius, J.	Amiens
32	Vermuyden	Holland
33	Vesalius	Brussels
34	Vlacq	Gouda
35	Wallis	Ashford
36	—	Cambridge

(*b*) *The West Country, chiefly in the 17th century.*

Number on Map	Name	Birthplace *
1	Bacon, R.	Ilchester
2	Bradley	Gloucestershire
3	Glisson	Rampisham
4	Graunt	Hampshire or Hants family
5	Harriot	Oxford
6	Hooke	Isle of Wight
7	Lower	Bodmin
8	Mayow	Cornwall (family)
9	Petty	Hampshire or Hants family
10	Sydenham	Dorchester
11	Willis	Great Bedwin
12	Wren	Tisbury

As to (*a*) Descartes and Spinoza worked, but were not born, in this region.

As to (*b*) In the realm of technology, Newcomen and Savery were born in the West Country. Later on Jenner was born in Gloucestershire and Davy in Penzance.

* In a few cases, work-places, or the nearest larger place, have been substituted. Numbers within a circle on the map indicate that only the region, not the town or village, is given. Cambridge is numbered and inserted for reference, owing to its importance; though it was not the birthplace of any of the scientists mentioned.

MAP II

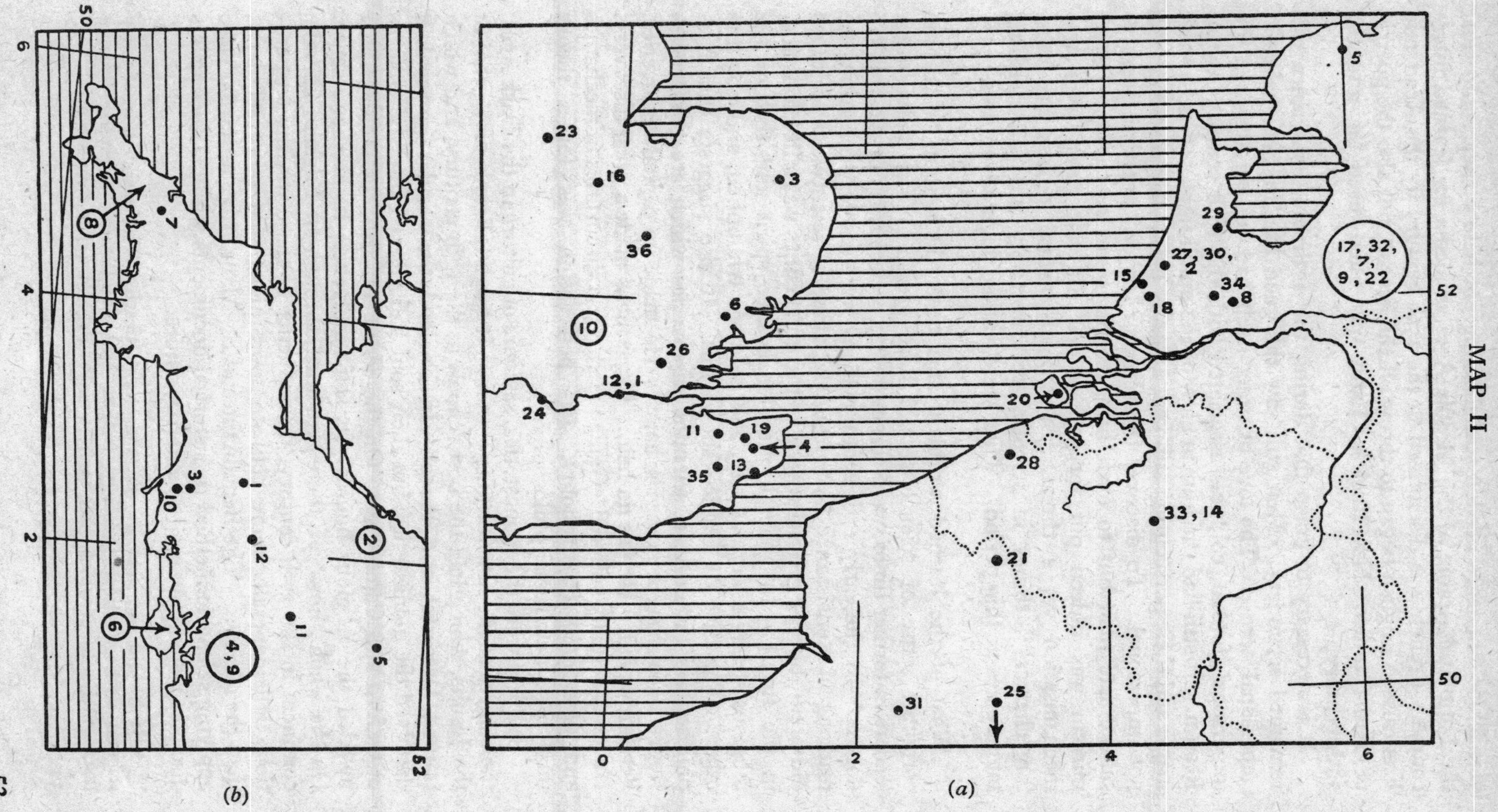

52
50
6
4
2
0
5
17, 32, 7, 9, 22
29
27, 30, 2
34
8
15
18
23
16
3
36
6
10
26
12, 1
24
20
11
19
4
35
13
28
33, 14
21
25
31
(a)
50
6
4
2
52
8
7
3
10
1
12
2
6
4, 9
11
5
(b)

vernacular languages needed straightening out before they could rival the Latin and Arabic of the Middle Ages as media for learned intercourse. French was disciplined in the 17th century by the Académie Française. English began to throw off the license, but also the poetry, of Tudor times. Nevertheless, Latin was often used far into the 18th century.

It is necessary to grasp that during the 17th century science still remained secondary. Men still saw the good life in personal, not impersonal, terms. The love of order itself had not yet acquired the impersonal, formal, coldness associated with the 18th century Age of Reason. It still had the human glow of the uphill fight. Witches and magic and their persecution still kept a deadly hold. Even science was not impersonal. To discover the reign of law did not then seem to subordinate man but to exalt him, its discoverer and user. Determinism was indeed put forward, but it did not rule. Nor was mechanisation yet a threat. Even mathematical tables, the most "mechanical" things in science, were certainly not soul-crushing monotonies at this period. They were, rather, high adventures.

What of the learned world at this time, the soil in which the corporate sense of science should have grown? The universities, leaders in former times, were committed to ecclesiastical or Aristotelian dogma. By the early 17th century, many were scarcely tolerating the teaching of Renaissance humanism, let alone of science. Exceptions did occur, but the general revival of the universities was still a long way in the future. In general, the Renaissance spirit and that of the religious struggle were at war, and now one triumphed, and now the other. The case of Galileo shows this. It also suggests that it is philosophies rather than scientific discoveries which are persecuted. Roger Bacon has a martyr's halo.* He made no capital discoveries. Peregrinus did. He has no halo. Copernicus made a great discovery. He never actually suffered. Galileo made great discoveries in mechanics and *observational* astronomy. His persecution was for a theory implying a new social philosophy.

Still less were the monasteries, saviours of culture in the Dark Ages, the centres of the new learning; though it was a religious body, that of the Jesuits, which had the best schools in the 17th century, and which trained Descartes and Pascal. In England the secular power had dissolved the monasteries, but their soul had entered into the colleges of Oxford and Cambridge. On the continent, also, the secular spirit invaded the religious houses—its self-indulgence, its aesthetic sense. On the whole, however, there remained, throughout England and the continent in the 17th century, the influence of these wealthy corporations, still generally more celibate, more pious, and more scholarly, than the average. The Benedictines of St. Maur (1632 onwards) were applying scientific method to historical documents while Galileo or his disciples were applying it to external nature.

* If he did not actually die a martyr's death. The association of early science with mystical theology, magic, and handicraft "mysteries" is a fascinating subject which such writers as W. Pagel are just opening up.

The real nurseries of science were the courts, the learned societies, and to some extent, also, the guilds of physicians and surgeons. In 1560, an "Academia Secretorum Naturae" had had a brief life in Naples before the Pope closed it on suspicion of the black arts. Then, in 1603, came the Accademia dei Lincei, which, again, was not permanent. In France the wealthy Peiresc (b. 1580) began acting as patron and liaison officer for the scattered scientists of the day.* In the second third of the century the friar Mersenne assumed a similar rôle. There were informal meetings in Paris of such men as Descartes (before he went to Holland), Desargues, Pascal's father, and Gassendi the philosopher and populariser. Mersenne also kept up a Europe-wide correspondence: with Galileo, Jean Rey the chemist, Hobbes the English philosopher, and others. For some five years prior to 1650 such men as Wallis, Boyle, Wren had been meeting fairly regularly at London or Oxford. In their activities the Royal Society was foreshadowed. In 1657 there was a transient Academy founded in Spain on the model of that at Naples; and in that year, also, some disciples of Galileo founded the Accademia del Cimento at Florence. Its single publication, the *Saggi* (1667), produced a great effect at the time. In these ways corporate self-consciousness was promoted; though all through the century, and after, individual exchange of letters retained a great importance.

Meanwhile, the kings of England and of France, like Rudolph earlier, had grown interested.† The Royal Society was founded in 1660 (chartered in 1662), the Académie des Sciences in 1666. Curiosity was a motive, at least in Charles II's case, as well as the hope of industrial advances. The latter was destined, in the main, to about a century's discouragement. Neither in France nor in England were the scientists purely speculative, but many technological problems are too difficult for the early stages of science. True to the derivation (see Chap. I) of the experimental method from the crafts, the new Royal Society undertook industrial and agricultural surveys of England, pointers to the industrial and agricultural revolutions of the 18th century. But the surveys soon languished. In 1665 began the great line of the *Philosophical Transactions* and also the transient French *Journal des Scavants*, first items of the now vast periodical literature of science. In 1672, Germany came in with Sturm's Collegium Curiosum, and in 1682 with the *Acta Eruditorum*, to which Leibniz contributed. In 1683 a short-lived society was founded in Dublin. Russia followed with the St. Petersburg Academy of Science in 1725, and Sweden at the instance of Linnaeus and others in 1739.

These societies had vicissitudes in their early days, and reflected the tempers of their times and nations as well as transcending them. By the end of the 17th century, the French Academy had developed that officialism which is the drawback of the high social esteem of intellect in France. Cartesian vortices became an orthodoxy. So did opposition to the calculus, so natural to the Latin mind, modern representa-

* For the extent of mutual knowledge or ignorance of the chief scientists of that day, see J. Pelseneer, *Isis*, Vol. 17, 1932, pp. 171–208.

† For the deeper social and economic causes, see works in the Bibliography.

tive of the logic of Greece. In England, the drawback was insularity. A worship of Newton, more devoted than intelligent, settled on us for a century. As for Germany, her day was not for a century yet.

Nationalism or earlier forms of localism are never, even by the impulse of science, more than just held in check. Readers of Picard's preface to the fine volumes of the history of science in France published under the name of Hanotaux, will be interested to learn that France holds the golden mean between the extremes of England and Germany. This is just the part for which we had cast ourselves ! So, doubtless, had the Germans. We cannot discuss this subject here ; but each nation's way of life evidently makes some of the needs of science easy to supply, others hard. The Frenchman has a clarity, the German a synthetic power and thoroughness, the Englishman an invention and practicality, which the others should admire rather than claim or imitate.

We turn now to the ideas, rather than the externals, of the 17th century atmosphere. Looking through the positivist eyes of Comte we may see each country reaching a climax in philosophy before it turns over to science. France's great philosopher is Descartes, England's is perhaps Bacon or Locke. Germany's climax in philosophy—Kant and Hegel—came over a century later, but again contemporaneously with her real start towards supremacy in science. It is this turn-over from philosophy to science which we have to study.

In all three countries its symptom was that the philosophers became critical of their own assumptions rather than constructive like the scholastics of the Middle Ages. For both Bacon and Descartes, the vital question was, by what method could sure knowledge be *founded*, not *built*. For both, experimental science as opposed to the now sterile scholasticism of the universities, was the principal hope. Yet both were typical philosophers in that they did not personally experiment, though Bacon, like the Greeks, was a great and cool *observer*, if chiefly of human nature. Descartes' typical contribution to the complex of science was mathematical, abstract, like his philosophy. Bacon's good fortune was that, while others like Nicholas of Cusa had sounded most of his notes before him, he was the first to sound them in the presence of a resonator ; for the British mind has resounded to his cry of " stick to the facts " all down the centuries since. It is a commonplace that Bacon tried to carry contempt for theory further than it can go. He substituted the " exhaustive " collection of facts for the true method of science, which has always lain in the use of hypotheses with facts as their constant check.

Several of Descartes' abstract tenets have lasted science almost ever since.* Mind as not extended in space, matter as extended, mind and

* Like Newton, Leibniz, and most scientists of that century, Descartes was no atheist ; was even religious. These men saw in the passiveness of matter (as expressed by Newton in the laws of motion) a need of a God to move it or to link it with the human will. In general, atheism has been associated with biologists rather than with physicists.

matter generally as mutually exclusive, animals as machines—all these have barely been modified for practical experiment even by the triumph of the quantum theory. The Cartesian theory of vortices banished from planetary theory such vagueness as Kepler's " spirits " ; although, as we shall see later, logic was by no means its sole progenitor. But then it is apt to be where Descartes' logic is weakest that his science is weakest also. When in the end he sapped his logical mind-body contrast by asserting a bodily link in the pineal gland, his assertion was of no effect on science.

It was perhaps Descartes who first brought into science abstraction as we now understand it. Copernicus had indeed given an impressive illustration of the fact that science cannot always trust to the senses, but must go behind them and assert that what we see is capable of more than one interpretation. But Descartes carried this distrust of the senses, like the distrust of mere *a priori* dogma, to its logical conclusion. Only the bare idea of space, of matter as extension, seemed to him undeniable and safe.* He thus prepared the way for that very bare universe of particles in themselves colourless, tasteless, odourless, which later on became a scientific orthodoxy.

But after all it is not in the philosophers, but in the scientists themselves, Galileo, Huygens, Newton, that we best see the change-over from philosophy to science. We will take one point only. When in the next chapter we describe the change-over from ancient to modern mechanics we shall see that it is often the idea of a *circular orbit* as the natural or perfect one which even the reformers give up last. Why should the circle have been preferred to the straight line, which seems to us so much simpler ?

The answer may be that men, conceiving themselves as the centre of the material universe, naturally took the orbits of the perfect (heavenly) bodies as circles, which are the simplest of central curves. But Menon has produced evidence that squares were at first preferred to circles. We seem nearer the point in saying that to us the orbits do not appear primarily as finished figures or patterns at all, but as the paths of a motion *still uncompleted*. In this last phrase the central contrast seems to lie. Is the world and our knowledge still uncompleted, is there room for progress ; or is there not ? Classical scholars all down the ages have tended to the assertion that the Greeks left us nothing to learn, that so-called Progress, like walking, is mere falling just arrested in time. Of this essential closure of view, this finite universe, the circle is as typical as the straight line is of the viewpoint of science, which Cusa, Bruno, Descartes, but not Bacon, heralded with the assertion of the infinitude of space. That viewpoint believes in step-by-step advance, distrusts the *a priori*, doubts ever reaching an end. The Greeks' comparative weakness in the sciences of continuity, of motion, and of experiment (dynamics and physics), is relevant here ; almost as if motion had been to them a slight error of taste.

* This is curious if there is truth in the suggestion in Chapter I that the idea of space as primarily elbow-room—extension—was then a fairly recent growth.

In fact this belief in infinity, this doubt of the cut-and-dried, at the back of the scientific viewpoint, is one of the rather few genuine elements which we derived from the " barbarians " rather than from the Greeks. The infinities, though not the pessimism, of the ancient Eastern religions, had been preserved by Christianity. For the step-by-step communion with the infinite by a series of sacraments, science has substituted the step-by-step communion with it by a series of hypotheses and experiments. But the infinity remains.

CHAPTER V

MECHANICS, ASTRONOMY, OPTICS IN THE 17TH CENTURY

THE story of this, the most important single episode in the history of science, is a complex one. Improvements in the science of mechanics are severely tested by difficulties in astronomy, brought to the front by advances in the manipulation of lenses. Other experiments made possible by skill in the handling of glass give definiteness to the concept of a gas. This, and advances in hydraulics, enrich the idea of the physical ; and the concept of mass crystallises out. With its aid, and with that of advances in mathematics, the astronomical difficulty is overcome. Doubly the advance depends upon the properties of glass ; as, in the next century, advances in more than one case depend on those of mercury. We leave the advances in mathematics to a separate chapter, and begin with those in mechanics.

To the Greeks, statics and hydrostatics, like mathematics, had been abstract and remote from practice. Thus, they had been apt to prove propositions in these subjects for both commensurable and incommensurable values of the weights involved. Their practical men must no doubt have roughly realised such maxims as " What is gained in power is lost in speed " ; but they worked in no fruitful correlation with the theorists. This phase of substantial isolation lasted right through the Middle Ages and beyond. Oresme and others, then (as we have seen) Leonardo da Vinci, foreshadowed many modern ideas. But they remained isolated.

Practical need started the truly continuous development. We have seen that in the Netherlands, hydromechanics had become urgently practical ; and it was the great Lowlander Stevinus (1548–1620) who in 1586 gave this subject and mechanics their real start. In contrast with the Greeks, he began from experiment. Stevinus was something of a universal genius, a military engineer, a designer of ingenious contrivances.* We first deal shortly with his hydrostatical work, and then pass on to the more extensive developments in mechanics.

The central difficulty in hydrostatics is that of fluid pressure. How can an entirely non-rigid body exert pressure ? Stevinus cleared the point up and calculated the pressure at a given depth and in a given direction. He made the assumption, then by no means universally admitted, that perpetual motion is impossible. Hence any given part

* Such as his wind-carriage which outstripped horses along the Dutch sands while carrying nearly thirty people. It is important to realise that it was not ingenuity which was lacking in those days but a settled habit of finding real uses for its products. For Stevinus, see a memoir by G. Sarton, *Isis*, Vol. 21, pp. 241–303, and Vol. 23, pp. 153–244.

of a liquid could not be supposed to be perpetually circulating under the pressures of the parts around it, and so preserved within a liquid its given place. This settling of the idea of liquid pressure prepared for the more difficult one of gaseous pressure cleared up by Boyle and Mariotte half a century or so later.

Mechanics and Astronomy

Turning now to mechanics, we find certain central difficulties holding up progress at every point. After collinear forces, parallel ones—levers—are the easiest to deal with. The Greeks had had a tolerable understanding of these. It was the next stage, non-parallel forces and their composition, which had held them up. Its first concrete form, the inclined plane, had been posed by Pappus. But it had not been solved. Stevinus gave an intuitive solution by observing that a uniform chain laid over a double incline must rest in equilibrium if its ends are on the same horizontal plane. His well-known diagram is illustrated in Plate VI. What the long part gains in weight it loses in that only a part or *component* of it is effective downwards. This was the first understanding of the difficult idea of what we now call a vector, or of the parallelogram or triangle of forces. Varo of Geneva had the idea at about the same time.

At about the same time, too, Galileo cleared up the allied difficulty of another vector, velocity. How, it was asked, can a body have two velocities at once ? The difficulty of conceiving this led to a confused notion that, for instance, a projectile sent off obliquely must first move straight along its initial direction until it has lost all its velocity along that line, and then drop straight down under gravity. This was evidently the theorising of the cloister, not the experimenting for which the battle-field now began to call ! It was just such cloistered things which Galileo (1564–1643) spent his life in thrusting aside. In him the mundane vitality of the Renaissance seems to flash forth as fresh as ever. He pointed out that there is no reason why a body may not have, at every point of its path, not indeed two velocities, but two components of velocity *varying independently*, and so a continuously curved path.

Mankind thus entered the 17th century with two " vectors," velocity and force, fairly clearly conceived by two men in different parts of Europe. Unlike velocities, the forces thus combined might be of totally different kinds, as, in the above instance, gravity and the reaction of the plane. What was asserted was in fact that these acted, when together, just as they did when apart. A new distinctness of conception had entered the subject.

Another not unreasonable idea was, that (earthly) bodies, when under no special force, slow down, instead of proceeding, as we now assume, with uniform velocity. A constant speed was thought to require a constant force. This enforced the obstructive distinction of earthly or corruptible and heavenly or incorruptible bodies. For these latter, when undisturbed, move uniformly and in a circle. So, at least, our senses seem to tell us. Here, again, Galileo, like Kepler and Benedetti, began to clear matters up, his instincts being generally

DE
BEGHINSELEN Elementa.
DES WATERWICHTS
BESCHREVEN DVER
SIMON STEVIN
van Brugghe.

TOT LEYDEN,
Inde Druckerye van Christoffel Plantijn,
By Françoys van Raphelinghen.
cIↃ. IↃ. LXXXVI.

FACSIMILE II. Title-page of work by STEVINUS, showing his method of expounding the triangle of forces. *Reproduced from a copy in the British Museum.*

PLATE VII

Sketch of GALILEO's proposed pendulum clock. *From a copy kindly presented to the Science Museum by the R. Ist. Studi Superiori, Florence.*

sound, though his enunciations were rarely water-tight. He preferred to think of undisturbed motion as uniform ; since the more the visible sources of disturbance were removed the more uniform even earthly motion became. It was not speed which he made constant under constant force, rather acceleration. This was a capital advance.

Galileo's views on the circularity question were more confused. He did not rid himself of the Aristotelian idea that a body might indefinitely pursue a circular path if acted on by no force.

It is with regard to the third muddle of contemporary thought, that of " gravity " and " levity," that Galileo is most clearly revealed as the first of an illustrious line, the line of mathematical physicists ; of men able to analyse their experiments by means of their mathematics. His experiments had begun in 1583. They reached a climax about 1588–91. Criticism now denies * his demonstrations from the Leaning Tower of Pisa, in which he was thought to have shown that bodies of different weights fall (apart from an obvious correction for air resistance) at the same rates. Whatever experiments he did perform seem to have been convincing enough to rouse the " Aristotelian " orthodoxy. This latter was committed to the view that bodies move faster in proportion to their weights. In his elevation of mathematical order, Galileo was in the Platonist current mentioned in Chapter III, but, as noted above, he himself also took ideas from Aristotle.

Galileo's mathematics came to the fore in a practical as well as theoretical sense in his use of the inclined plane and the pendulum to give greater precision to his proofs ; since freely falling bodies move too fast to be directly observable. He rediscovered the fact that a reasonably wide pendulum swing takes no longer than a very narrow one. He then showed mathematically how this implies that " gravity " is increasing the speed of the bob by equal amounts in equal times. This, also, was contrary to the orthodoxy of the time. We leave the analysis of the conclusions drawn from these experiments to a later section and proceed to astronomy, the subject of Galileo's next great experiments.

These require a brief divergence into optical history. We have seen in the first chapter that, in spite of the spread of spectacles, the advantage of using lenses in surveying and astronomical instruments was not realised in the 16th century. It was perhaps another case of the separation of craftsman's knowledge and theorist's knowledge. Eventually, however, experiments began. Between 1590 and 1608 Lippershey (and perhaps two other Dutchmen independently) put together long and short focus convex lenses in both the usual combinations ; that is, as both compound microscopes and telescopes. The microscopes depended for their superiority over single lenses on a much closer knowledge of theory than was possessed for another century. Hence it was the telescopes which gave the immediate results. In 1609 Galileo, in Italy, heard of them. Soon he was making much better ones, with magnifications up to about thirty. With these he quickly made a series of revolutionary discoveries in astronomy.

* For recent discussions see *Nature*, 4 January, 1936, and *Isis*, Vol. 24, 1935, p. 164. Stevinus and De Groot did a similar experiment about 1590.

They revealed the planets, hitherto structureless points of light, as finite, physical bodies, full of detail and " corruption " like the earth. Further, they revealed that the sun has spots. These facts took the heart out of the old idea of incorruptible heavenly bodies, in definite contrast with earthly ones ; and another great obstacle to advance was under attack. Universal curiosity fastened on the phenomena, plainly visible to anyone who had access to one of Galileo's telescopes. The mysterious rings of Saturn were revealed, and above all a miniature Copernican solar system in the moons of Jupiter. At last the eyes of man were fairly turned towards the skies, and towards the Copernican question which they raised. The Aristotelian universities moved the Church to action, with results to Galileo too often discussed to need further comment here.

It will be recalled that not only Galileo but Kepler was just then at his height ; and it is typical of those days before scientific cooperation that these two men ignored each other. Galileo ignored Kepler's elliptic orbits, Kepler drew no nourishment from Galileo's mechanics, although he had some of its ideas himself. Kepler thought, as we do, that the moon causes tides. But his idea was muddled and astrological. Galileo rejected it, but his own was no better. We have seen that Kepler chose magnetism rather than Aristotelian gravity as the force on the planets. But for us the chief point is that neither he nor Galileo conceived either of these forces as universal ; that is, as acting everywhere if they act anywhere.*

All such forces, in fact, were highly speculative ; and the next great theorist in astronomy, Descartes, conceived that his chief service lay in making them unnecessary. His vortex theory accounted for the remote heavenly motions by a common-sense force, familiar on earth ; that of a swirling fluid vortex. Unfortunately, it lacked the necessary feature of a good scientific theory : quantitative predictions exactly comparable with experiment. The hydrodynamical theory of vortices is even now difficult. In those days it did not exist. Descartes' theory was thus a mere vague suggestion, not specially related even to such already-known facts as Kepler's laws.

We must now return to mechanics ; for of course every such astronomical superstructure implies a basis of axioms in mechanics. True to his logic, Descartes was not content to leave his axioms implicit, but stated them. One of them repudiated circular in favour of rectilinear motion as natural and fundamental. Indeed, two of them would show him approaching Newton's clarity of perception if the rest did not destroy the pretension. Descartes' notion of matter itself, however, was in some ways more modern than Newton's. He eliminated, as part of his repudiation of vacua and of action at a distance, any fundamental distinction between occupied and empty space. A piece of iron, for him, was almost a piece of space which transmits a push but does not transmit light, " empty space," almost another portion which transmits light but not a push.

* Still, it was a gain to have compared these forces ; and especially to have put forward, as Kepler did, the profoundly novel idea that gravitating bodies come together at an *intermediate* point.

Galileo had left the best minds with experimental evidence that the acceleration produced by gravity is independent of the mass or weight of the body concerned. We say " mass or weight " in order to stress that it was just a foolproof distinction of these which was the task of the theorists. Lack of one had led to the age-long confusion from which these men were slowly drawing clear; but this was perhaps not the only reason why gravity and inertia had to be distinguished. A great primary fact of the time was the growth of quantitative knowledge of other forces besides gravity. All these forces contended with inertia, while gravity was irrelevant to them. They thus forced inertia, for men like Descartes, Gassendi, and Baliani, into the primary place. Yet drawing this distinction faced these workers (about 1650) with the extraordinary fact that the gravity or weight, and mass, thus distinguished, were yet *always proportional*. Why? It was a question which might well have diverted the best minds from the easier, but at this stage more profitable, concepts of inertia, force, etc.; for it was only " solved " by the difficult concepts of the general theory of relativity. But it was coming to be a mark of the scientific spirit to evade the hypnosis of the insoluble problem.*

Before entering on the final definitions of mechanical quantities, we briefly describe the " other forces " above-mentioned, which may thus have exerted such a crucial influence in directing men's minds. In particular, we consider two of them in which the advances were most definite: the expansive tendency of gases, and the elasticity of solids. In doing so, we should remember that, though we are here concerned with mechanics, these advances were the very ones the lack of which had been holding up the diverse sciences of chemistry and physiology.

We have, in fact, had several examples of the confusion which existed in 1600 on the subject of gases. The gases, for instance, in flames were taken as examples of " levity." Galileo brushed this aside, and substituted the modern notion of relative density. Meanwhile, as we have seen, Stevinus was clearing up the concept of fluid pressure so far as liquids were concerned.

The metaphysical principle that nature abhors a vacuum was not then " explained " by the weight or pressure of gases; and it was not until Galileo's time that † it was realised that nature only abhors a vacuum up to a certain definite point; in fact only up to thirty feet of water. These thirty feet were quite enough to quench the metaphysical principle into a physical fact. When (in 1643) Viviani at Torricelli's suggestion proved that the length of a column of mercury " supported by a vacuum " is less just in proportion to its greater density, the simple mechanical explanation that the weight of the outside air was involved came irresistibly to the front.

This explanation was tested by Pascal in 1646. Having procured

* Almost all the fundamental concepts of practical science, from that of action at a distance to that of species, enshrine " insoluble " philosophical problems. It is indeed their very impenetrability which fits them for a pivotal rôle.

† Apparently in the course of pumping operations.

glass tubes from Italy, he showed that the " abhorrence " was less on the top of a mountain, obviously because the superincumbent air was less. The familiar U-tube barometer began to come in.

This facility in the manipulation of glass tubes was of vital importance.. It had been growing here and there since before 1600. Perhaps as early as 1593, Galileo had constructed a thermoscope, parent alike of the barometer and the thermometer. It was simply a bulb, with a stem in which water had been caused to rise by immersion of the open end when the bulb was warm. Van Helmont also used thermoscopes to measure body temperature. Ingenious persons were using the variations of atmospheric temperature and pressure in such instruments to give what was still called perpetual motion. The separation of thermal and pressure effects was in fact a difficulty throughout the century, but we defer its discussion,* and confine ourselves to the pressure aspect in its development into air pumps.

This was an expensive matter, demanding workmanship of an accuracy perhaps then commoner in Germany than elsewhere. The wealthy Guericke (1602–86) of Magdeburg made (1654) the first air pump, also his well-known hemispheres. He showed that a vacuum cuts off sound but not light, and destroys life and combustion. Using Guericke's experience, Hooke (1659) constructed for Boyle (1627–91) a less clumsy pump. In the receiver of this, Boyle placed a barometer, and thereby proved that it is not here the (negligible) superincumbent weight of the air which depresses the mercury in the U-tube, but its pressure. Thus he completed the process of substituting for the vague idea of " spirits " a perfectly definite entity, a " gas," with a perfectly definite pressure, like that of Stevinus for liquids. This pressure he, or Hooke, showed (1660) to be inversely as the volume ; so that pv for any given body of gas is as definite as v for a liquid or solid.

We must dismiss Hooke's (1635–1703) work on the elasticity of solids even more briefly. He showed that compressed or stretched solids develop, like gases, a perfectly definite pressure or tension proportional to the change of length involved, though one which, unlike that for gases, varies with the solid concerned.

Hooke was a man of febrile energy, and he and others were constantly engaged at this time on every sort of mechanical, optical, electrical experiment. The total effect was a permanent enrichment of the concept of the physical, and an enrichment, also, of the stock of instruments. Holland, with its great sea trade, was an important centre in this work. Huygens (1629–85), in fact, began his career with navigation in mind. For this and astronomy, he started to improve the telescope and the clock. He (1657) coupled pendulums with the gears of time-pieces to regulate their rate of running. Hooke (possibly in 1658) suggested balance springs as an alternative : it was with this in mind that he investigated the elasticity of solids.

Huygens' suggestion under this head was a part of his theoretical work on the *compound* pendulum. This was the first success of the crystallising science of mechanics in dealing with bodies not of negligible

* See Chap. VIII.

size. It is also the point at which we return to the central process of defining concepts which was soon to culminate in Newton.

This work on the compound pendulum was a triumph of that abstract viewpoint which, for the future, was to mark continental as opposed to English scientists. Through Van Schooten, Huygens had felt the abstract influence of Descartes, in sharp contrast to the concrete activities of Boyle and Hooke. It was an abstract principle which this continental group began to develop.

We have seen the growth of the idea of inertia or mass, the essence of which was that something was conserved through changes of volume, state, etc. Descartes had called attention to the importance of the *work* * done by a force ; in fact, of what we should call energy. It was the conservation of one form of energy, namely the kinetic form, which Huygens assumed as the basis of his calculation. We shall have to note several more such conservation principles in the history of science, each of great importance. Their central value, already present in Huygens' example, was that of enabling us to find out certain general features of a change without first calculating all its details.

We now know that kinetic energy is not always conserved, so that Huygens' method would not have applied to much more complicated problems than he chose. Granted it, however, he did succeed in showing (1673) that every solid swings precisely in time with some simple pendulum or other, and that for strict isochronism a cycloidal, not a circular, path is necessary. This familiarity with vibratory motion was of incalculable importance in the future of science.

Although it is not clear that he always understood Kepler, his work in astronomy dovetails closely with the main history of mechanics. Assuming, like Descartes, that, for heavenly and earthly bodies alike, the "undisturbed orbit" is a straight line, he found that a central force of $\frac{mv^2}{r}$ (m the mass, v the velocity, r the orbital radius) is needed to keep a planet in a circular orbit. From this he further concluded that, had the planetary orbits been circles and Kepler's other laws still true, the central force must vary as $\frac{1}{r^2}$, as suggested without proof by Boulliau (1645). This was also realised in the Royal Society in London, by Wren (1632–1723), Hooke, and Halley (1656–1742). It has been remarked by Pelseneer (*Isis*, Vol. 17, from p. 171) that the history of Kepler's Laws up to Newton is obscure. In England, the youthful Horrocks (1619–41) had applied them to the *lunar* orbit. Horrocks thus preserved the Keplerian tradition and adumbrated a *universal* force.†

The question of the hour was thus : would the inverse-square law account for the elliptic orbits demonstrated by Kepler ? If so, the whole tremendous system of the planets would have been shown to be governed by a single universal force. It was a question involving very difficult mathematics, and it defeated all these men.

* Concept used earlier (1615) by S. de Caus.

† In a different way, Descartes' vortex idea was also universal.

Meanwhile, since the middle sixties and his own early twenties, Newton (1642–1727) had been meditating this idea of one single universal force. He had distinguished the three types of case with which anyone attempting to prove it would be required to deal. These were, sun and planet (two points, but an elliptic orbit), moon and earth (two finite spheres) and earth and small body near its surface (requiring, like the second, a knowledge of the attraction of a sphere not of negligible size). We shall consider in the next chapter the peculiar type of mathematics which these problems demanded. Newton was one of the two first effectively to provide it.

This enormous work was but a fraction of his service to science, and his whole scientific work was but a fraction of his activities. Nor was it the fraction which always impressed him as the worthiest. Over-sensitive to attack, distracted by theology and also by worldly ambition, he frequently laid his researches aside.

In 1682 he proved the point about the elliptic orbits, but lost his proof. In despair upon the point, Halley called upon him, and Newton gave a proof again. Halley pressed him to go on, and, under this stimulus, Newton also overcame the difficulty of summing up the attractions of all the parts of a finite sphere, and thus of calculating its effect at an outside point.

Swiftly proceeding to formulate the underlying axioms of his own and a century's effort, he completed in 1687 the decisive " Philosophiae Naturalis Principia Mathematica." This gave a theory of the planets and of the moon, and a lunar theory of the tides. The laws of motion exactly defined the concepts of mass, inertia, force, and their relation to velocity and acceleration. A synthesis was effected of the various approaches to the idea of mass, and it was assumed (wrongly if Mach in 1883 was right) that they afforded a measure of it independently of acceleration and force.*

Newton established for the first time in any subject more concrete than pure mathematics the strictly unadorned expository style of Euclid, an important advance on the cumbrous and pretentious habits until then current. His precise enunciations outmoded such arguments as that of Leibniz against the Cartesians : does the " efficacy " of a moving body depend on its velocity or on its $(\text{velocity})^2$? These arguments, however, went on for many years afterwards.†

Leibniz and others noticed that the laws relating to uniform velocity and acceleration, while implying some kind of " absolute " space, or frame of reference as we should now say, specified none which applied to all possible cases, terrestrial or astronomical. Thus the question of relativity, which we have seen raised (Chap. III) in the 15th and 16th centuries, came up in a more concrete but perhaps more limited shape in the 17th. Galileo had assumed that " uniform velocity " has a sufficiently definite meaning, and Newton's development of the idea was

* The late L. N. G. Filon criticised (1937) some of Mach's conclusions. Rosenberger (1895) and Jourdain (1914) believed that, for Newton, mass was measured by number of identical constituent particles. Were present-day ideas perfectly clear, it would be easier to settle the status of past ones.

† Such lag was usual in the early days of science. It still occurs.

found to work very well. Hence the question, though raised, was again dropped—for two hunded years.

Thus in the course of this century the points (as to gravity and as to relativity of velocity), of both the general and the special Einstein theories came up for decision and were decided in a sense different from that now adopted.*

As to action at a distance, the position is not simple. We suggested that Descartes could logically have maintained that matter's transmission of a push *is* an action at a distance—matter being only a kind of space. But he chose a different view, vortices as both transmitting force and filling up the inadmissible vacua. As against these vortices, Newton's editor Cotes identified his master with action at a distance; and in this he has been followed by nearly everyone since. Yet Newton wrote an explicit repudiation of the idea.

By this work in mechanics, Newton reinforced Descartes' critical process of stripping matter of its more sensual attributes, leaving only a bare universe of hard particles to the 18th century. But such a formalistic view is alien to the 17th century in general. To that century, the world was full of light and colour. We shall now turn to trace the optical side of its work.†

Optics

Two main phenomena had drawn attention to the question of colour, in addition to its physiological aspects (for which see Chap. IX). One was old, the rainbow, the other new, the chromatic aberration of the lenses then coming into serious scientific use.

In the early 14th century, two writers gave the refractions and reflections by which the primary and secondary rainbows are formed, but the colours were left unexplained. Then the Dutchman Snell—in 1621, first published (by Descartes) 1638—discovered the law $\frac{\sin i}{\sin r}=n$, a constant for any one material. Thus the problem of the lens-section to give a true point-focus with parallel light became, theoretically, solvable. The section, however, was obviously not a circular one, nor even, as with mirrors, a parabola. Its calculation was not carried out. It lost importance not only because non-circular lenses could not easily be ground, but because an imperfect focus was not the worst fault in powerful lenses such as those used in the 17th century for microscopical purposes. The worst fault was the coloration of the image.

The explanation of the rainbow by Descartes (1637) on the analogy with a prism depended on the variation of n from colour to colour (dispersion). This evidently explained also the coloured images with lenses, if it did not explain how they could be remedied. The "origin" of the colours (whether they are "contained in" the original white light

* For the chemical, thermal, and other aspects, see Chap. VIII.

† In Barrow's day, space and time were often compared, time being likened to a line. Leibniz had a clear grasp of them as merely orders of coexistence and succession.

or are " due to " the lens), is, in one sense, a question of the type that science can never answer ; although early scientists were often motived by the belief that it can. Marci (1595–1667), however, did succeed in defining a sense in which white light is always composite while coloured light may be simple. He dispersed it into a ribbon of colours on a screen by means of a prism. When he then passed a monochromatic portion of this ribbon through a further prism there was no further dispersion. Newton, taking up the subject in 1666, freed it of a number of obscurities and, in performing this experiment, completed it in the vital particular of resynthesising the white from the coloured light.

We turn now to theories of the subject, and in the first instance to abstract rather than physical theories. We begin where the merely geometrical theory of perspective ended ; and we also leave behind the old, pre-scientific, " theory of tentacles."

Descartes had deduced Snell's law from a metaphysical theory, which Fermat criticised. Fermat's was just as metaphysical, but in our eyes it has two merits. As against that of Descartes it maintained that the velocity of light is less in a " denser " medium (that is, where $n>1$). It also led to the view that the path taken by light is such as to make its time of passage a minimum. This was an improved form of the ancient idea of Hero of Alexandria that the *length* of an optical path is a minimum. For neither of these merits was there experimental evidence at the time ; but the first happens to be true, and the second, modified, has proved to be the first of a highly significant type of scientific principle.

These " minimum principles " should be thought of in connection with the conservation principles already discussed, since each completes the other and they grew contemporaneously in this 17th century. They share also the important peculiarity of enabling us to ignore the details, or the physical mechanism, of a change, and yet to calculate some of its features. Each is a crystallised, quantitative, form of instincts almost universal to man, such as that of the Economy of Nature and " What we gain on the swings. . . ."

As we shall see in Chapter VI, Fermat was at that time developing also the purely mathematical theory of maxima and minima—of stationary values in general. This last phrase was later found to be the correct general form for Fermat's Principle ; since light sometimes follows a *maximal* path. The true technical use of the principle, however, still lay in the future, and here it is more important to note how metaphysical it was in Fermat's hands than to know exactly what its details were. It bore, in fact, a teleological tinge ; as if each ray cast about before settling its best route to a desired and predetermined goal.

We now turn to more physical theories. Both the obvious mechanisms for light, projectiles, and waves in an ether, seem to have occurred very early to speculators. Hooke gave a crude wave theory in 1664. Newton was as familiar as Huygens with waves and ether, which he explicitly recognised as part of the mechanism of light. He made them, however, subsidiary ; thinking them more akin to heat.

Both he and Huygens seem to have had their wave ideas dominated by waves of sound.* The result was that Newton thought an emission theory necessary to explain why light, unlike sound, does not bend round corners; while Huygens on the same analogy adopted longitudinal waves in his theory.

When in 1665 there was published Grimaldi's (1618–63) observation that light does bend round corners (diffraction), Newton used the ether, but not waves, to explain it.† For definitely periodic phenomena like the "Newton's" rings observed by Boyle and Hooke, he recognised the need of a periodic element in the explanatory suppositions; but it was "fits of easy reflection and refraction," not precisely waves, which he suggested. Hooke tried to explain colour, but it was Newton who first correlated it with differing periodicities.

Hooke, too, had the idea of a wave-front, but it was Huygens (1678, published 1690) who made the capital advance of analysing a wave-front's effect at a further point into wavelets; thus, again, making mathematical development possible.‡ By the theory thus reached, he was able brilliantly to analyse the double refraction observed by Bartholinus of Copenhagen in crystals of Iceland spar (1669). But because he had only longitudinal waves in mind, he was denied the characteristic triumph of the wave theory—its ability to explain why in these cases the rays have different properties in different planes containing them. This "polarisation" was announced by Huygens himself in 1690 and interpreted in 1717 by Newton, who pointed out that the (longitudinal) wave theory could not explain it.

There was a crucial test between these theories, in the fact that the wave theory must suppose a refraction towards the normal to be due to a slowing up of the wave, while the particle theory, as then held, must suppose it due to a speeding up, and that in the ratio of the refractive index. But this test could not be applied. In those days, it was by no means certain that light had a finite velocity. Descartes thought not. Galileo, about 1600, had failed to find one by crude shutter-and-distant-observer methods. Römer's attribution (1676) to light's finite velocity of certain irregularities in the apparent motion of Jupiter's satellites as the earth moved towards and away from them was not at once accepted.

Halley's belief, however, in a finite velocity led to the work of Bradley (1692–1762) in 1729. In this, another astronomical phenomenon was explained by a finite velocity of light, and gave a value roughly concordant with Römer's. This was aberration, the slightly different angular positions of the "fixed" stars according to which way the earth, in its orbit, is moving across the rays from them. This led in the 18th century to a general acceptance of a definite finite

* Newton worked out a value of the velocity of sound, the incorrectness of which was to be crucial later.

† As a superficial refraction or attraction, the ether being denser in free space.

‡ He did not, however, envisage a *train* of waves of definite wave-length; rather a single "pulse." None of these men seized the wave-idea in all its length and breadth. See Whittaker, E. T., "A history of theories of the aether and of electricity . . ." 1910.

velocity for light. But terrestrial methods powerful enough to compare the velocities in different media had to await the middle of the 19th century. Hence the decision on this ground between the two fundamental optical theories remained suspended.

Meanwhile, optical instrument-making went on; and indeed professional instrument makers, especially opticians and clockmakers, had never been wanting throughout the 17th century. It is impossible to gain a true idea of the astronomers of this time without realising that they were keen practical observers and constructors. Astronomical and other observatories, indeed, constitute one entire aspect of science of which this book can give no adequate history. Improvement of the means of navigation was a main motive in all this development. In

GRAPH I

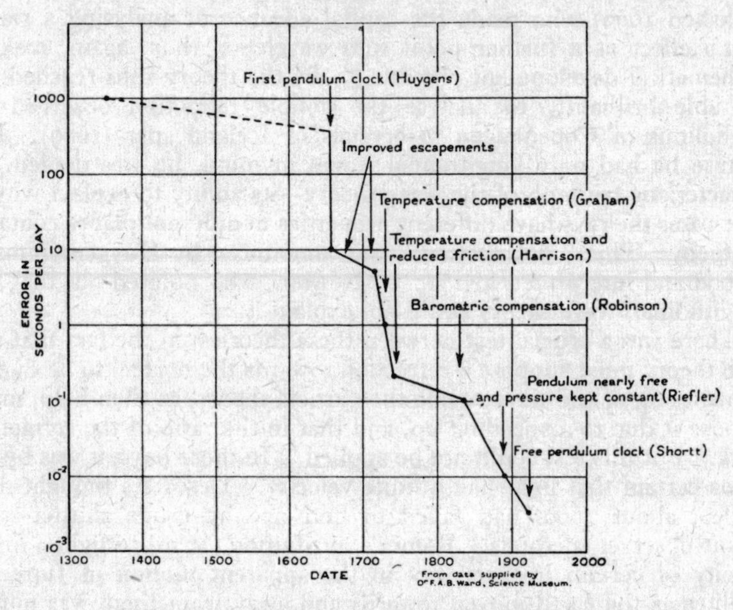

ACCURACY OF TIME MEASUREMENT.

particular, there was constant effort to discover means of finding longitude at sea. Galileo had tried and failed. Hooke and Huygens had this need in mind in their improvements in the clock. So had Newton when he made those lunar tables by which, given an accurate timepiece, longitude at last became calculable.

The accurate time-piece was longer in coming. But, again, it was the overseas trade which brought it forth. Harrison's chronometer (1761–2) received a government prize. It turned out that not the regulator alone (balance-spring or pendulum) but its temperature compensator, was the crucial improvement (see Graph I). Le Roy, in France, and others, should share the credit with Harrison.

The idea of compensation seems to have been strong just then, for it is the leading one in the contemporary production of achromatic lenses for telescopes.* Huygens in the 1680's, Gregory in 1695, much later Euler, and Klingenstierna of Upsala, had considered this matter. Actual achromatic lenses were made privately by Hall in 1729; but it was 1758 before the London Huguenot optician Dollond made an achromatic telescope and with it a European sensation.

The absence of such instruments had not held up the progress of astronomy. Despairing of their possibility, Newton (1668) had taken up an idea of Gregory (1663) and of the Jesuit Zucchi, and made a *reflecting* telescope. For the largest sizes, this type has been used ever since. Hadley made another in 1723. The astronomer W. Herschel (1738–1822) made larger and larger mirrors (e.g. in 1789, one of 40 ft. focal length, 4 ft. across) and many other improvements. He was a Hanoverian who, under the dynastic internationalism of pre-democratic times, lived in England.

Eighteenth Century

We turn, in concluding this chapter, to the further history of gravitational science. This now becomes an enormous subject, and it is impossible to follow it in the detail which its importance demands. On the whole, it was pursued in the 18th century on lines laid down in Newton's day. Newton and Huygens, for example, had been aware that the earth is not a sphere, but an ellipsoid of revolution. The consequent variation of gravity and of the period of a pendulum from place to place had been noticed by Richer in French Guiana in 1671. In the 18th century it became the subject of more and more accurate surveys (e.g. Clairaut, 1741–3).

Maclaurin and others took up the general question of the attraction of an ellipsoid at an external point. The mathematics of this proved very complex, and caused Lagrange long after (1773) to introduce the idea of potential (see Chap. VI).

Laplace (1749–1827) was the supreme mathematician of Newton's planetary theory. The greatest single missing link—and a great one it was—which he supplied in Newton's work was his partial proof that the system would be a stable one; but it was his prodigious power in dealing with both the detail and the general features of the subject which gave him his characteristic place in scientific history. Some notes on the fate of his nebular theory are given in Chapter XXII.

Another aspect of Newton's law had meanwhile been attacked. The law gives the force between two masses m_1 and m_2 as $\frac{Gm_1m_2}{r^2}$ where G is a constant universal for all matter, the second of its kind.† Newton gave indications of a method by which G could be experimentally determined, but it was necessary to know m_1 and m_2, and this is impossible as long as at least one m is of astronomical dimensions. The attraction of small weighable terrestrial bodies would be very

* Not to be confused with the much more difficult, because more powerful, achromatic *microscope* lenses, which were a crucial advance of the *nineteenth* century.

† If we take the early measures of the velocity of light in free space as falling into this category.

minute. Even that of a mountain would be small, but about 1740 it was proved measurable (Bouguer, in work published 1749). In 1774–6 it was measured, not entirely inaccurately, by Maskelyne on Mount Schiehallion. The difficulties, of course, included that of estimating the size and density of the mountain.

In 1797–8, Cavendish, a wealthy aristocrat who had made science a life-work, took over from J. Michell (1724–93) a delicate torsion balance for measuring the attraction of two *small* bodies, an apparatus of the same type as we shall see used at about the same time for finding the laws of electrostatic and magnetic force. We cannot further describe any of these methods, nor the many 19th century improvements of them. The chief significance of that of Cavendish for the history of science is perhaps not the values it gave of G and of the density of the earth, but its being a brilliant essay in the enormously important business of eliminating even the minutest disturbing forces from experimental determinations. Even more than the accurate geodetic surveys of the time, it required extraordinary precautions against slight variations of temperature, pressure, and the like, and even against human activity in its neighbourhood. It might be taken as a beginning of the movement for rigorous accuracy which marked the 19th century in every branch of science.

We have now surveyed the effective births of mechanics, of hydraulics, of physical optics and of the theory of the elasticity of gases and solids. Other branches of physics, such as heat, came into line so much later that their discussion is deferred. Light has at least one simple property, straight line propagation, and so has elasticity, Hooke's and Boyle's laws. But until Boyle's property had been thoroughly digested it was hard to find thermal effects with even approximately simple features on which to seize.

CHAPTER VI

MATHEMATICS, 1600–1800

RENAISSANCE mathematicians had added little to the Greeks in geometry. On the other hand, great advances had been made in algebra ; and algebra was one great element in the mathematics of the 17th century. Another was the idea of infinity, to which the Renaissance, like the Greeks, had in the end been led. We treat the mathematics of the 17th century in these main sections :

(*a*) The classification of curves.
(*b*) Kepler's " Principle of Continuity," joining with Desargues' projective geometry.
(*c*) Descartes' algebraic geometry.
(*d*) Infinite series and the calculus.

These make the 17th century one of the greatest periods of mathematics so far as ideas go. But ideas are not all. There are other things : for instance, elegance and rigour. In these latter, Europe, with its Gothic element, was longer in recovering the standard of the Greeks. We have seen that another tone of mind, the physical tone, fought with the pure abstractions of mathematics for the mind of Europe. Each blurred the other until the time of Newton, when they began at last to settle their frontiers. Thereafter, in the 18th century, we recovered the elegance of the Greeks. For their rigour, we had to await the 19th century.

(*a*) CLASSIFICATION

We have seen that mathematics has a purely empirical part. It has also, no less than biology, parts where classification is the only approach. One such region is that of curves, of which any child can draw a vast variety. To understand 17th-century development, we must sketch the way in which the Greeks had approached this vast region.

The straight line and the circle, and plane combinations of them, early separated themselves, and were classed together by the Greeks as " plane loci."* The next idea was to combine straight lines and circles into the non-plane figure, or solid, which we call a cone, pursuing it both ways from its vertex. The (" conic ") sections of this by a plane, they grouped together as " solid loci." These include the parabola, hyperbola, ellipse, circle, and pair of straight lines. Next came higher curves, which the Greeks found too difficult to pursue very far, and which in fact showed up how fatally the absence of symbolism hampered them in developing a really suggestive classification. For the Greek name for these curves, derived from their mode

* English translation.

of construction, was simply " linear loci," which throws no clear light at all upon their relation to the " plane " and " solid " loci. Had symbols been used in ancient mathematics, the Greek recognition that conics are connected with quadratic equations would have led to the very classification which we consider in (*c*).

There is another type of classification which is just as vital: that arising not from curves as individuals but from the sets of curves of which they are taken, explicitly or by implication, as instances. We have seen in Chap. IV that the Greeks were much interested in cases of what we should call topology (when the set of curves is that obtained from any one member by bending or stretching it*), and that their conic sections and perspective might have suggested the projective viewpoint to which we proceed in section (*b*). But these viewpoints were only implicit in the Greek approach, and the essence of mathematical advance is to make explicit and basic, ideas which have hitherto been only incidental. Neither of these ideas was made fully explicit until the 19th century, largely because, like the theory of numbers, they are either too difficult or too easy.

(*b*) Kepler and Desargues

The Greeks had realised, from their mode of viewing the conic sections, that these are continuous with one another in the sense that the continuous rotation of the plane of section can give us all of them in succession: circle, ellipse, hyperbola, parabola, straight lines. But Kepler suggested that we could validly go further and deduce that therefore (for instance) the two arms of the hyperbola are " really " continuous with one another, and form, with a " branch of infinity," one curve no less than an ellipse. Here was an entry of an infinity distinctly different from that of the Greeks, or from that implied (see (*d*)) in Viète's infinite product for π; one with, certainly, some very strange consequences. For example, this view would suggest that parallel straight lines, although they never bend, may be viewed as joined at infinity. It should be noted that this can be quite distinct from non-Euclidean geometry and is now treated as such. But we have also noted in Chap. III the struggle in Kepler's mind of the idea of a pattern and the idea of the physical. Had he a confused non-Euclidean notion in which geometrical space itself, no less than the " physical space " of the surface of the earth, would ultimately come back to its starting point? If so, the struggle of physical and realistic tones of thought was not infertile.

In Desargues (who dared to take this kind of infinity seriously) another case of this struggle brought projective geometry prematurely to birth. He wrote (1636) on perspective; and it was from this, with its pencils of rays and its shadows, that he derived the projective idea (1639). Map projections implied a similar idea. Using the invariant of a pencil of lines (known to the Greeks) he was able to prove great numbers of complex theorems with an ease which was bewildering and highly suspicious to his contemporaries. Difficult theorems in

* Under certain limitations.

conics were deduced from easy ones on circles, difficult theorems on polygons in general from easy ones on regular polygons. The key lay in the Keplerian processes, in which Desargues indulged very freely, of regarding opposite ends of a straight line as coincident, and straight lines as circles with centres at infinity. It was partly suspicion of these which led to the fate awaiting his work. For, apart from Pascal (who used the method to reach some astounding results) so few mathematicians took the subject up that it not only made no advance but was forgotten for nearly two hundred years. Desargues' book was in fact only rediscovered in 1845, by which time others had rediscovered his methods and rectified his numerous questionable applications of them (Chap. XIII). For, as has been remarked, rigorous accuracy was not a 17th-century feature. It may be added that Desargues never realised the vital point that in his geometry the conception of rigid *length* is irrelevant.

Another reason for the complete obscuration of the method was the development of the calculus and of Descartes' algebraic geometry at about the same time. These were much better than Desargues' methods when it was a question of getting, not some new result, but solutions of specific problems. This was especially the case when the specific problems were those of the new mechanics and physics, in which (unlike projective geometry) the concept of rigid length was vital.

(*c*) Descartes

Classical surveyors and astronomers, such as Hipparchus, had specified the position of a place on the earth's surface by the two " coordinates " latitude and longitude. Further, historians have detected the idea of a graph as early as the 10th century. But none of these contain the essence of Cartesian geometry. This lies in representing curves or graphs by coordinates *connected by a formula* (1637).*

It must be remembered that, while the 16th century had made progress in the theory of equations, these had only been (as we should say) in one variable. The idea of a relation of two current variables, of a function, was in effect as new in algebra as was its application in geometry. It was an idea, and an application, the unfolding of the riches of which is still not finished ; and Descartes made only a beginning. He stated, for instance, that if any curve could be mechanically described, it could be represented by an equation. He saw that the axes could be either perpendicular or oblique, but he did not realise, as did Newton later, the immense usefulness of the negative extensions of the axes.

The new method was not widely known until the annotated edition of 1659. Wallis (1616–1703), however, systematised and extended it, and took the step of identifying conics with curves having equations of the second degree. This clinched the classificatory primacy of degree both for equations and for curves, and made it a natural step when Newton went on to examine cubic curves.

* Fermat and Harriot had the idea, in a more cautious, less suggestive, form at about the same time.

Coordinate methods were long in becoming, what they are now the standard elementary resource not only in geometry but in mechanics and physics. Newton, at the end of the century, while using them, was conventional enough to translate his working back into geometric form. The element of standardisation has been the most notable feature of the coordinate method, and its great advantage over others. It was not until the 19th century that it was realised that, in pure mathematics at least, its benefits have been purchased at a price. The Greeks would have recoiled from Descartes' assumption of the equivalence of length and number. This had to be disentangled in the 19th century. The obliteration, too, of projective methods probably delayed the 19th century realisation that several kinds of geometry are possible. More recently still it has been realised that there are regions in applied mathematics where grave confusion may arise from the tendency of Cartesian methods to confuse properties due solely to the choice of coordinates with those intrinsic to the figure treated (see Chap. XXII).

On the other hand, sufficient attention is perhaps not paid to the benefit of analysis from another kind of geometrical conception which, as mentioned in Chap. IV, made an early appearance in the East and may be said to have been introduced into Europe with two-dimensional paper calculation. This was the *configuration* of quantities (Pascal's triangle, the determinant, the matrix) with its rich formal suggestiveness.*

Wallis' practice of tabulating rather than graphing functions had a very similar sort of suggestiveness : for instance that arising from interpolation where the latter involved extension of existing definitions of the quantities involved. It suggested, in fact, a new view of functionality.

(*d*) Infinite series and the Calculus

We have already noted one way, Kepler's, in which Europe came to its first attack on the idea of infinity. But for the time the two main ones were infinite series and the calculus, which, so far as they were distinct, we take in that order.

One of the "insoluble" problems left by the Greeks, that of squaring the circle, led to an infinite product for π (given by Viète in 1593). As the 17th century wore on, it was gradually realised, also, that no finite series of terms could provide the values needed in tabulating the all-important circular functions and their logarithms, or the logarithms of numbers. In 1668, Mercator (*c.* 1620–87), and also Brouncker, gave :

$$\mathrm{Log}(1+x)=x-\tfrac{1}{2}x^2+\tfrac{1}{3}x^3- \ . \ . \ . \rightarrow \infty$$

Newton seems to have had the series for sin x and e^x by about 1666 (printed 1685 in Wallis' Algebra), and in 1671 J. Gregory gave that for arc tan x.

* Determinants remind us that finite algebra did not halt after its Cartesian junction with geometry. They were posed by Leibniz in 1693, and taken up by Vandermonde in 1771 as entities in themselves, not necessarily associated with simultaneous equations.

Wallis was already aware that the use of infinite series is tricky. For instance, three has a finite logarithm, but the above formula would not give it. The first question about such a series is, does it " converge " to a finite, definite, limit ; for all, or for what specific range of, values of x ? Newton was among the earliest to see the full importance of the question. He gave a rough test, but it was not adequate to the extreme difficulty of the subject ; as indeed he realised. For instance, in 1676 he enunciated his binomial theorem for fractional and negative indices (for positive integral ones it amounts to Pascal's triangle), but was thereby confronted with infinite series for which his proof, like Vandermonde's in 1764 and Euler's in 1770, broke down.* While this was realised, stress on it held, as we shall see again later, but little interest for these adventurous generations.

The approach to the calculus was twofold. There were questions of tangency and gradients on the one hand, questions of lengths, areas, and volumes on the other. As regards the first, a typical case had confronted Galileo when he had had to find the speed of a body at a given instant, knowing that it falls a distance proportional to the square of the time. This is a question of the gradient of the space-time graph. The ancients had been interested in the similar problem of the maxima and minima of curves, that is, of finding when their gradient becomes zero. We have already noted in Chap. V Fermat's revival of interest in this question. We give later the further history of minimum principles as applied to external nature. Here we confine ourselves to the mathematical point.

Some of the logical difficulties of the subject were raised by differing ideas of tangency, derived mainly from the Greeks ; touching as opposed to cutting ; cutting in two coincident points ; and the tangent as giving the current direction of movement of a particle moving along the curve in question. These caused controversy, but the remarkable fact which began to emerge with Fermat (*c.* 1628 onwards) was that the controversy did not seem to affect a certain number of simple rules for finding gradients and tangents in practice. Thus, the zeros of a certain " derived " function gave the maxima and minima of the original function. For instance, for $y=ax^n$ the derivative was nax^{n-1} for all integral values of n. For such algebraic functions the derivative was found without resort to suspect notions such as limits. For higher functions, this was no longer true. The notion of *movement* was used, and this involved all the logical difficulties mentioned.

At this point we must take up the second type of problem, areas, lengths, etc., where very similar logical difficulties had appeared. For such problems the best of the ancients, such as Archimedes, had reached the " method of exhaustions." This avoided the treacherous notions of (for instance) lines made up of points, areas of lines, which

* It was also slowly realised that for indices like (e.g.) $\frac{1}{n}$, n integral, other difficulties arose. These were never adequately attacked until the 19th century. There must, for instance, be n values, some involving $\sqrt{-1}$; while irrational and complex indices are also possible, as Wallis saw in 1655.

Zeno had exploded. It lay roughly in finding (say) curvilinear areas by finding the areas of inscribed and circumscribed polygons, and then showing that the difference between their areas could be indefinitely reduced by increasing the numbers of sides. An infinite series is involved here, and in but few cases does this lead to a finite expression for the area in terms of the dimensions of the curve. The ancients proved only certain simple expressions for areas by its means.

When the subject was resumed in the 17th century it was in a more rough and ready atmosphere. Men like Kepler, with his law as to elliptical areas for planetary motion, were more anxious to get results than to be pedantic over proof. They did not always avoid even the more obvious of Zeno's pitfalls. The important Italian Cavalieri (1598–1647) did so by basing himself on Archimedes * and developing the method of exhaustions into that of "indivisibles" (1635, esp. 1647 and 1653). In this, recourse to infinite series was minimised and to some extent standardised, an important step towards the complete (formal) standardisation called the calculus.

It will be remembered that the Greeks had been led to abandon mere mensuration in terms of a fixed unit, and to concentrate on relations, or comparisons. True to this tradition, Cavalieri and the other early European workers avoided some pitfalls and jumped others by confining themselves to the comparison of corresponding infinitesimal segments of known and unknown areas.

We think of integration problems (like areas) as harder than differentiation, or gradient, problems; but this is due to a realisation which we are about to describe. Before the emergence of this, the opposite was the case, for the simple reason that a sum is simpler than the quotient of two differences. This realisation constituted the hardening out of a tendency which, only implicit in ancient times, had been enormously strengthened by symbolism, the tendency, namely, to view mathematics as primarily concerned with formulae, with manipulation. This particular instance of it was a rule connecting the operations of integration and differentiation.

To take the above case as an instance, it was realised that the area under the derivative curve $y=nax^{n-1}$ is none other than ax^n+b, b being given a suitable value; that is, it is in essence given by the equation *of the original curve*.† This can be shown to be true for all simple cases such as were in question at the time. As we should say, each formula for a differential coefficient or derivative gives, by inversion, an integral; and although this still leaves many functions unprovided with integrals, it standardises a large number of problems in both parts of the subject. Barrow's proof of it, the first, was geometrical; but the full development of the implied point of view is associated with a new and most gifted generation, and notably with Newton in England and Leibniz on the continent.

* So far as he was known then. Recent discoveries show that Archimedes himself went further than Cavalieri knew.

† Until the late 1650's, the lengths, as opposed to the areas, of curves were supposed not to be calculable. The reason for this was to be significant later. It was, that the ellipse and hyperbola had been the curves attempted.

Between these two men an unfortunate controversy as to priority broke out, which for over a century deprived English mathematicians of the benefit of continental advances. Here we only remark that the whole matter was but a symptom of the isolation which still ruled in science. Learned societies were starting; they were not yet a settled habit. Newton used his method in 1666, but gave no printed account until 1693. Leibniz (1646–1716) was in possession of the differential calculus in 1675 and published it in 1684.

Newton's extraordinary gifts were in one way against him, for, given his stress on applications, they led to a certain casualness about presenting methods or results in their most elegant or general shape. The same phenomenon seems to be discernible in other mathematicians, such as Laplace. Leibniz, on the other hand, cosmopolitan haunter of courts, was sensitive to formal beauty and to the extreme methodic importance of notation. For thirty years he maintained a Europe-wide correspondence on notational matters; and his $\frac{dy}{dx}$ was much more suggestive, if more cumbrous at first sight, than Fermat's $f'(x)$ for derivatives, or Newton's $\dot{y}$ for "fluxions."

But with all aids, the calculus was so hard to grasp that it was well into the 18th century before the Bernoullis and a few other highly gifted mathematicians had so far systematised it that it was widely used by those of only moderate gifts.

Even in the case of Newton and Leibniz themselves, the manipulatory gain was achieved by ignoring some of the basic difficulties already mentioned, such as those involved in motion and flow. Had the French Academicians who banned the calculus been motived less by conservatism and more by these difficulties, their stand would have been admirable. But it might have remained disastrous. For it was these difficulties which had so delayed the Greeks.

Eighteenth Century

We now turn to the 18th century, but we give no continuous history of it, for mathematics now becomes technical and voluminous quite beyond the scope of a short history. We have mentioned the isolation into which a blind worship of Newton thrust the English school. We use this here to justify separate treatment; for, partly because of it, some characteristics of our rather limited mathematical faculty make an obtrusive appearance. These perhaps include originality and suggestiveness coupled with inadequacy of development. Three lines scarcely pursued further for periods of 100 to 150 years were started by the generation immediately following Newton, and then were mostly dropped. Taylor (1685–1731), before Arbogast (1759), foreshadowed the operational calculus which, 150 years later, was developed * by another Englishman, Heaviside. Taylor also considered the results of changing the variables in functional expressions. A century later, this suggested invariants.

Cotes by 1710, the French-born De Moivre (1667–1754 ?) even

* Again, without due rigour.

earlier, noticed a few isolated cases of another phenomenon not fully discussed for a century, but destined then to become central: namely, the way in which the "mysterious" $\sqrt{-1}$ furnishes connections between the logarithmic and trigonometrical functions. Besides these three cases, we must remember that Taylor's (1715) and Maclaurin's theorems on the expansions of $f(x+h)$ and $f(x)$ in powers and derivatives were not proved in any serious sense (although they were much used) until nearly a century later.* Maclaurin's (1698–1746) work on astronomy, Taylor's on vibrating strings, illustrate another English tendency—stress on applications. But none of these were kept up. After this first generation, English mathematics decayed until after 1800.

The rest of this chapter will mostly be concerned with the continental school. Into its work we are led in the wake of the Bernoullis, a family which, for a hundred years after Newton's time, was almost as rich in mathematicians as were their contemporaries the Bachs in music. There were, for instance, James (1654–1705) who systematised the calculus, John (1667–1748), and Daniel (1700–82) who systematised hydrodynamics (1738).

Our few notes centre mainly round applied mathematics, for not only were the applications studied on the continent even more brilliantly than in England, but they gave rise, as we shall see, to vital developments in pure mathematics, especially in differential equations.

In contrast to England, the continental workers treated the applications from a characteristically general and theoretical standpoint, almost (as D'Alembert and Lagrange actually said) like Euclid with time as a fourth dimension. Every problem was reduced to a differential equation. Hence the central position of this subject.

They generally selected as their basis some particular interpretation of Newton's laws such as the "Conservation of Mechanical Energy" (Bernoulli), or "D'Alembert's Principle" (1743, from Newton's second reading of his third law) or "virtual work" or "least action" (Lagrange). From these as axioms, they deduced immense superstructures of formulae. The physical background was disproportionately small, but the prior existence of these great bodies of theory accelerated the development of mathematical physics when it bloomed in the 19th century. In particular, the great Lagrange pointed out (1780) that there is no need for the data about a mechanical system to take any one specified form, such as the conventional one of initial positions, velocities, masses, and forces—that *any* independent coordinates, sufficient in number, would do. He gave equations into which these "generalised coordinates" could be directly inserted. This is a great help in physics, where the inner mechanism is often known only in part.

Least action, the most powerful of these doctrines, was a matured form of the minimum principles of Chap. V. One of these was Fermat's optical principle, which John Bernoulli had put in a form

* H. W. Turnbull believes (Nature, 9 July 1938) that J. Gregory had Taylor's theorem forty-five years earlier.

valid for refraction also. Proposed (1744) in metaphysical shape by Maupertuis (1698–1759), least action was proved from Newton's laws by Lagrange in 1760, for a very general type of case. Action was the space integral of the momentum, or the time integral of the *vis viva* ($\int mv^2 dt$).

To deduce the path of a mechanical system from this involved a new type of problem in pure mathematics. The " pure " and " applied " fields thus advanced together, the new problem being the discovery of the path or function which made a certain other function of the same variables a minimum. This is clearly not a matter of the simple maxima and minima of the differential calculus, where the function is *given*, but of a new and very much more difficult subject. This subject came to be called the Calculus of Variations. The ancients had known the solution of certain simple problems in it ; for example, the sphere as the minimum surface of a given volume.

The use of the word " given," here, reminds us of the important point that there must be a fixed—conserved—quantity in a system for a minimum principle to be applicable : that conservation principles and minimum principles are correlative, and must not be thought of apart. For the 18th century the quantity conserved was that which we have linked above with the name of Bernoulli—mechanical energy. Lagrange's methods applied only to mechanics, to the massy, rigid particles of 18th-century science. They were not at once applied to physics and chemistry—to light rays or to the " force " of affinity. The doctrine of the conservation of *all* energy was still in the future.

The 17th century had found the solution of certain problems in the calculus of variations, notably that the cycloid is the brachistochrone or curve of minimum time of fall between two points not in a vertical line. Euler (1744) found necessary conditions for the case of one variable, Lagrange, refounding the subject (e.g. 1755 and 1759), for two. As with the simpler case of a given function, these did not discriminate between maxima and minima—they were necessary but not sufficient ; and in 1786 Legendre, somewhat on the analogy of the simpler case, went on to develop criteria for this purpose. But when Weierstrass, nearly a century later, returned to the subject, he found it still in its infancy. It is, in fact, such a difficult one that we cannot go further into it. We merely note that the solutions of Euler and the rest reduced it to differential equations. It was thus on these that everything turned. Unfortunately these, too, are so difficult that we can only give superficial notes upon them. For us their chief significance lies in certain further problems to which they drew attention.

Newton (1676, published 1693) solved a differential equation in infinite series ; and by about 1693, also, Leibniz and the Bernoullis began to work out most of the isolated " dodges " by which equations finitely solvable can be brought under control. Euler joined in this pursuit. He probably knew of integrating factors by 1735.* By about 1750 this " formal " development of the subject was nearly over, and

* " Singular " solutions not given by the general formulae were known to Leibniz. Lagrange pointed out (1774) that, geometrically, such a solution is the envelope of the family of curves given by the general formula.

it had begun to be realised that such development usually leaves the worst of the problems still to be solved.* In one simple sense this is indeed obvious, for these men reckoned that they had solved an equation when they had reduced it to the form $y=\int f(x)dx$, and, as we have seen, it was only for a few functions that integrals could be found. But it was true in a much more general sense.

It was *partial* differential equations—those with more than one independent variable—which raised the peculiar problem destined to have such important effects in 19th-century mathematics and physics. The physical connection should be noted. Indeed it was a physical problem, that of the vibrating string in acoustics, which gave rise to the partial differential equation in question (Euler, D'Alembert, 1747).

Before turning to this, we may cite a case in which 18th-century mathematicians reduced a complex physical problem to a differential equation. The idea of potential was first use by Lagrange in 1773 (not the name, see Chap. X). At that time the special problem which most attracted applied mathematicians was that of finding the gravitational attraction of various bodies at an external point. The convenience of the potential is that its differential coefficient at a point with respect to any direction gives the force there in that direction. It is the sum over the attracting system of all terms like $\frac{m}{r}$, where r is the distance of the point from the mass m. In 1784, Laplace showed that this potential, for empty space, satisfied the now ubiquitous differential equation :

$$\frac{\delta^2 V}{\delta x^2}+\frac{\delta^2 V}{\delta y^2}+\frac{\delta^2 V}{\delta z^2}=0$$

which he gave at first in its vastly more complex form in polar co-ordinates. Suitable solutions of this equation when the value of V is known on concentric spheres were called spherical harmonics. Legendre had given a two-dimensional form of them in the preceding year. By substituting $-4\pi\rho$ for zero on the right-hand side of the equation, Poisson (1813) made it give also the potential in space occupied by gravitating matter of density ρ. It was long (see Chap. X) before the equation was applied to the electrical case where it is now so familiar. We cannot discuss the formulation of its solution, but it will be clear that to reduce the gravitation problem to this equation, and to solve the equation, are two decidedly different things.

We now consider in conclusion the mathematical question arising out of the acoustical equation mentioned. A differential equation with one independent variable has an arbitrary constant in the solution, to be settled by reference to the boundary conditions in each particular case. One with two independent variables has an arbitrary *function* to be settled (for instance, in this case, by the initial form of the vibrating string) and it was this function which was the trouble. What, Euler asked in effect, are we to do if at the start the string is (say) partly in a straight line, partly in a circle ? Can we use the expression

* The realisation had never been wholly lacking. Huygens, who was never lost in formalism, had proved an existence theorem for a solution.

for a circle part of the way, that for a straight line the rest of the way? D'Alembert said no. Then in 1753 D. Bernoulli added a fresh piquancy to the position by an extraordinary assertion (made largely on physical grounds), which would have made Euler's question needless. This was, that we can always represent *any* function *or set of such* by an infinite series of sine and cosine terms with properly chosen coefficients; so that only one expression is needed. Euler retorted that any such series is periodic, while not all functions are.* Even if it referred only to a finite stretch, Bernoulli's assertion seemed to comprehend under a single infinity of coefficients all the "infinitely infinite" variety of possible curves.

These men had no idea that two functions might coincide for a continuous infinity of values and then diverge. The problem was, in fact, too difficult for them, and was not understood until the great 19th-century clarification of the idea of a function to which we come in Chap. XII. For, with the advantage of this clarification, we can see that this problem confronted the men of the 18th century with all the peculiar weakness of their point of view. By function they meant, with Euler, expression in x, y, etc.; and it did not occur to them that a functional relation could exist apart from a formula. The formulae for $\sin x$, $\log(1+x)$, etc., had suggested the ambition to bring every relation into an explicit form, with the variables, or variable and constants, wholly separate. Taylor's theorem seemed to be a supreme triumph here. So did the formula connecting $\sin x$, $\cos x$, e^x via $\sqrt{-1}$, or Euler's view (1748) of logarithms as indices. And yet, as the century went on, certain apparently quite simple problems continued to defeat this ambition. Prominent among these were the quintic equation, and the "elliptic integral" $\int \frac{d\theta}{\sqrt{1-e^2\sin^2\theta}}$ (or similar forms) involved in finding the arc of the ellipse.

Defeat over these prompted the nineteenth century to a humbler or more critical frame of mind, to a realisation that formulation, whether in equations or in the apparently more satisfactory form of a solution of them in terms of imaginaries or infinite power series, can leave most of the problems still to be faced, can (like so much else in the 18th century) be mere empty formality unless we know exactly what (say), the operations $+$, $-$, etc., *mean* as applied to infinity or $\sqrt{-1}$—unless we know the range over which an infinite series ceases to converge. For a century after Huygens, the last of an older tradition, the continental mathematicians were almost as much drugged by an elegant symbolism as the Englishmen were by an inelegant one. In their opium dream they largely forgot the old Greek doubts about such matters as the infinite. But by 1800 those doubts were coming back.

* See, for all this, *Isis*, Vol. I, 1913, pp. 661 onwards. Article by P. Jourdain.

This was another type of applied mathematics, destined to be of immense importance. Here we can only mention that it became a subject in itself from the time of the solution by Fermat and Pascal (1623–62) of the problem of the unfinished game. This problem had halted its earlier workers. As earlier, the impulse here came from actual gamblers. At the same period, economists or administrators, like Sir W. Petty (e.g. 1662) in England, De Witt (1671) in Holland, Vauban in France, initiated the empirical side of the subject and its connection with annuities, insurance, and the like. Deaths had long been (very imperfectly) recorded. Graunt (1662) and, more important, Halley (1693) constructed mortality tables. The subject made great advances in the 18th century on several fronts. Its finite side was substantially completed by such brilliant mathematicians as J. Bernoulli, who also raised the thorny question of inverse probability. Another side of the subject was raised by Lagrange (1770), namely the theory of errors, of how to obtain the best value from a set of non-concordant observations of the same physical quantity. The method of least squares was used in astronomy by Euler and Gauss, but, apart from an uninfluential early exposition by De Moivre (1733), it was Laplace who first (1778) explicitly gave the normal error law $y=ae^{-cx^2}$. Here the subject leaves the bounds of the finite.

Laplace drew together into one whole all the lines which made up the theory of probability. As with mechanics, the existence of this ready-made body of theory had great influence on the development of physics and other sciences in the 19th century.

CHAPTER VII

MICROSCOPY, CLASSIFICATION, GEOLOGY

THE subjects of this chapter are not as miscellaneous as they may seem. The first two represent tendencies which may be contrasted in descriptive biology since Vesalius turned inwards and divided up the human body, and the early botanists turned outwards and set about collecting specimens of all the forms of life. The first seeks light in cutting things into smaller and smaller parts, and may be called analytic. The second seeks it in seeing things as parts of wider and wider groups, and may be called synthetic. The two interact, and their separation is purely for convenience. Microscopy continued the analytic process, classification the synthetic. Nowadays, classification would perhaps be better called analytic, and microscopy synthetic ; but this is a change which has occurred in the last century.

Geology shares with them a mainly descriptive, as opposed to abstract, character, which makes it convenient to treat all three together. But, further, it had, and has, great intrinsic connection with biology. Its picture of wide natural processes laying down strata, embalming life's past, gave to the bloodless categories of the classifiers the living warmth of a background, a continuity, which led up to evolution.

This chapter does not represent all sides of 17th and 18th-century biology. It leaves physiology for the most part to the next chapters. But it sketches some of the chief lines of preparation for Darwin.

MICROSCOPY

In microscopy, we turn to the second of those effects of the Galilean concentration on lens systems of which 17th-century astronomy was the first. It has few landmarks until the end of the Thirty Years' War (1618–48). But from quite early in the century scattered men, chiefly in Italy and the Netherlands, had been working with magnifying lenses, submitting to them every type of small object on which they could lay their hands. They developed the uncanny skill of early craftsmen in every medium. Then, towards 1650, they began to make decisive improvements in technique. They found out how to keep animal specimens long enough for minute observation, by preserving them in alcohol. They began to get single lenses of great strength. These, and not the still unsatisfactory compound systems, were the instruments of most of their discoveries. They experimented with " preparations " in which the parts should have the attitudes of life : filling them with air, with liquids, and finally with fusible solids (Boyle, 1663 ; especially Swammerdam, 1672).

Swammerdam's father had collected, from the Dutch tropical trade, the largest private museum of exotic animals and plants in the then leading city of Amsterdam. We are thus reminded that these microscopists form an early fruition of what might be called the Museum Movement. With the Jesuit Kircher (1602–80), Swammerdam (1637–80) was the first to see the red blood corpuscles (1658, published only 1738). Thus the now complex study of blood chemistry was made possible. Then (1660) Malpighi saw the blood moving in the capillary vessels of the frog's lung, and thereby supplied the missing link in Harvey's proof of the circulation of the blood. His discoveries on the lung made possible the physiology of respiration developed in the next century. His work on embryology revealed a fact later to be of great evolutionary interest. Like Hooke earlier, he was quite familiar with *cells* as features of plant tissue, without seeing in them the universal significance given them in the 19th century. With Leeuwenhoek (1632–1723) he was an early foreign correspondent of the Royal Society, and one of those who set the important fashion of short papers as opposed to bulky, over-systematic treatises.

Leeuwenhoek was the first man who set eyes on bacteria and (with his pupil Ham) on spermatozoa (1674) ; thus making possible all the 19th-century knowledge of bacteriology and of reproduction. Similarly his discovery of the striations which distinguish voluntary from involuntary muscle made him a founder of histology.* But it should be specially noted what long intervals elapsed before the germination of the seeds sown by these men, intervals commonly of a century or more. An adequate theoretical outlook had to be developed first. To this, the early microscopists made no contribution. Their work often illustrates the contrast between a child's and a scientist's curiosity—the fertility, but yet inconclusiveness, of the former. So largely did they exhaust the structures visible with single lenses that with one main exception (and physiology apart) there was a gap in analytic biology until the 19th century.

The exception was the subject in which Harvey had been halted precisely by the lack of microscopy : embryology, with its observations of the chick and its theories of spontaneous generation. Malpighi had tested inconclusively whether the earth itself could bring forth plants, but in those days men had not our clear-cut distinction of living and non-living to apply automatically to such doubtful cases as fossils or the humus of the soil. Tournefort, one of the most considerable botanists of his time, thought that crystals were plants ; " self-nourishing " as they were, and growing (as we might say) from " seeding." It was all part of the time's still grave imperfection in the concept of the physical. Redi (1627–97), another of the microscopists, had shown, with his new weapon, that one class of cases then supposed to show spontaneous generation did not show it, namely, grubs and other small creatures. But plenty of other classes remained. Spontaneous generation as then held now seems an absurdity, but the name is equally applicable to the view that life must have arisen from the non-living in some past geological age ; while the analogy of crystal growth is still used by mechanistic theorists, and recent work on viruses should be remembered.

* Baglivi (1700) made a similar discovery.

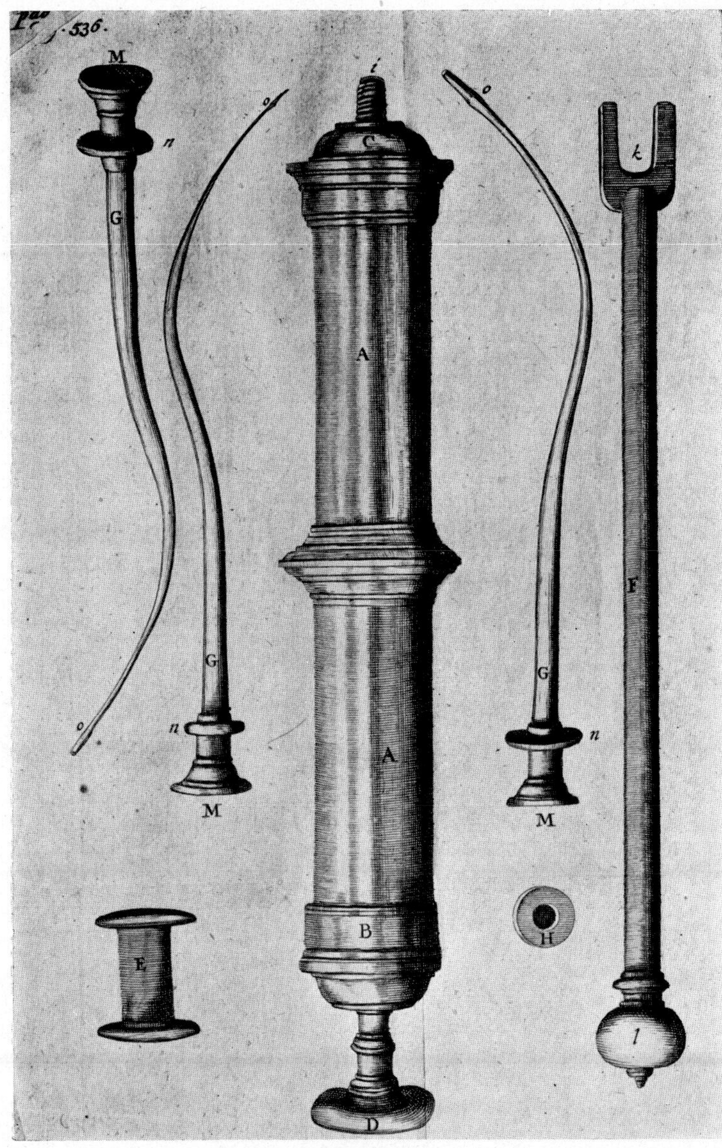

De Graaf's injection syringe.

PLATE IX

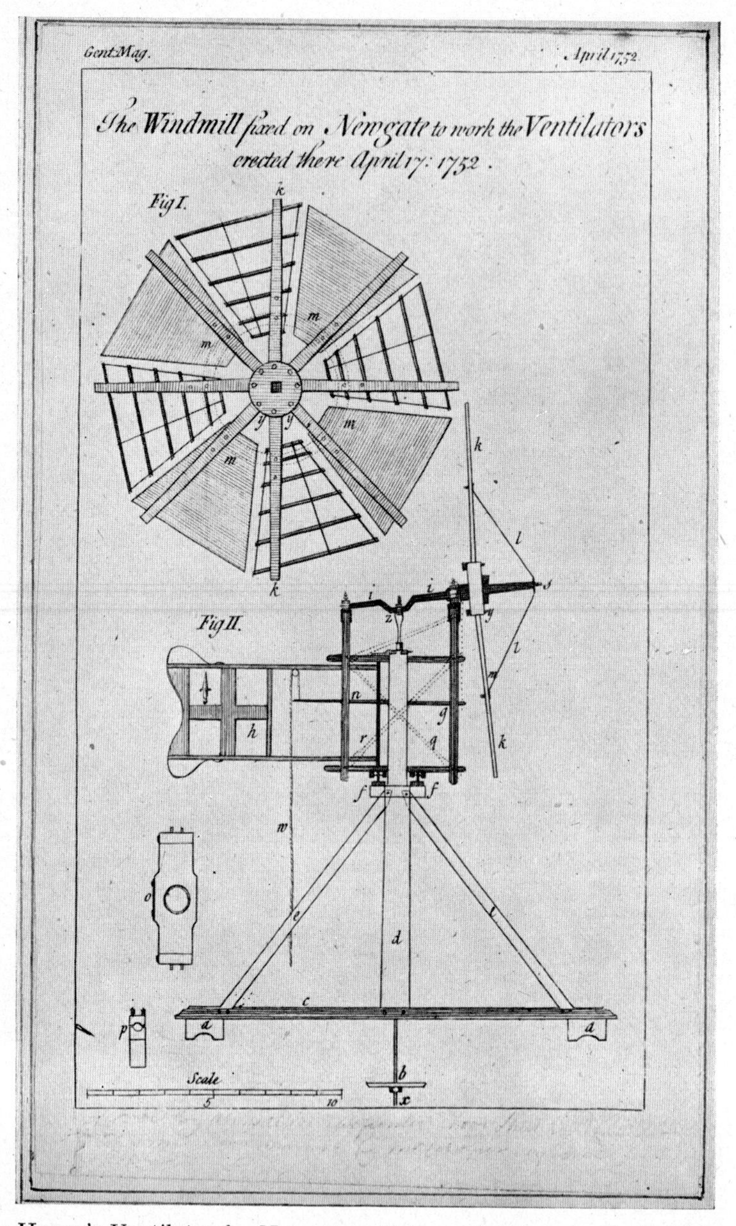

Hales's Ventilator for Newgate. *From a drawing in the British Museum.*

Leeuwenhoek's discovery of the spermatozoa suggested to the 18th-century mind a much greater refinement of absurdity than spontaneous generation. This was "emboîtement," or preformationism carried to its logical limit in the following way. Harvey had failed to find in the deer a mammalian "egg"; and although De Graaf, with lenses, had (1672) seen the resemblance between the ovaries of a bird and the ovarian follicle of a rabbit, the denial of a second partner in the embryo's origin became widely orthodox in the 18th century. The orthodoxy took two forms, one maintaining that the male, and the other that the female, was the only factor in generation. The great Haller (1708–77) adopted the view of the womb as a mere incubator for the spermatozoon. Now, according to emboîtement, the sperm (thus become the sole factor) must contain every detail of the mature creature ready formed, though on a much smaller scale. For, if, say, the surface of the spermatozoon presently splits, there must have been some *pre-existent* reason why it should do so at one point rather than another. But if it (therefore) already contains every detail of the mature creature, it must, *inter alia*, contain spermatozoa; and each of these, another complete creature in little. And so *ad infinitum*.

The Aristotelian doctrine had been that the male contributes form, the female only substance. But, as the alchemists degraded Aristotle's non-material "element" into a crudely material entity, so by Harvey's time "form" had been degraded into a sort of aura surrounding the embryo. Such was the hampering side of the greater richness of the European mind. Harvey supported the degraded idea and yet his ultimate effect was against them both. This is a typical and recurrent phenomenon among the great biologists. Their observational work, and the discoveries for which they are memorable, often seem to us directly to controvert their own theoretical convictions. Aristotle himself, an excellent observer, is another case in point. We meet others over the evolution question. It is perhaps because in biology generalisations or speculations which merely *cover* the facts are fatally easy to arrive at, while true hypotheses, which also go beyond them (and that in ways admitting of definite test), are very hard to reach.

This particular Preformationist doctrine lasted until Wolff (1733–94), of Berlin, described (1759) the formation of the chick's intestine from the layers the discovery of which, in the 19th century, proved to be such a turning point. These layers appear in the formerly undifferentiated ball of cells into which the egg divides.

But while it began to be evident that what is sensible is merely a question of the power of the lenses used, it began to be equally evident that the purely logical argument becomes inapplicable for this very reason. Moreover chemical differentiation was ignored. This type of lesson has had to be learnt by each branch of science in turn.

Classification

It is the accumulation of vast masses of data which gives the impulse to the ordering, classifying spirit; and we have seen that as early as the 16th century there had gone with these great accumulations other things which they imply, cooperation and communication. Gesner had had correspondents in several countries. L'Obel had culled specimens from Asia and America. In the 17th century Ray (1627–1705) had helpers in East Anglia, in Teesdale, and elsewhere. In a classic partnership with his patron Willughby, he travelled England and Europe. Tournefort (1656–1708) traversed Europe and the Near East. Thus by the 18th century much experience had been gained and the subject itself had grown well-defined.

Most of this section will be concerned with botany, for, as we have seen, it was here that early workers met their greatest difficulties and their greatest triumphs. Most of our notes on animals will be coupled with comparative anatomy and palaeontology at the end of the chapter.

Bauhin is regarded as the first man who set out not only to classify, but to do so on a *system* (" systematist "), that is, so to organise his pigeon-holes that their symmetry or subordination was an intellectual satisfaction in itself and a stimulus to further search. He began (1620) to organise species into genera, and so, like Jung, made a faltering approach to the now standard binomial nomenclature. He seems to have influenced Linnaeus, as Jung seems to have influenced Ray. The latter line, which perhaps began with L'Obel, moved towards a " natural " classification (see later).

Ray marks, with Willughby and Tournefort, a great step forward in many matters besides classification. He recognised, for instance, what Jung did not, the sexual nature of flowers; while his idea of the biological status of species was much more modern than that of the Linnaeans of a century later. For him fixity was no dogma,* and in many matters he had that breadth of outlook which has made Englishmen such invaluable early workers in many sciences.

Ray was primarily a botanist, Willughby a zoologist. The former's " Historia Generalis Plantarum " appeared in 1686–1704. In this, the basis of classification was the fruit, coupled with leaf and flower. There was over-emphasis, as we think, on the distinction of herbaceous and arboreal habit; but, like us, Ray divided both into mono- and dicotyledons. He applied to " herbs " the division of flowering and non-flowering. Some of his groups correspond with ours. Into others he imported many incongruities, even sometimes animals! For the animals and fishes which Willughby had collected, Ray adopted a classification by toes and teeth which places many species as they are placed now.

Tournefort's spirit, as became the country of Louis XIV, was more formal. It denied, for instance, the sexuality of flowers. But it led to the further development of the binomial nomenclature and to many of the detailed features which were used by Linnaeus.

We must pause upon the question of sex, for it illustrates a marked feature in the development of classification, its dependence upon physiological advances. We have already remarked (Chap. II) how much harder classification and nomenclature were in plants than in animals, because function—the creature's working or physiology—was not clear in the former case. Hence, when any new element in a plant's

* As we shall see in the next section, on geology, he was not alone in such views. Hooke held similar ones.

working was cleared up, advance in classification could follow. One major one was that of sex; while we shall see presently that another physiologically important feature, the vascular system of plants, was later on made the basis of further advances in classification.

As to plant sex itself, Tournefort was supported in his denial by Malpighi. The ancient Assyrians had known and used the sexuality of the date-palm, of which the trees themselves are male and female. Soon after 1690, Buonanni and Camerarius began to accumulate evidence on the subject. When men like Linnaeus (1735) began to consider horticultural practice in hybridisation, another case of the isolation of craft and theory was ended, with stimulus to at least one party. Holland, which Linnaeus visited, had been a centre in this respect for a century or so. The similar case of plant and animal breeding was not long in following (see Chap. XI). The sex of plants was thus in the air in the early 18th century, but has proved very complex.

In Linnaeus (1707–78) the idea of classification as an end in itself reaches the zenith of its influence. After him it gradually fell to the status of a convenience, gaining vastly, in the process, in breadth and naturalness. Linnaeus used, as we have hinted, the sexual organs—flowers, and especially their stamens—as his basis. His book "Systema Naturae" (1735–58*), like Newton's in another realm, was a great advance in business-like brevity and system. He standardised the binomial naming—perhaps his greatest service. A regular series of points was stated for each species. The prolixity with which the old writers had covered the gaps in their information was excluded. A complete hierarchy, classes, orders, genera, species, was erected. His placements under the first two have rarely, those under the last two have often, survived in our present schemes. Linnaeus was the first man to be able to cope with the flood of extra-European species. Like all work until long after this, his system strikes us as very weak on the "lower" plants—algae, fungi, etc. These were, of course, a knotty subject.

The Swede was no armchair theorist, drawing abstract distinctions between living forms in which he took no further interest. He was, on the contrary, a tremendous personal force and a life long lover of wild nature. From the age of twenty-five he was constantly exploring and inspiring pupils to explore. He or his agents travelled Lapland, Arabia, Egypt, Palestine, the Cape, the Antarctic, Java, Ceylon, N. America. From 1732 to 1749, he was often on the road himself. His correspondence, like his fame, was Europe-wide. The stimulus to field-work was immense. It became usual for every expedition to have its naturalist, such as Joseph Banks and Linnaeus' pupil Solander with Cook's first voyage (1768–71).

If the Linnaean system froze up progress for a time, it was, as with Newton's, by the immensity of the founder's influence. Linnaeus perfectly realised the artificiality of fixing on certain characters by which to classify, and of sticking to them. Before his death, he started on a more "natural" system; that is, one in which more and more

* Many editions.

characters are taken into account. But this would not have been an evolutionary one : Linnaeus, though he discussed hybridisation as a source of species, had scant sense of continuity. It would have been based, like modern systems, on balancing many criteria ; but it would have been a system, a *finished* thing, all the same.

Like all the systems of the 18th century (in medical practice, for instance, but also outside science) this of Linnaeus has for us an empty, formal, look. The arrays of species lack the connective tissue, the life-blood, of the evolutionary doctrine. But the point to grasp is that to Linnaeus it had very much the kind of attraction, and the supreme status, that evolution has for us. It was emphatically a great biological synthesis, a victory of mind over matter. Its very imperialism shows this—its sweeping inclusion not only of plants and animals but of minerals and even of diseases.

The inclusion of minerals was an old tradition of the natural history cabinet.* That of diseases reminds us that the great Englishman Sydenham (1624–89) had been the first to treat epidemics as objects for study and comparison in precisely the same sense as any of these other objects of descriptive natural history.

Whatever the contrast of spirit between Linnaeus and ourselves, it remains true that classification had with him reached a stage when most reforms amounted only to moving existing divisions in blocks. Had this not been so, the mass of field work which Linnaeus provoked would have become unmanageable.

Two men, A. L. de Jussieu (1748–1836) and A. P. de Candolle (1778–1841), somewhat in the tradition of Ray, are especially associated with the changes which enabled the systematics of 1859 to absorb the Darwinian viewpoint with so little friction an ‹ to survive in many essentials to the present day. Both these men worked in France, though de Candolle was a Swiss. Both were members of families which included many other botanists. Bernard de Jussieu (1899–1777) was in fact responsible for the basic ideas of the system brought out by his nephew (1789). This added acotyledons to Ray's two classes, but abandoned Ray's primary division into woody and herbaceous plants. It made much use of stamens and petals. To it de Candolle (1819) added the important perception of the systematic value of the vascular bundles in the stems. These are absent or much simplified in the lower plants, on which de Candolle was the first to make a serious attack. He went further than de Jussieu towards a natural classification in which every important character should have due weight. But this major advance was not the work of any one man, but of the whole spirit of a new century, to which we turn in Chap. XI. It involved the recognition that *no* basis of classification is more than a convenience.

* Linnaeus made geological tours, and in his last edition (1768) of "Systema Naturae" he gave a complete list of fossils in the order of their occurrence in his system of rock succession. The real classifier of rocks, however, was Bergman (1735–84), another Swede ; while it was Werner who, adopting Bergman's divisions, carried the systematic spirit of the 18th century to its greatest extreme where minerals were concerned. See later sections of this chapter.

GEOLOGY

Unusually late, geology remained throttled by philosophy and theology. In spite of Copernicus, men found it hard in practice to imagine that the earth did not form the major part of the universe. Thus the Record of the Rocks became the Record of Creation, and geology was linked with cosmology, a branch of philosophy. Positive facts remained few and local. Also, like living forms, the rocks are too complex to be among the earliest subjects for hypothesis as opposed to generalisation. It was not until the late 18th century that hypotheses capable of test began to be proposed in true scientific fashion. Speculation on geology had, however, by them been rife for several generations. Indeed, as early as 1571 the Dane Severinus had called for facts to balance it.

Early observations were greatly conditioned by man's rather few means of access to the interior of the earth. In Italy notes had been made on volcanoes from Leonardo onwards; in Germany, notes on mines from Agricola onwards; in most countries, notes on fossils. In France about 1750, *extinct* volcanoes began the train of decisive discoveries. In Germany, not much later, mining provoked the serious classification of rocks. In Great Britain the surface cuttings and canals of the agricultural and industrial revolutions aroused the curiosity of Hutton and W. Smith. In the early work, partly palaeontological as it was, botanists bore a large share.

Leonardo da Vinci, Giordano Bruno, Stevinus, and other men of the Renaissance, with their vivid freedom from tradition, at once reached such common-sense conclusions as, that the rivers had sculptured the valleys, that some rocks had once been molten, that the sea, which is (Bruno) always near active volcanoes, must have some connection with them, that the land and the sea may not always have been in their present positions. But the clarity of these early views was soon lost. It was unsupported by work in the field, and put side by side with pre-scientific ideas. Fossils, for example, were Sports of Nature played on mankind, or the earth's imperfect successes in its efforts at the (spontaneous) generation of life.

We have mentioned in the first chapter that Agricola (1494–1558) was a recognised mining expert. He made observations on crystals, their cleavage, lustre, colour, hardness, and was among those whose imagination was struck by those columnar formations of basalt (such as the Giant's Causeway in Ireland) which were to prove crucial later on. He did not, however, eschew speculation.

Regularly stratified rocks and their faults have always been a central theme for geological curiosity. Steno (1638–86), a Dane like Severinus, was an early worker on this (Italy, 1669). Men began to dream of geological maps, though ordinary ones were still of the crudest. Matters, in fact, were beginning to move, though isolation of effort was still the rule. In 1674 the Frenchman Perrault produced evidence (against the Aristotelian view) that rainfall is more than enough to explain existing springs and rivers.

By this time, the subject's few serious students fully realised that fossils were organic remains. So did the new " natural philosophers," who saw nothing impossible in wide crustal upheaval or climatic change. Such were, for instance, Hooke, Ray, Woodward in England, Vallisnieri and Moro (*c.* 1700) in Italy. Hooke used for fossils the metaphor of a coin. He also suggested extinct species, as opposed merely to dead individuals, to account for these relics. While this last view was strongly opposed, it could itself issue in fallacy and speculation. The Flood (asserted the " Diluvialists ") had destroyed previous creations, of which the fossils were relics. When the 18th century revealed several very different series of fossils, a whole series of " Catastrophes " was asserted, without evidence, by the school of that name.

To us, the essential point in this is that the evolutionary connection was denied, each catastrophe entailing a fresh creation. To the scriptural idea of the Flood is perhaps traceable, also, the contemporary notion that the whole surface of the earth was primitively an ocean. This fell in with the idea that the past was essentially *simpler* than the present.* Emboîtement was an instance of the opposite idea that, logically, we must suppose a constant " complexity." Both ideas still provoke *a priori* reasoning.

But by the early 18th century truly scientific bodies of theory were being built up, not indeed in geology, but in a subject related to its cosmological side, namely astronomy. It was not the field-workers of the time, earth-bound and non-mathematical, who realised this, but the theorists : Leibniz, Buffon, and later Kant. They brought Newtonian astronomy into the service of nascent geophysics.

Leibniz's theory (1691) was developed from one of Descartes (1644). A molten earth, solidifying, had crinkled into mountain ranges, while the interior became cellular. The cells, now and then collapsing, engulfed some of the surface water which was condensing from the surrounding vapours. In such commotions the strata which had already solidified sometimes broke down into the " sedimentary rocks," whose heterogeneous structure had already been contrasted with the homogeneity of other rocks.

In some points, such as the inward leakage of the Universal Ocean, Buffon's theory (1749 onwards) was the same. This leakage explained the dry land. But Buffon's service was his emphasis on the great periods of time which it all implied ; though his actual periods remained absurdly short. He also made the significant observation that the smaller fossils, with the least resemblance to living forms, were the deepest, actual bones very shallow.

Buffon's geological theory was an integral part of a typical 18th-century system, this time in natural history—a partly popular exposition of most of the science then known, including Newtonian astronomy. Like most such systems, it eked out fact with fancy. The planets, for instance, were produced from the sun by cometary shock. But it remained a fine achievement.

These speculations as to earthly and planetary origins are linked via Kant's nebular hypothesis with the more definite theory of Laplace.

* Hooke has, however, been claimed as the first Uniformitarian (see later).

But we must turn to Buffon's successors in geology proper, which was now beginning its definitive advances.

In Guettard (1715–86), France found a man with a sobriety of outlook worthy of Locke or Sydenham * or anyone of the English school. A passion for botany brought this son of a provincial apothecary into touch with established science in the persons of the brothers Jussieu at the Jardin des Plantes in Paris. Guettard kept his botanical—and so his palaeontological—interest when he became fascinated by geology. He noticed how rivers, rain, and weathering gradually change the surface of the earth, and how new surfaces are formed by the detritus brought down from such action elsewhere.

By field work around Paris, Guettard came to realise that superficial minerals generally lie along bands. When some of these bands ended at the English Channel, he inferred, from their general curvature, that they would reappear on the English side. He found that they did so. Thus, about the middle of the century, true hypotheses had begun to appear.

Guettard never realised clearly the definiteness of the *succession* of his bands, that side by side on the surface meant above and below underground, and that this in turn meant after and before. In this he was forestalled in 1719 by the Englishman J. Strachey. It was, however, the time of the appearance of the first truly surveyed map of France ; and the French Academy saw that a new field had opened. Under these official auspices, Guettard, with the help of no less a young man than Lavoisier, wore himself out beginning a geological map (published 1780).†

On the meaning of stratification two Germans, Lehmann (1756 †) and Fuchsel (1762 ‡), were ahead of Guettard ; though otherwise their observations were much more local and their theories more obscure. Each did work in the German mountains and on its basis drew up schemes of successive strata due to the action of different physical agents in successive epochs, strata which, generally horizontal, had tilted to form ranges. Somewhat similar work had been published in Italy by Arduino and in England (1760) by the same J. Michell who had been responsible for Cavendish's apparatus. The idea of succession began to spread.

It drew strength from the first exploration of a mountain range of first-rate magnitude. This was made by Pallas in the Urals, as part of the huge survey (1768–84) of the natural and human resources of Russia and Siberia ordered by the westernising Empress Catherine II. In spite of theories of crinkling already instanced, mountains had generally been regarded as solitary relics of entire strata which had once covered the earth and had themselves been covered by the primeval ocean. Pallas showed that on the contrary mountain strata bear every evidence of having been tilted or wrinkled up, the older, harder, non-fossili-

* Though, of course, Sydenham allowed himself some *a priori* concepts, such as that disease is nature's effort to eliminate morbific matter from the system.

† First proposed by Lister in 1683.

‡ Dates of publication.

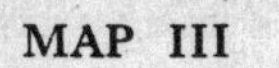

MAP III

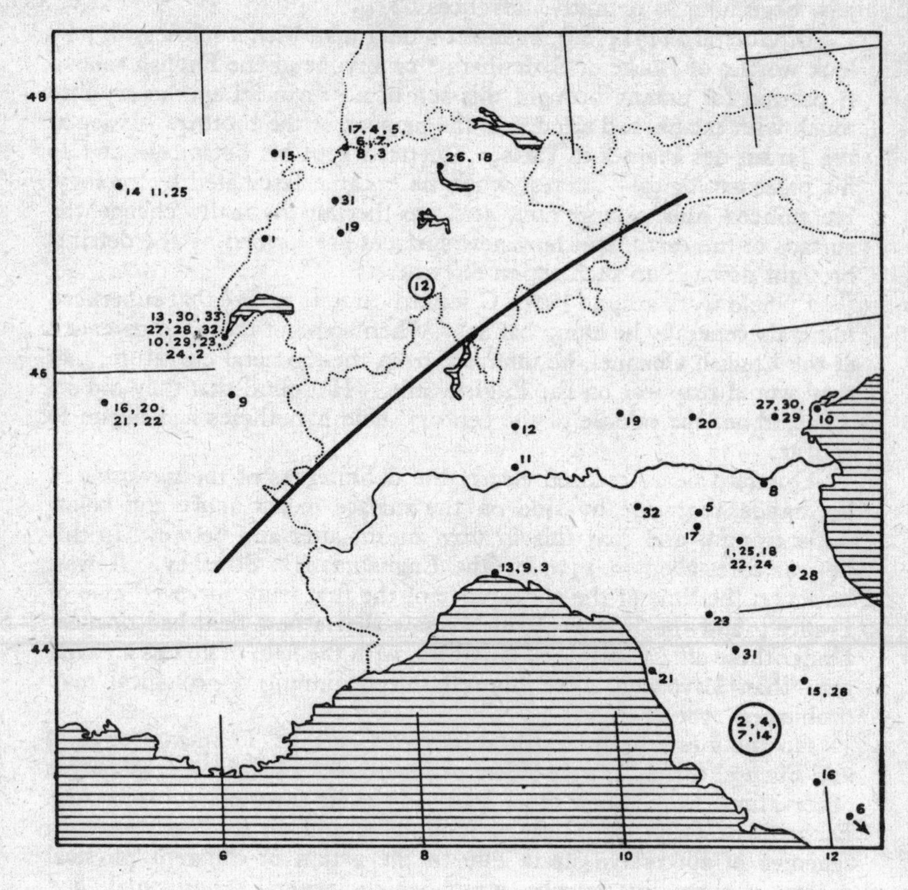

BIRTHPLACES OF SCIENTISTS

Please consult Explanation on p. 10 before studying the Map

North Italy up to about 1700

Number on Map	Name	Birthplace *	Number on Map	Name	Birthplace *
1	Aldrovandi	Bologna	17	Fallopius	Modena
2	Baglivi	Italy	18	Ferraro	Bologna
3	Baliani	Genoa	19	Fontana	Brescia
4	Benedetti	Italy	20	Fracastoro	Verona
5	Berengario	Carpi	21	Galileo	Pisa
6	Bruno	Nola	22	Grimaldi	Bologna
7	Buonnani	Italy	23	Leonardo	Vinci
8	Cabaeus	Ferrara	24	Malpighi	Bologna
9	Cabot	Genoa	25	Novara	Bologna
10	Cadamosto	Venice	26	Redi	Arezzo
11	Cardan	Pavia	27	Sanctorius	Padua
12	Cavalieri	Milan	28	Torricelli	Faenza
13	Columbus	Genoa	29	Vallisnieri	Padua
14	Cotugno	Italy	30	Vesalius	Padua (worked)
15	Cesalpino	Arezzo	31	Viviani	Florence
16	Fabricius	Acquapendente	32	Zucchi	Parma

BIRTHPLACES OF SCIENTISTS—*contd.*

Switzerland with S.E. France, chiefly about 1700–1800

Number on Map	Name	Birthplace *
1	Agassiz	Môtiers
2	Argand	Geneva
3	Bauhin	Basle (16th century)
4	Bernoulli, D.	Basle
5	Bernoulli, James	Basle
6	Bernoulli, John I	Basle
7	Bernoulli, John II	Basle
8	Bernoulli, N.	Basle
9	Berthollet	Talloire
10	Bonnet	Geneva
11	Buffon	Dijon
12	Bürgi	Switzerland (16th century)
13	Candolle, A. P. de	Geneva
14	Clément	Dijon
15	Cuvier	Mömpelgardt
16	Desargues	Lyons (17th century)
17	Euler	Basle
18	Gesner	Zurich (16th century)
19	Haller	Bern
20	Jussieu, A. de	Lyons
21	Jussieu, A. L. de	Lyons
22	Jussieu, B. de	Lyons
23	Le Sage	Geneva
24	Luc, de	Geneva
25	Morveau	Dijon
26	Paracelsus	Zurich (16th century)
27	Pictet	Geneva
28	Prévost	Geneva
29	Rive, de la	Geneva
30	Saussure, de	Geneva
31	Steiner	Utzensdorf (19th century)
32	Trembley	Geneva
33	Varo	Geneva

Eustachius and Borelli are among the rather few great Italian scientists born south of this map.

The Swiss towns were cities of refuge ; though in the 16th century, Servetus had been burnt at the stake at Geneva. The " Alpine System," in Switzerland and the Massif Central of France, gave us much of our early natural history and geology. Freiberg (see Chap. VII) in Saxony is another instance of this. Other Pictets, de Candolles, and de la Rives might have been listed. Mathematics was also important. Lagrange was born in Turin (not shown) in the 18th century. In wider matters the names of Rousseau, Voltaire, and others are much associated with this district.

* In a few cases, work-places, or the nearest larger place, have been substituted. Numbers within a circle on the map indicate that only the region, not the town or village, is given.

ferous rocks emerging at the centre of the range, where erosion has bared them of the later, fossiliferous ones outside.

Guettard had made another discovery destined to have decisive results, that of volcanic stone in the Auvergne and of a whole set of extinct volcanoes from which it came. These were now the centre of interest, especially from the work of Desmarets (1725–1815). Desmarets made a detailed survey, and a fine map, of the Auvergne volcanoes, with their lava streams deeply carved by later streams of water.

Agricola's " basalt " is found in the district. Now to explain its peculiarly definite forms was already an established ambition of geologists when Desmarets insisted on its connection with volcanic action (1763, published 1774). The evidence for this seemed to many insufficient ; for basalt is also found remote from volcanoes, and at that time it was supposed that rocks once molten must remain glassy. Desmarets himself had grave misconceptions which were unrealised at the time. He thought that granite, also, was a form of lava, an even older one than basalt. Only a much improved chemistry could remove this error, but another geologist, H. B. de Saussure (1740–99), tried to put the matter to laboratory test. He attempted unsuccessfully to obtain basalt from granite by fusion.

De Saussure started, like Guettard, as a botanist. A Swiss and, like Gesner and Haller before him, a great lover of mountains, he spent his life in the Alps, exploring the strata and collecting rock specimens on a great scale. All the time he believed that the central granite was a deposition of the primeval ocean.

This aqueous origin, in fact, as opposed to an igneous one, for these rocks, was the theory developed by Desmarets' opponents in the " Vulcanist " controversy which followed. It was a violent, most unscientific, controversy, which Desmarets himself avoided. Against his supporters was ranged the most powerful geological orthodoxy of the last quarter of the 18th century. This was the Wernerian school at Freiberg.

We have mentioned Werner as a Linnaeus of the rocks. Werner did good service to nomenclature ; and his pupils' observations, though not his own, spread over the known world. From him modern German geology takes its rise, and in himself he is no bad symbol of the German mining tradition, with which his family had been associated for three centuries. But at the crucial points, his influence, pontifically exercised, was against the truth. He failed to see the systematic importance of fossils, or of the crystalline forms of rocks ; and on the basalt question he indulged in an orgy of apriorism only to be expected from one who took overmuch pride in his freedom from speculation. For it is a fact very noticeable in the descriptive sciences, biological or geological, that too much theory in the wrong place is the nemesis of too little in the right place. It is a similar point to that mentioned under embryology.

In the end, some of Werner's own disciples, Von Buch and others, supplied their master's best refutation ; but by then the igneous origin of basalt had been agreed, and the whole subject of geology had been put on a new footing. In this result the Scot J. Hutton (1726–97) was the chief agent. Hutton, with a view to farming his paternal acres, had gone to study agriculture in the best school of his time, the Norfolk of the agricultural revolution. There the flints embedded in the local chalk attracted his curiosity, and he took up geology. He was already a student of chemistry, and a personal friend of Black, one of the greatest chemists of the time.

Hutton read and travelled widely. In him met the tradition of the great cosmological minds, Leibniz and the rest, and the less ambitious tradition of the field workers. Many agencies had been suggested by his time to explain the stratification of the rocks and the other phenomena of the earth: deposition by the oceans, transport and attrition by streams; melting, pressure; tilting and breaking of strata. But few had thought that even all these together could produce the observed effects. "Catastrophes" without everyday parallel were still thought necessary.*

Hutton added to the number of these normal agencies, but, more important, he first thought of testing whether they really could and did produce the observed effects. He concluded that vast periods of time must have been needed, and that there was no vestige of a beginning, no prospect of an end. But he persisted that substantially the same agents were operating now as had operated all through the past. This was "Uniformitarianism." Among the factors which he reckoned proven were the internal heat of the earth, with extrusions of molten granite, and volcanoes as safety valves. The rocks once extruded, there came the chemical, hydraulic, atmospheric, glacial, and other attrition and weathering factors on which he lovingly dwelt.

He published these views in 1785, more fully in 1795; but in neither case with any great clarity or attractiveness. It was a time when the sound and fury of controversial "Theories of the Earth" had put these in discredit. Hutton's work was not widely appreciated until Playfair expounded it (1802); and then it was keenly contested.

His chemistry, moreover, was not always sound, and on one point he was happily too pessimistic. He had contended that the processes of stratum-formation were too vast to be tested in the laboratory. But his friend Sir J. Hall, in the line of de Saussure earlier (1762–1831), disproved this, and disposed of an old point against the Vulcanists, by showing that we can generally cool molten "glasses" into solid ones, or into crystals, at choice, by varying the conditions. It was the beginning of a number of similar investigations which have led to modern geochemistry and geophysics.

Lyell (1797–1875) carried Uniformitarianism to an extreme which, while it became a dogma of almost Wernerian proportions in England, never commanded much respect on the Continent. But Hutton's essential service remained. He had provided geology with a theory at once covering the whole ground and critically founded.

We now turn to palaeontology. In France especially, the interest in the mapping of strata had grown with the realisation of fossils as guides to the chronological succession. This was an extension of an

* E.g. by the Wernerians.

observation by Buffon already mentioned. By about 1780 an abbé, Giraud-Soulavie, had begun to realise that strata can be divided in a definite way. Earliest, and typically deepest, are those containing only fossils with no living analogues (ammonites, etc.), then the shallower ones containing these, mixed with others having living connections; next, those with the latter alone; and finally, and shallowest, those containing trees and higher animal remains. With this realisation of a succession (though not yet an evolution) of simpler, then more complex, forms, geology yields up its full significance for biology. Thereafter palaeontology becomes a matter of huge schemes of detailed work in the field. For instance, the number, as well as the nature, of the fossils in successive layers were put under contribution in the work of Brongniart (1770–1847) and the great Cuvier.

To understand the work of Cuvier (1769–1832) it is necessary to realise that while the broad classification of the higher animals had never been a primary difficulty, the study of their comparative anatomy had suffered from the extreme practical importance of the allied study of human anatomy. Spigelius (1578–1625), an unworthy successor of Vesalius at Padua, had separated the two; and until Cuvier's time only that isolated medical genius John Hunter (1728–93) had fully realised the importance of comparative anatomy even for clearing up that of the human body.

Hunter went even further, and studied comparative physiology; but here we must confine ourselves to noting the few 17th and 18th-century landmarks in the study of anatomy. One important effort, started in 1645 by Severino (1580–1656), was the tracing of analogies of construction between more and more widely different forms. To this, as Severino had foreseen, the microscopists made important contributions. From about 1670 began the flow of monographs on individual organs or groups of animals: for instance, Grew's * on Stomachs (1681) and Trunks (1675). In 1681 G. Blaes (Blasius) published the first general summary of the comparative anatomy of vertebrates. As we have noted, the "classical" microscopists somewhat exhausted their subjects; and, in consequence, in the 18th century speculation tended to oust observation.

We must not pause over any further comparative anatomists before Cuvier. The palaeontological importance of being able to deduce, from a small part of an animal, a general idea of the rest of it (at least its systematic place), need not be insisted on. This was one of Cuvier's most sensational lines of work during what has been described as his dictatorship of biology. It was founded on a prodigious number of dissections, and upon a detailed classification of the animal species (1816).

Cuvier recognised four parallel "branches" in the animal kingdom: vertebrata, mollusca, articulata (e.g. insects), and radiata. His personal triumphs were in the mollusca (and in the fishes); his failure was the radiata, the radially symmetrical creatures, which (inevitably, in the

* Grew and Malpighi also studied the comparative anatomy of plants, fixing (e.g.) the distinction of root and shoot.

absence of powerful microscopes) were a mere collection of unclassifiables. None of his last three classes entirely correspond to ours: nor does his division of functions and organs into vegetative (respiration, circulation) and animal (muscles, nerves, brain). But his concentration on the axial skeleton, the two axial nerve chords of the articulata, the discontinuous nerve-masses of the mollusca, the tracheal system, and many other points, has remained significant.

The vast number of his detailed investigations entitled him to be—what he was—the first man to enjoy the full luxury of a bird's-eye view of the whole of life spread out backwards in time as well as around in space. For his work had convinced him that most fossils represent extinct species. But in spite of this great four-dimensional survey, Cuvier never doubted the fixity of species, and was an 18th-century Catastrophist in general outlook. He lent his authority against the work of his speculative contemporary Lamarck (1744–1829), who asserted the sensational doctrine that the species had changed into one another, the extinct ones having given rise to those now living. We pursue this line of thought in Chap. XI. In detail, Lamarck's ideas were too simple, for he envisaged the old single scale or ladder of nature, simply transformed into an evolution in time. Here Cuvier's "branches" were much nearer the facts, and, even without evolutionary theory, gave palaeontology the necessary animal classification to correspond with the geologists' succession of the rocks.

Geology is not a subject of which we can give a continuous history in this book, and we must leave it at the close of this, its formative epoch. From about 1808–11 the older tertiary rocks began to be mapped in detail in France (where their investigation is most natural). The very geology of France, that land of classicism, is classic in its richness and simplicity. Older again, the secondaries are more easily recognised in England, where they were attacked by William Smith (1769–1839). Smith was a civil engineer in an age of canal-making, and in the cuttings which he met or made he found his impulse to the production (1815) of his wonderful geological map.

Since then stratigraphy, palaeontology, petrography, and all the descriptive sides of the subject, have reached gigantic proportions, and are a matter for government surveys in most countries of the world. We cannot follow them.

Agassiz (1807–73), in a less merely descriptive direction, found in his native Switzerland his first evidence for an altogether more sensational importance of *ice* than had hitherto been suspected. Glaciers had been recognised by the Huttonians as breaking up and transporting rocks, but by the work of Agassiz (from 1837) it gradually emerged that entire ice ages had occurred, when large portions of the globe had for long and recent periods been entirely covered with ice. But this leaves the 18th century far behind.

One might have expected, after Newton, a great burst of discoveries. What came was a long period of slightly stunned assimilation, due in part to extraneous factors such as war. An observer born early in the century, and making the Grand Tour, would have been an old man before he came across, in the Paris of Lavoisier, anyone worthy of Newton. It was a century of long ripening of ideas ; in field physics ; in heat and chemistry and physiology ; in mathematics ; in geology. It was a century of growing specialism and concentration : no longer did men of the stature of Malpighi hope to make contributions to every science.

The results began to come at the century's very end. All through the period, science grew in externals such as learned societies and periodicals. It became popular in society. Men like Clairaut and D'Alembert were seduced from it by social favours. But the gospel of reasonableness, as if influenced by this courtly connection, had lost its 17th-century attractiveness and become the formalism of the Age of Reason, the age of the classifiers like Linnaeus and of the system-makers in all subjects. The old mediaeval encyclopaedic tradition, which we noticed in botany in the 16th century, had never lost its hold. In the 18th century it began to split up, into forms some dangerous and some benign. The old, often uncritical, compilations of earlier days began to harden into bibliographies as we now understand them, like Schurig's in embryology (1732). Haller is usually taken as the founder of scientific bibliography—as the first man who realised, and tried to cope with, the flood of scientific literature.

Equally benign were the cooperative encyclopaedias which existed largely for information and did not pretend to expound a closed symmetrical system of completed knowledge. The most famous of these is the French Encyclopédie of Diderot and his helpers. But unfortunately the Napoleonic pride of single individuals remained more conspicuous than cooperation. It is perhaps in medicine that system-making, with its inevitable slurring of points still unknown, became most notorious, but it was present in most non-mathematical subjects. In every age, medical ideas with a certain undoubted scope and a certain ingenuity about them have been erected into panaceas. In the 18th century this tendency reached its climax so far as qualified professional doctors go, being thereafter increasingly the province of quacks.

One of these ideas was the very simple one of John Brown of Edinburgh, that all ills are either depressions or excitements ; to be treated, respectively, with alcohol or opium. His own ills were depressions and he died of the cure. The matter, however, was not a joke. In Göttingen, for instance, this " Brunonian " doctrine was still causing students' riots at the end of the century.

Far more interesting was another of these ideas, that what kills cures—in extremely small doses. It is puzzling how often this is true ; until we recall that many diseases are due to animate enemies which succumb to the same agents as ourselves. Not only drugs, but the surgical knife itself, may be lethal or curative according to dose.* Much newer to Europe was another example of the "small dose," preventive inoculation. This, long practised in the East, was brought from Turkey in the 18th century by that eccentric female traveller Lady Mary Wortley Montagu (1689–1762). It began to be used for small-pox in Europe from about 1750. Of the development of this into vaccination, we shall speak briefly in Chap. XI.

The too-ambitious schemes of this century naturally failed, in every subject, to hold the ground they claimed. They became but a part of the obstructive crust of elaboration and formality, which even touched the art of war. Beneath this crust burned the passions of Pope and Swift, of Voltaire and Rousseau, heralding the French Revolution and its burst of activity. The word revolution recalls that the best energies were still not (any more than in mediaeval or Renaissance times) in scientific work. In France they were in politics ; and in 1700 France was still three times as populous as England and gave the tone to the Continent. In England more of the best energies went into the development of natural resources, partly because England's were larger and better-disposed than France's. The landed and moneyed interests, fresh from their 17th-century triumph, took up native and foreign technical advances. The great East Anglian landlords systematised and extended home and continental practices in crops and stock, manures and implements. The inventors and mill owners, putting in automatic machinery, and engines to drive it, accelerated the industrialisation which had long been taking place in certain areas. Large factories were not new, nor were simple machines. It was automatism and power which made the crucial difference. In varying degrees, all these changes spread in the 19th century over most of the white world.

In "pure" science France led throughout the 18th century. England's sponsoring of the Machine, however, was significant. It was to give her science, in the next century especially, a potency, but also a peculiar twist, influential far beyond her shores. This was the love of mechanical models as ultimate explanations of the universe.

It should be noted that, during the first two-thirds of the 18th century, "applied science," as we understand it, was undergoing a period of discouragement after the bright promise of the previous century. The great primary advances of the time, in textiles, in the metallurgy of iron and steel, in power, were inventions, not scientific discoveries. Even Watt was as much a mechanic as a scientist. There were many engineers, especially in France, of keenly scientific bent. Men like Desmarets and Lavoisier were consulted by the French government. But then it was not in France that the primary inventions were made.

It seems to be true that, even within science itself, invention is primary, research (in the sense of intensive application of already-systematised theory) only improves it or prepares its ground. But by the third quarter of the 19th century this improvement increasingly

* The "cure-all" system erected on this did not, however, interpret the idea so widely. It was confined to drugs, where it failed in many cases.

made the difference between commercial success and failure; and, at least in certain industries, systematic technical research, as opposed to brilliant forays, was becoming the business of undoubted professional scientists. In the 18th, many key industrial improvements had little systematic connection with the corporate body of science. And yet the interconnection and the small scale of everything in those days must be remembered. When Dr. Beddoes, of Clifton, wanted to exploit oxygen as a curative stimulant, the chemical assistant he found was the young Davy, and the instrument-maker, James Watt.

What of Germany? Even at the end of the century, the day of her scientific (and political) greatness was still ahead. And yet, with Frederick the Great and the "Sturm und Drang," the sleepy old ducal university towns were blossoming out of their pedantry; and by 1800 Göttingen and Halle were ahead of French and British universities in their general touch with life. By that date, too, three supreme men were simultaneously at work within the Empire: Goethe in letters, Beethoven in music, Gauss in science. With one characteristic German subject, philosophy, this section may fitly close.

With the "philosophy" of Industrialism, indeed—with the doctrine of Progress—France was more concerned than Germany. By her early 18th-century argument as to the relative value of Ancients and Moderns in literature, she brought in the doctrine of the increasing perfection, of the ultimate perfectibility, of man. But, as has been remarked, in another sense, "philosophy" is opposed to Progress. We contrasted this classical view of it with the unclosed progressiveness of science. The contrast remains even when, as with Berkeley and Kant, philosophy, in the name of science, takes a sceptical turn. For in that, as in the other case, there is no asymptotic, historical approximation to truth. Truth has already been found—to be unattainable.

It remained for a 19th-century German, Hegel, to try to unite the two points of view. His falling beyond our century illustrates two points. One is that our 18th century lacked, even in its doctrine of progress,* the sense of historical continuity, the sense of evolution, which blossomed in Hegel and in the German schools of history.†

The other point is that Hegel's extreme weakness in physics reminds us that it is after all in its *emancipation* from philosophy that science emerges, in spite of the present scientific and mathematical interest in some aspects of Kant and Leibniz. In the 18th century this emancipation went on both in physics and in biology. We have spoken of the colourless, soundless universe of masses and forces which this century adapted from Newton. This involved whittling away the merely fantastic and "spiritual" elements in a concept which was now coming to serious grips with experiment, that of the ether. Much use was made of this, and of "effluvia," in, for instance, 18th-century electricity and magnetism. Further instances of isolation, in this case of experiment and speculation, were thus ended; and whole branches of science were prepared for their definitive, 19th century, advances.

* Exceptions, such as perhaps Vico and Montesquieu, only prove the rule.

† Earlier than in Darwin, though without crucial features which Darwin added.

CHAPTER VIII

EXPERIMENTAL SCIENCE I

In several chapters of this book we have had to emphasise the dependence of chemistry and physiology on the development of the idea of the physical. This dependence appeared first in the ancient symbolic correspondences of planets, elements, humours, etc. It developed, with Harvey's analysis of the circulation on physical lines, into a truly scientific connection. Far apart as are mathematical physics and observational biology, the *experimental* sides of the subjects are inextricably entangled, with chemistry as a further strand.* To show how this tangle was straightened out is one of the main preoccupations of this book. This chapter and the next are devoted exclusively to it.

Two physical ideas will especially concern us, those of heat and of gases, with the chemical idea of combustion as a middle term. In the 17th century these ideas remained in a confusion which we shall not attempt artificially to clarify. Their gradual separation brought out another middle term, this time physiological, namely, respiration. The thread where we take it up is primarily chemical.

Seventeenth Century

We have said in Chap. I that even in the 16th century, chemistry, if feeble as a science, was strong as a trade. The 17th century was a time of further trade development. The glass of Venice and Murano made possible the crucial experiments of Torricelli and others on gases. Pottery and dyeing and agriculture were being improved. Pharmacy in Holland and Germany, metallurgy in England and Germany, were rich and ancient trades. Glauber (1604–70) and Becher (1635–82) attempted, though without success, great extensions of the German industry, so that its later predominance was no new idea. For the moment, the Netherlands and the British Isles were prominent, the former with Van Helmont, Sylvius, and later Boerhaave, the latter with Lower, Glisson, Boyle, Mayow, and later Hales.†

Van Helmont (1577 or 80 to 1644), of Brussels, was distinctly less mystical and alchemical than Paracelsus a century earlier. Already he illustrates the preoccupation with heat and with gases. It was his great service to concentrate chemical attention on these latter, and to realise definitely that there are several kinds; though he could not usually give clear tests or definitions for them. He also knew the three chief

* We have seen in Chapter VII, however, how physiology influences another biological subject, classification. Physiology, in fact, is historically a science of central importance.

† Oxford was a prominent centre of this activity in the one country, Leyden in the other.

mineral acids, which we call sulphuric, hydrochloric, and nitric, and knew that metals dissolved in them can be recovered.

Fludd in 1617 described enclosed combustion in an inverted glass vessel over water—a valuable device for manipulation, and one which informed man that after a certain volume-diminution has occurred, the flames go out. Van Helmont knew of this latter fact. He distinguished gases from the more easily liquefied vapours. He denied that fire is a separate " element "—merely gas at a high temperature. He is less modern in professing to have effected transmutation of metals and in particularising the " archaei " * which in his view presided over the various bodily systems, nervous, digestive, respiratory, etc., and in arranging them in a hierarchy. Yet even here his ideas had a new definiteness which ensured their ultimate criticism. Thus, he tried to identify digestion with a definite if uncomprehended process, fermentation, on which he had modern-sounding views. He assimilated other processes of this sort with the type-case of the leavening of bread.

His quantitative use of the balance—his stress on the importance of weight-changes—was long in getting the imitation it deserved. It did not become a habit of chemists until well into the 18th century.

Yet he was still concerned with the old alchemical elements, and with transmutation. With Thales, and against Paracelsus, he reduced the four to two : air and water, with stress on the latter. One of his experiments in support of this well illustrates the enterprise, yet the quaintness, of the science of the time. For five years he watered a weighed willow in a weighed quantity of earth. He compared weights before and after. The weight of the earth was unchanged. From this he concluded, wrongly but not unreasonably, that all the willow's increase was due to the (all-important) " element " water. Clearly, the full part played by gases was still unrealised when his work was posthumously published in 1648 !

In fact, the needed clarification of that subject was still in the future so far as chemistry was concerned. It will be recalled that the *physics* of gases was about then being cleared up by Boyle (Chap. V) ; though neither Boyle himself, nor Mariotte (1620–84) its rediscoverer, realised the significance of the law pv=constant. Mayow showed that this law was true not only of air but of what we call nitric oxide and of hydrogen. Mariotte also discovered that the sap of plants rises with truly remarkable force. This fact may seem unconnected with Boyle's laws, but it was not, for (true to the keynote of this chapter) the experimental technique was similar in the two cases. It is vital to remember that at this time the use of barometer-like glass tubes, with liquids in their bend, was rapidly extending. For it was only through the facility this gave in handling definite bodies of gas at definite temperatures and pressures that the Lavoisierian revolution in chemistry became possible.

The Englishman Hales (1677–1761) went further than Mariotte. He used a true manometer, and with it actually measured the pressure of sap, and blood-pressure in man. No doubt, like Leeuwenhoek with the microscope, he was looking round for things on which to experiment

* Spirits. But the confused implications of all such terms at the time must be remembered.

with his new toy. He found that blood-pressure varies during each heart-beat, and from health to sickness, as well as from species to species. He was, in fact, a founder of quantitative physiology.

He measured the rate at which water enters roots and evaporates from leaves. Like Grew, he was interested in locating regions of most active growth in plants. He made pricks at different points of growing stems and followed their movements. He realised, too, that plants derive something from the air. But chemistry was not yet capable of identifying this as the product of ordinary combustion and of human respiration (our " carbon dioxide "). Hales, in fact, was in advance of his time, and had no immediate successors. We must return half a century to chemistry proper and to animal—medical—physiology.

The great medical teacher F. Sylvius (1614–72) stands for many things. First, we may claim him as an example of a recurrent feature in the history of physiology, the importance in it of great personalities and of teachers, who inspire hundreds of brilliant students. Later examples will be Boerhaave, Hunter, Müller, down to men like Sherrington in our own time. With Paracelsus, Sylvius stands for the view of the body as primarily chemical. He tried to identify what we should call organic and inorganic chemistry. He stands, too, for one of the first chemical theories not concerned with fire and metallurgy : the " affinity " of acids and bases, their union in " salts " which are neither acid nor base. This was a conception made clearer later on by Lémery (1645–1715) and then by Macquer (1718–84). For Sylvius it had a medical significance. Acidic and alkaline replaced bilious, phlegmatic and the rest in the old scheme of human constitutions. With Boyle, Sylvius stands for a great, though not nearly successful, effort to banish mysticism from chemistry.

Boyle's step in this direction was necessary but not sufficient. He adumbrated the contrast of element, compound, and mixture. He began to stress the idea of *pure substances* and of others as composed of these both by mixture and by combination. He tried to obtain criteria of purity, but he did not succeed. He called for quantitative statements of how much of A combined with how much of B. But this quantitative point of view received little attention. Van Helmont indeed had implied the constancy of weight throughout a reaction. But he was working at a time when even in the simpler matters of mechanics, the constancy, indeed the very meaning, of mass was far from clear.

Like him in stressing weight was the Frenchman Rey, who realised (1630) the importance of a fact mentioned here and there in the 16th century. This was that, at any rate in the case of lead, there is an *increase* in weight when earths are formed from metals in calcination. But this promising start was not followed up, although Rey was in touch with the " liaison officer " Mersenne. It could not have had its present meaning ; for the earths were generally assumed to be the elements, the metals (as less stable) being the compounds. Rey denied this ; but his assertion that earths are formed by fixation of air on metals is spoilt for us by his comparing it with the absorption of water by sand.

His best point was perhaps his perception of an essential similarity of calcination and combustion. All these subjects seemed complicated

CHART II

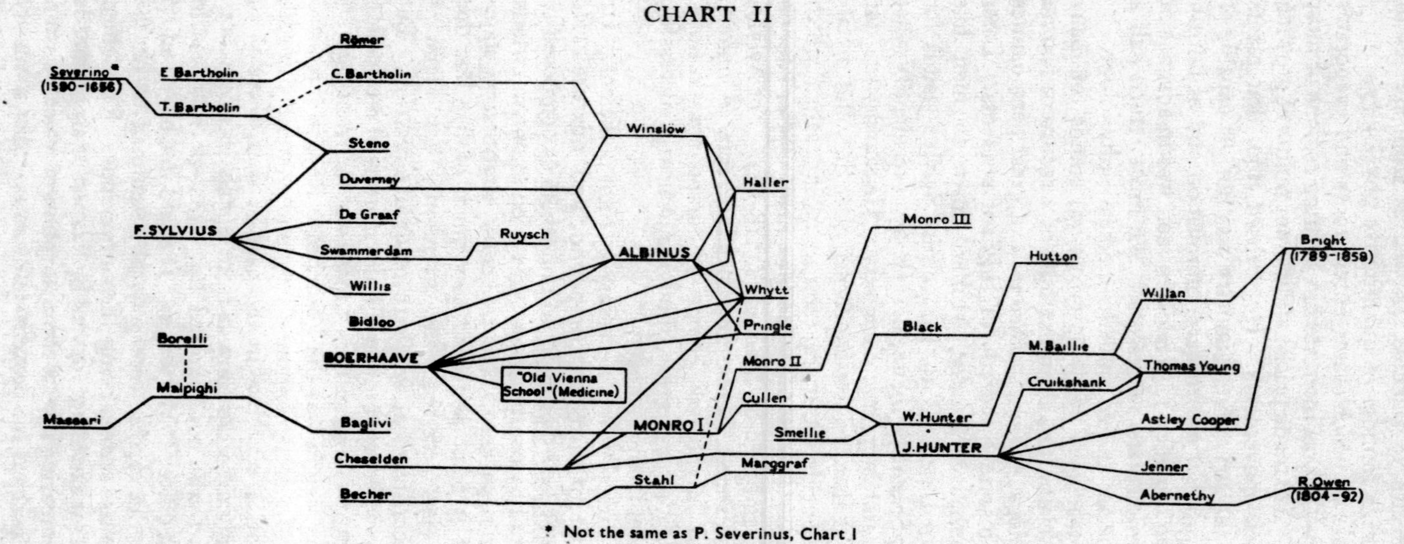

* Not the same as P. Severinus, Chart I

EXPLANATION

(BUT SEE "SCHEME OF CHARTS" AT BEGINNING OF BOOK)

A —\ B Means A taught B

A -- B Means A influenced B indirectly

CAPITALS for conspicuous teachers

Contemporaries are roughly, but only roughly, one above the other

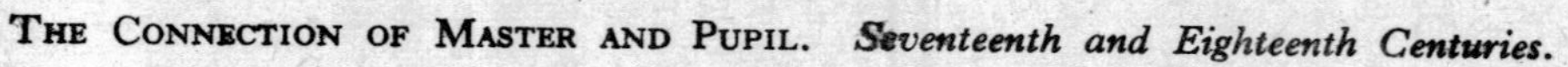

THE CONNECTION OF MASTER AND PUPIL. *Seventeenth and Eighteenth Centuries.*

to early workers because, lacking the distinctions drawn later, they were obliged to treat together aspects (physical, chemical, physiological) which we can take one at a time.

We have spoken of the confusion engendered by the idea of spirits. "Air," to the 17th century generally, did not mean what it does to us. It was, of course, one of the elements in the old schemes. As such, it could not be a mixture; though it had to be admitted that "effluvia" could accumulate in it. Only Hooke (e.g. 1663, 1682) seems to have had the idea that air might contain two gaseous parts, one supporting combustion, the other not. Mayow (1643–79) and others seem to have held that it was the elasticity rather than the volume which diminished in the experiment over water. This was not unnatural in view of their interest in the current experiments on "the spring of air." *

The sensational case of gunpowder, then much studied for reasons of state, had acquainted man with the fact that air is *not* always necessary to combustion. It was inferred by Hooke that a principle in common between nitre and air (called by Mayow "nitro-aerial particles") was the supporter of combustion. Flame, to Mayow, consisted in these particles being torn from their partners. It was as much a disruption as a combination. This was in line with the traditional view of fire as the "sulphureous" (that is, fiery) element manifesting itself or *escaping*. It also, as we shall see, contained roots of the phlogiston theory. Becher (1669) adopted Rey's parallel, which Mayow denied, of combustion and calcination, the traditional parallel of the calx being ash (residue, not product). In another respect, also, in its all-too-great ingenuity, this nitrous principle contained sinister premonitions of phlogiston. Thus, when agitated, the nitrous particles were responsible for heat; but (on the basis of the cooling produced when nitre is dissolved in water!) when fixed, they were responsible for cold (e.g. for cold winds), and indeed for rigidity and elasticity also. In this rich confusion there thus lurked the beginnings of a kinetic, as well as of a material, theory of heat.† And indeed both theories were already held.

Combustion was, simultaneously, associated with respiration. In the vacuum of his air pump Boyle (1660) had seen animals die and flames cease to burn. Soon after, Lower (1631–91) showed that venous becomes arterial blood by absorption of something from the air.

But any strokes of insight which these workers made were apt to be drowned in mere speculation and lost to subsequent workers. Thus, Mayow does not seem to have held that nitro-aerial particles *alone* could support life or flame. For this and other reasons, it is very doubtful if his views were a genuine premonition of oxygen—rather, of phlogiston. But, if they were, the important point to notice is that they were forgotten. Communication was still weak among chemists, not only because of the evil individualism of the alchemist, but because chemists had as yet no perfectly clear technical terms with which to

* Mayow used the pneumatic trough brought into general use by Priestley.

† The full discussion of the confused writings of the 17th century is beyond our capacity or space. Mayow's merits, especially, are controversial. They seem to have been seriously over-estimated. See *Isis*, 1931, Vol. 15, pp. 47–96, and pp. 504–46.

communicate had they wished to do so. It was such technical terms, such "units of thought," which Newton was just then giving to mechanics.

Iatromechanics and Iatrochemistry

Modesty is not a characteristic of early stages; and already, in the 17th century, the many ways in which living creatures resemble inanimate systems had led to the view that the dominion of mechanics and chemistry is complete—in fact, to a first crisis of mechanistic philosophy.

The reader will recall Descartes in connection with this "iatromechanics" and "iatrochemistry." The mere identification of muscles and bones with levers had been made by Galen and carried into considerable detail by Fabricius (1537–1619). Galileo had recognised that interesting principle which we call the "scale effect," and had applied it both to inanimate structures and to living ones. Weight goes up as the cube, but supporting area only as the square, of the linear dimensions. Hence the slender flea, and the thick-limbed elephant; also, the enormous size of some marine animals, whose weight is supported by the water. Hence, too, the practical stimulus to the subject, in connection with the launching of ships. The flea is not only slender but immensely agile, and this fact raises the question of the way in which muscular power varies with muscle size, and so the further one which had obsessed Leonardo, of the possibility of unaided human flight. This and the whole subject were treated by Borelli (1608–79) in "De Motu Animalium" * (1680–81).

Long before this (1614), Sanctorius had made his much-illustrated balance-chair for comparing weights before and after meals. He had also taken up Galileo's idea (of the middle 1590's) of using thermoscope and pendulum as what we should now call clinical thermometer and pulse-watch. Neither of these ideas was destined to reach regular bedside practice until the 19th century. They were taken up and dropped at long intervals. Pulse-counting, indeed, had already been suggested by Nicholas of Cusa in the middle of the 15th century.

All these were genuine achievements, as was much that was done under iatrochemical inspiration by Sylvius. But unfortunately these movements went to excess and lost all connection with experimental verification, as when the stomach was explained as a grinding machine. The opposition which they called up from vitalists and animists, however, made much noise. It also brought to the front two men, Stahl and Hoffmann, both of Halle, who were destined to hold up the revolution in chemistry and in heat, to which we now turn.

Stahl was the anti-mechanist. His "spirits" were like Van Helmont's archaei of a hundred years before. Vitalists did not then confine "life" to the narrow bounds now usual, and Stahl, a true

* A much later writer, Le Sage (1724–1803), made further beautiful applications of the scale idea. These form a link between Galileo and the modern surface-physics of colloids. Le Sage pointed out that the surface area per unit volume increases, shape for shape, with decreasing size, and therefore that tiny creatures like insects would dry up by evaporation if their skins were as porous as those of mammals.

18th-century system-maker, felt bound to include chemistry under the hierarchic dominion of his spiritual powers. A more modest and useful service, that of assembling information, was done (for medicine and chemistry separately, 1708 and 1732), by the great Leyden medical teacher, Boerhaave (1668–1738).* But something more than collecting and classifying was needed to set heat and chemistry on their way.

Heat

Though many had long held the material theory of heat, both it and the phlogiston theory crystallised out of the matrix which we have already sketched. It is consequently hard to separate them without violence to chronology. Phlogiston, especially, tried to keep thermal aspects (of chemistry) under its dominion. There was a simpler region of "pure" heat, however, in which decisive advances came, on the whole, sooner than in the chemical subject.† Hence we treat it first; but before doing so we must note that light, like heat, was often included in 18th-century lists of the chemical elements. Both were often held to be etheric, but the word ether had a rather indefinite sense. Though contrasted with matter, it was very like it. It "pervaded the minute pores of bodies."

It is vital to remember that most 18th-century chemists had no adequate sense of the advances which the 17th-century physicists had made. Otherwise their point of view is inexplicable. We shall notice several other instances of this isolation from the general stream of science which marked chemistry as still pre-scientific.

We have noted in our Newtonian chapter that Galileo and Van Helmont had, before the opening of the 17th century, experimented with the expansion of air. Galileo had used a bulb with an open-ended stem inverted over water. In the same chapter we briefly described the stages by which gas pressure became a definite physical concept, definitely related to volume under isothermal conditions. We now turn to the effects of heat—to non-isothermal conditions. By virtue of the temperature-sense, men were saved, in these early days, some of the confusion felt later between "level" and "quantity" of heat. We can thus defer the consideration of the latter.

In Galileo's original instrument these effects were confused with the effects of barometric pressure; and both thermometry and barometry progressed throughout the 17th century largely by separating these two factors. In 1632 Rey filled the *bulb* of Galileo's appliance with water, thus getting a crude water thermometer. But the end of the tube was still left open. A quarter of a century later, instruments with sealed tubes were being made in Florence; but the decisive steps of progress were still well in the future, and the stage had by no means been reached when each worker automatically used the results of the last. Long after this (for instance, in the constant volume air thermo-

* His works had a long vogue. The great Haller superseded the medical treatise in 1759–66. See next chapter.

† Stahl propounded phlogiston ideas in 1703 (and later); but they only acquired full importance about half a century afterwards, when he was dead.

meter of Amontons, in 1702) no care was taken to compensate for variations of atmospheric pressure.

One step forward was taken when the non-volatile mercury, already used for barometers, was extended to thermometers. Amontons managed with its aid to arrive at the law, rediscovered about 1787 by Charles (see later), as to the effect of temperature on the volume of gases. But he was only using mercury as an indicator. In using it as the thermometric substance, inaccurate results still sprang from not using it pure.

To make satisfactory thermometers had by then, however, become a settled ambition, partly because they had acquired a practical use (for instance, in meteorology). It was a celebrated meteorological instrument-maker, Fahrenheit (1686–1736), a German living in Amsterdam, who took the crucial step of purifying the mercury by filtration through leather. With his much superior instruments, Fahrenheit was able to show (1724) not only that pure liquids boil away at a constant temperature (a fact roughly realised for water by Newton, Huygens, and others) but that this temperature is changed by alterations in the atmospheric pressure.* Thus it came slowly to be realised that for significant measure of any one of the three quantities, p, v, and T, the other two must be specified. Thus, too, a further item of definiteness, boiling-point, and a further enrichment of the idea of the physical, was arrived at. It will be noted, however, that the only *measurable* entity was still temperature. Richmann may (1744–7) have begun to view it as a factor of *level* or intensity, presumably of a second factor. What this second factor was to be, we now examine.

Heat, by then, had begun to have another and sensational practical use. Various engines for raising water out of mines by its use had been tried in the 17th century. For mines were ever getting deeper and therefore wetter. Newcomen engines began to spread over England from 1705 onwards. For all their glaring faults, they were in quite wide use when Watt started out in 1760, to improve them.† Now Watt's impulse came from Black, the man who, besides his work on heat, began the chemical revolution. We have thus reached the very centre of things.

Indeed we are at the centre in more ways than one, for the Scotland, and especially the Edinburgh, of that time, was, after Paris, the intellectual centre of the world. We have spoken of Black's friend Hutton. We must also remember Hume and Robertson and Adam Smith, Galt and Sir Walter Scott. Birmingham was another important contemporary town.

Black (1728–99), like Priestley, Pasteur, and Phillips of the contact process for sulphuric acid, seems to have received some of his interest

* Boyle had, however, noted the ebullition of liquids in a vacuum. The process of deciding convenient fixed points for these thermometers, which we cannot pursue in detail, had begun in the 17th century. Réaumur (1683–1757) worked in France independently of Fahrenheit. Centigrade scales began to come in about the middle of the 18th century.

† They were not in our sense steam engines, but engines in which the condensation of steam was used to bring atmospheric pressure into play.

in science from the fermentation industries, brewing, tanning, and the like. Distilleries formed another practical use of heat. They were booming. In the early 18th century, gin engulfed Britain in an overwhelming wave of drunkenness. We may, fancifully, see in these new sciences of chemistry and of heat one reaction to this ! For not only is distillation one of the best methods of preparing pure substances, but distillers, Black found, were aware, as users of fuel, of a quantitative aspect of heat other than mere temperature. Two branches of technology were thus exchanging influences, at the very start, with this new theoretical science of heat. Black, in fact, ended another case of the isolation of craft knowledge and theorist's knowledge ; and the union of traditions gave him the doctrine of the latent heat of freezing and of vaporisation (1757–62). This was of special interest to Watt, but for brevity we pass straight on to specific heats.

Boerhaave (1732) had presumed, from the identical final readings of thermometers in two connected hot bodies, that equal volumes of all substances have equal capacities for heat. There was little logic in his special conclusion, but it is clear that with him the second factor, quantity, was beginning to enter the measurable region.

Krafft (1744), and also Richmann, considered another point, the final temperature when substances at different temperatures mix. They calculated this, but their results were wrong. This was partly due to a failure of ideas, partly to a fundamental difficulty which makes heat harder to deal with than light, the fact, namely, that heat cannot be even approximately confined for long in a given space. In other words, the containing vessel must be taken into account.

On both these points, Black * succeeded ; and by 1760 he was able to formulate a correct and applicable measure of quantity of heat, and thus, also, of specific heat. He thereby added yet another element of numerical definiteness to the concept of the physical.

Lavoisier and Laplace in 1783 determined a large number of specific heats. The amount of ice melted was found to give the most accurate measure of heat. At the same time, the arbitrariness of the mercury-in-glass scale was occasionally being realised, in preparation for the thermodynamic viewpoint and the accurate gas thermometry of the mid-19th century.

Black tried to avoid all theories of the nature of heat, but seems to have held, like most of the experimentalists, that it was a self-repellent fluid which he was measuring. We have noted how both this view and the kinetic theory could arise out of one 17th-century notion. An effort at a kinetic theory had been made by D. Bernoulli as early as 1738. But it remained isolated. Not until a very definite type of atomic theory—the theory of *identical* atoms—had sunk deeply into consciousness could even the greatest mathematicians see that " a commotion among the invisibly small parts of bodies " could lead to the simple definiteness of Black's and Charles' and Boyle's laws. For

* His priority over De Luc and Wilcke seems clear, also their independence. See McKie and Heathcote, " Discovery of Specific and Latent Heats," 1935.

the moment the material theory was nearer to experiment and was therefore perhaps the more profitable. Experiments, for instance, were made to find whether the fluid was strictly weightless. From Boerhaave and Buffon onwards, the difficulty of weighing hot bodies led to conflicting views. But the decision on this point had to await the fall of phlogiston. We turn, therefore, to this latter, and to chemistry in general.

PHLOGISTON

It was not chiefly Stahl * who made the phlogiston theory into the notorious doctrine of an element of negative weight. He was still in the stage when weight questions were thought of small importance. What views he propounded were not self-consistent. His theory was in the main a case of the realist attitude (Chap. I): all metals, for instance, still had, for him, a "metallising principle." The fiery phosphorus would be made up of an element (the dull phosphoric acid) plus the fiery principle, phlogiston. It was only gradually that these principles became corporeal enough for their weight to be considered. The idea of actual levity only came to the front with Venel (*c.* 1750). The more important theory of Morveau (1772) was rather that phlogiston is the lightest of bodies or media. "Levity" never spread to the majority of phlogistonians, some of whom, by dint of ignoring these particular implications of the theory, contrived to make many of the definitive advances to which we come presently.

The phlogiston theory was an advance in definiteness on the old alchemical views, but when it came to be armed with negative weight its power of covering facts remained excessive. The best theories are precisely those which, because they *cannot* be stretched to cover *any* facts, can predict new ones. In fact, phlogiston never inspired much experimental work. Its service was rather the characteristic service of "principles," from Aristotle onwards: the intellectual one of enabling us to arrange our thoughts and give them some sort of unity. In this it falls in, too, with the systematising tendency of its century. That negative weight ("levity") could still be held in the days of the theory of universal gravitation was an instance of the isolation of chemistry, though the relation of the two theories did not go undiscussed (e.g. Chardenon, 1763). The theory, too, always took insufficient account of the gases which were then coming into chemical prominence, and which ultimately upset it. Yet purely from the qualitative chemical point of view it had a certain naturalness. For flame always rushed upwards, and lime lost weight as it became "fiery" or "quick," that is, as it gained phlogiston. "Air" † remained necessary for combustion, since the phlogiston combined with it as it escaped from the burning materials. The kinship with 17th-century ideas is very evident here.

* The subject is complex. See White, J. H., "History of the Phlogiston Theory," London, 1932; but especially Partington, J. R., and McKie, D., *Annals of Science*, Vol. 2, pp. 361–404, and Vol. 3, pp. 1–58.

† The uncertain meaning of the word should be remembered.

As in flames, so elsewhere, phlogiston could take the form of light as well as of heat—in what we should call photography, for instance. The essential reaction of photography was brought forward in support of phlogiston after Stahl's time, by Scheele in 1777. Silver chloride can derive from light the phlogiston necessary to convert it into silver !

It should be noted that we always try to explain phenomena by *one* principle before we call in more. We shall see that, even when phlogiston was abolished, men still tried to keep chemistry one-complex. After that they had a *dualistic* theory, and only later recognised a number of principles of combination as well as a still greater number of combining elements.

Chemical Technique

It is necessary to glance at the general condition of chemical apparatus and manipulation in these times. In the 16th century Agricola had begun the serious assaying of metals. In the 17th, Boyle had considerably improved what we should call qualitative analysis, beginning the systematic use of indicators. He had tests for ammonia, vitriol, and the silver test for hydrochloric acid (our name). Kunckel (1630–1703), a chemist connected with the glass industry, had invented blow-pipe analysis. Marggraf (1709–82) used flame-colour to tell soda from potash, and the microscope to distinguish the sugar in beet.

According to Meyer, relations with the mineral and heavy chemical industries became as important in the 18th century as relations with pharmacy were in the 17th. With the coming of the industrial revolution in England, the manufacture of fundamental reagents, and especially of sulphuric acid, took a sharp upward turn. By then, too, one of the exigencies of commerce—coinage and goldsmith's work—had occasioned the construction of balances less convenient, but not incomparably less accurate, than many of ours (see Graph II). As isolation broke down, these became available for theoretical chemists. It is needless to stress the importance of this.

Stahl's opponent Hoffmann, and others, examined mineral waters, which had far greater importance in those days of imperfect waterworks than they have now. In doing this they developed analytical methods for distinguishing sulphates and nitrates, and also a number of bases. After Stahl's time, Bergman (1735–84) the great Swedish rock classifier, began that systematisation of all these methods of analysis with which every student is now familiar. Quantitative analysis was still in the future, but already analysis was becoming the key to all serious chemistry.

The Chemical Revolution

We may say in general that men like Marggraf had begun to realise that most acids can combine with most bases. Rouelle (1703–70), who taught Lavoisier and improved nitric acid manufacture, took a difficult step, that of recognising that salts themselves can be acidic or basic. As early as 1718 Geoffroy (1672–1731), of the French Jardin du Roi, had prepared tables of the relative affinities of different acids and bases from the consideration of the violence of their reactions and of which displaces which. Much later, Bergman really set the fashion in these tables.

We have noted that Boerhaave and others made experiments on the conservation of mass through a variety of changes. This was an undoubted service in the slightly phantasmagoric atmosphere which still prevailed in their day. But it was left to Black to make the first really decisive experiments (1754) in chemistry proper; and there is no doubt that he was able to do this because he knew enough physics to hold the physical and the chemical before his mind, together and yet distinct.

These experiments concerned the relation of what we should call the carbonates to the oxides (or hydroxides) of the alkali metals (magnesium, potassium, and calcium). From the point of view of modern chemical theory these carbonates appear to be just two or three salts among hosts of others. But at that time the last two, and the corresponding oxides, bulked much larger in chemical and industrial consciousness. Their metals were not known, so that they did not at all obviously fall into the class of substances obtained (with increase of weight) by burning metals in air. It was thought that the highly caustic oxides of the last two were obtained from the mild carbonates * by *addition* of the fiery element phlogiston, instead of, as we should say, by the removal of a ponderable definite gas, carbon dioxide.

Black showed that magnesium carbonate loses weight on heating alone, to give a substance which, when put into acids, gives the same salt as does the acidified carbonate itself. There is, however, an absence of the bubbling so marked in the latter case, and Black inferred that the substance which bubbles off has the weight which the carbonate loses. It is a gas with properties distinct from those of air.

The choice of magnesium was crucial. The existence of slaked lime complicated the experiment for calcium, the (then) impossibility of obtaining the dry oxide at all complicated it for potash. But once the relation was cleared up for one case, Black soon succeeded with the others.

In none of these experiments had "phlogiston" been needed. Instead, a gas, carbon dioxide,* had begun to assume the definiteness so often foreshadowed in this book. Boyle had collected hydrogen.* By Boerhaave's time, chemists could distinguish carbon dioxide, hydrogen, marsh gas, etc.† But, for all that, these facts did not mean to them what they mean to us. In the absence of the idea of purity, gases were only "differently modified forms of air."

After Black's work, other gases were soon as definitely known, though still in terms of the phlogiston theory. But the latter, at the hands of men like the Swede, Scheele (1742–86), and Cavendish (1731–1810), was reaching the climax which goes before the fall. In describing hydrogen fully for the first time (1766), Cavendish, indeed, took it that he was actually describing phlogiston itself.

* In our words.

† For simplicity, modern names are used.

Among new gases was chlorine, discovered by Scheele, 1774, and independently by Priestley (1733–1804). This was to be a crux later on.* About 1774, Priestley was searching systematically for new gases, using the vital improvement of collecting all of them over mercury, which dissolves or attacks far fewer than water. He thus isolated sulphur dioxide, " spirits of salt " (HCl), and ammonia.

Finally, there came oxygen. Scheele, and perhaps Priestley himself, had isolated this as early as 1771 ; but publication came later, with priority for the Englishman (1774 onwards). By 1772 Priestley had shown that a gas, necessary to animal life, is given off by plants. Finally, in 1774, he proved that the red oxide of mercury, when heated, gives off a gas in which both mice and candles show much increased activity. It was the vital constituent, evidently, of the atmosphere, with which the emitted phlogiston unites. In itself, therefore, it must be " dephlogisticated air "—for Priestley was a phlogistonian.

This (oxygen) soon acquired an extreme importance at the hands of one who, like Black, united the manipulative and observational skill of a Scheele with the logical and quantitative instincts of a physicist. This was the great Frenchman Lavoisier (1743–94).

Lavoisier was a man of varied practical experience. He had worked with Desmarets, the geologist. He had been consulted by the French government. It was to design lanterns for lighting Paris that he investigated air. It was mineral waters which led him to investigate water. By 1772 he was already urging, from analogy with the cases of phosphorus and sulphur, that the gain in weight on burning is due to the air " fixed " thereby, not to the loss of phlogiston. In 1774, he calcined tin in a closed flask, using weighed quantities, made a hole in the flask and observed that the weight of the new air which rushes in is the same as the gain in weight of the tin. Finally, also in 1774, he hit, like Priestley, on the red oxide of mercury, which is convenient because at different temperatures it turns into mercury and back again. By a now standard classroom experiment, he conclusively demonstrated the rôle of the oxygen, and in 1777 published a full theory of combustion in which phlogiston was abandoned.

About this, there are several points to be noted. It was the end of the previous hope of accounting for the thermal and chemical sides of reaction by the same principle. The (material) theory of heat was now necessary in good earnest for thermal questions, and the name of " caloric " was invented for it. In Lavoisier's great " Traité Elémentaire de la Chimie " (1789) heat was included with light in the table of the elements. Fusion and vaporisation were chemical unions, of solid and liquid respectively, with caloric. We shall soon see, however, that another point of view was also being reached.

* At every stage of science there were of course many such discoveries, isolated at the time but crucial later. For a true picture, it is vital to remember the existence of these, even when space is lacking for their enumeration. Thus, Cavendish (1784) gave the properties of nitrogen and the composition of the atmosphere. Scheele was an extremely notable discoverer of new substances : compounds of manganese, arsenic, and barium, the organic substances, glycerine, and uric, oxalic, tartaric, citric, lactic, prussic, and other acids.

Finally, it was not yet modern chemistry which Lavoisier brought in. As Mme. Metzger has pointed out, oxygen was to him a principle, not precisely a gas. The realist point of view clung on in the man whose experiments made it irrelevant. Gases, for him, must have in common a "gasifying" principle, and each gas was a combination of this with something else; that is, was not an element. As of old, combustion was a double decomposition. The conception, however, was a great advance over "differently modified forms of air." For, as neither element occurred separately, the vital idea of *purity* was applicable to the substance formed by their union.

In another way Lavoisier was not yet modern. No less than the phlogistonians, he saw chemistry as organised monistically round *one* principle, only oxygen instead of phlogiston. Combustion and, as we shall note later, respiration, he had accounted for by oxygen. "Earths," or bases, were of course metals plus the "principle" oxygen. Their opposites, acids, he now proceeded to argue, were non-metals plus oxygen.*

Now oxygen is certainly present in most acids; while the hydrogen which acids give off with metals was explained as due to the water. To this latter explanation we should now, in our own sense, give assent; but at that time the crucial point remaining to be discovered was that some of the most important acids (for instance, hydrochloric acid) have no oxygen.† We defer this discovery to Chap. IX, while noting that it was what gave chlorine the importance foreshadowed above.

The French chemical world did not abandon phlogiston until about 1787; while Priestley, Scheele, Cavendish, and many others never did abandon it. The supersession of phlogiston by oxygen, together with the accumulation of what were now seen to be antique errors, drew attention to a matter of primary importance in chemistry as in every science, nomenclature. The technical terms of the time made accurate thinking almost impossible. In 1782 Guyton de Morveau began to urge reform. In 1787, with Lavoisier and Berthollet, he published a new system, with caloric as one item. This was a great improvement, but it had not yet the almost algebraic suggestiveness of modern chemical symbolism. For this we await Chap. IX.

Lavoisier's many accurately weighed reactions made the law of the conservation of mass a definite and final possession of chemists. Its immediate consequence, gravimetric analysis, could now follow. Of this, Lavoisier laid the foundation-stone when in 1782–3 he quantitatively decomposed steam by passing it over hot iron, and quantitatively burned hydrogen, and also carbon monoxide, in oxygen. He thus established the composition of carbon dioxide by weight.

* It will be seen that, while there was the old monism here, there was, though secondarily, the old dualism also. All compounds, bases and acids included, were, like the salts formed by their union, made of *two* entities, of which oxygen was always one. (See next chapter.) The difference between metals and non-metals was another thing which the new theory, unlike phlogiston, did not try to "explain."

† On the further point that all contain hydrogen, Lavoisier was less at fault; for by acid was then meant what we should call anhydride.

This raised the question of *equivalent* weights of reacting substances, compound or elementary. The tacit assumption was made that combining ratios are constant. In 1799 Berthollet (1748–1822), in attacking the old tables of affinity, attacked this assumption also. Unknown to the dominant Paris school, Richter in Germany had begun, from 1792, publishing serially a number of determinations of equivalent weights for salts. These determinations were later collected together by E. G. Fischer, but they did not touch Berthollet's real point, which was that the yield of a reaction was rarely (as we should say) quantitative, but depended not on theoretical combining weights or on the ancient figment "affinity," but on factors like initial quantities, and on the relative volatility or solubility of the reactants. In fact, he asked, was the idea of a compound really of much use? Were solutions, alloys, glasses, etc., in that category? The distinction of physical and chemical change began to assume its present form.

In a classic eight years' argument, Berthollet was refuted by another Frenchman, Proust (1755–1826), who succeeded in showing that in all Berthollet's examples, perfectly definite compounds were involved. His success was perhaps too complete for the good of science. It obscured for half a century the fact that Berthollet had founded thermal chemistry and discovered reversible reactions.

This chapter is strung on a thread of gas properties, and to this thread we now return; for the *volume* relations of reacting gases led up to the atomic theory. Cavendish in 1783 sparked hydrogen and oxygen, and, noticing that only water was produced, saw also the simple integer proportions of the volumes involved. It was but the first instance of what was gradually seen to be a general fact. The two carbon oxides formed further cases, and also cases of the similar "law of multiple proportions." The combining *weights*, of gases or of solids, though fixed, were not *simple*.

What was the explanation of this volume simplicity? Dalton (1766–1844) found it in atoms. As a mere idea, not within the range of experimental test, these were nothing new. Found in Democritus, atomism has been traced in a thin but continuous line even through the Dark and Middle Ages (Bede *c*. 600, Isidore of Seville *c*. 700) when Aristotle's authority was against it. It was used, from the Renaissance onwards, with foresight but with vagueness, even in biology ("seeds of disease," foreshadowing bacteria). It was popularised by Gassendi, accepted by Galileo and Newton. Boyle, on mechanistic lines, saw in it the best means of explaining chemical and physical changes. But it was not applied to any really *specific* set of facts. Dalton gave it definiteness (informally 1803, finally 1808) by saying that the atoms run in classes, those within each class being identical, each class constituting a chemical element. Then he used it on specific facts, the simple volume relations of gases.

But this Cumberland handloom weaver, typical Englishman, needed this concrete model primarily to aid his imagination, and he by no means fitted it unaided for its gigantic rôle in science. With curiosity and insight for every subject which came to his hand, from colour-blindness to the weather, he made his first approach to atoms from the physical side. He was concerned to explain atmospheric diffusion, and for that purpose he supposed that the atoms of different gases occupy different volumes.* Now his atomic theory had been eagerly taken up by the French chemist Gay-Lussac (1778–1850) as implying, on the contrary, that they all, in gases, occupy the *same* volume. Indefinitely adjustable volumes would have no similar suggestiveness. In 1805, Gay-Lussac, with the great German explorer and scientist Von Humboldt (1769–1859), began to test this hypothesis by the most extensive and exact investigations of gas reactions yet made. By 1808 they had the volume ratios for ammonia, hydrochloric acid, nitrous and nitric oxides, and sulphur dioxide. The ratios were always simple, and often implied equal volumes for different atoms. But to the latter proposition there were fatal exceptions, as Dalton pointed out. One was the reaction which Gay-Lussac would have had to write

$$N + O = NO,$$

implying that it halved the volume ; whereas it leaves it unchanged.

It was a fundamental difficulty ; but in 1811 Avogadro (1776–1856) of Turin, and in 1814 Ampère independently, suggested that it could be got over by supposing that the atoms of some, but not all, gases went about in pairs :

$$N_2 + O_2 = 2NO \; ;$$

in other words, that there might be " molecules " of elements as well as of compounds ; and even multiple molecules of the latter.

Here was another theory which, starting with only one category, was found to need two. Two brought a vast simplification, but at the time the uppermost feeling was that molecules savoured of the over-easy theories of the immediate past. It was pointed out that no one had ever *seen* either atoms or molecules. Many, like Wollaston (1766–1828), saw no reason for committing themselves to any such " figments." The facts of combining weight were enough for them.

This distrust of theory by chemists must be noted by anyone wishing to gain a clear picture of chemical history. Even very recently, the great Ostwald took up Wollaston's position. In the latter's day there was a specific excuse, for in 1815 and 1816, Dalton's theory had been given a discreditably speculative extension. Prout, observing that most combining weights were nearly integers, suggested that all atoms were made up of hydrogen atoms. This stimulated the researches of Stas and others, which showed that it did violence to the facts. But its fate might possibly have been similar had this not been so. At all events it was the new chemists' complete lack of sympathy with such ideas which led to the fate of Avogadro's hypothesis, its being dropped for nearly fifty years.† The atomic theory itself, however, did not share this eclipse. Although, as we shall see in the next chapter, it could not by itself give any final clue to the relation of combining to atomic weights, the determination of the latter as well as of the former became the word of ambition for a new school of chemists.

* We will not pursue this side of his work. Priestley was another product of the prevalent economy of home industries.

† A strong attempt to use it was, however, made by Dumas in 1832.

We turn, in conclusion, to the great advance in physiology, the first or second since Harvey, made possible by these advances in chemistry. Lavoisier and Laplace showed that carbon dioxide and water are the products of animal and human respiration; and by 1779, the Dutch engineer Ingenhousz had demonstrated the first great cycle of mutual dependence between plants and animals. He showed that, in the light, green plants absorb carbon dioxide, thereby building up their tissues and restoring oxygen to the air; while in darkness they, like animals, give off carbon dioxide. He failed to discover that they also normally inhale a small quantity of oxygen.

Lavoisier and Laplace had established that animal life really does go on as an oxidation of carbon and hydrogen. They even tried to establish quantitatively that an animal and a candle produce the same quantity of heat in producing the same quantity of water and carbon dioxide. Though their experiment failed through the complexity of the case, it founded an important branch of physiology. They were at fault in assigning the lungs as the place of oxidation, as were others in assigning the blood. It must be remembered that none of this work was properly appreciated until the days of the cell theory.

In giving this as the first great triumph of chemistry in physiology, we must bear in mind that physiology made a reciprocal contribution in that physiological effect was an early test for oxygen. A much more striking reciprocal contribution occurs in the next chapter.

CHAPTER IX

EXPERIMENTAL SCIENCE II

WITH this chapter, we conclude the 18th century, and advance far into the 19th. Our guiding thread is still the close relation of physics, chemistry and physiology ; and we start from an especially sensational instance of this connection, which our general observer of science would certainly have emphasised in giving a contemporary account of the period. This was electricity and magnetism, galvanism and electrolysis, to give the physical, physiological, and chemical connections respectively. After this instance, we revert to chemistry, first inorganic and ultimately physiological. The chapter ends on physiology proper.

In our chapter on Copernicus, we noted the early work of Peregrinus and Gilbert on magnetism. Magnetism may be said to have become an autonomous line of research by the middle of the 17th century ; but a hundred years later it was still often a name of magic rather than of science, as the words animal magnetism, and the name of Mesmer (1734–1815) will suggest. Gilbert had likened magnetism to gravity and had called in an ether to convey it. He had also regarded it, and electrification, as " effluvia " of " souls " in magnetised bodies and in " electrics " * respectively.

In these subjects, theories exact enough for mathematical development only came a hundred years later than in the case of gravity, a sound reflection of the relative complexities of the cases. To understand early difficulties, we must remember the lack of the vital distinction of charge (or in magnetism, pole) and field, a distinction which exists already drawn in the case of gravitation.

Facts, however, began to accumulate. About 1630 the Italian Jesuit Cabaeus (1585–1650) noticed that a particle electrically attracted to another is often repelled after contact. About 1660–70 Guericke made a first (frictional) machine for the *continuous* production of electricity ; while Boyle discovered that electric attraction, like light but unlike sound, traverses a vacuum. The real development began about 1730 ; and by the forties electricity was popular science.

In 1729 the Englishman S. Gray discovered that the very substances—chiefly metals—which cannot be induced to give the electric effect by friction, can convey it hundreds of feet when it is produced in electrics in contact with them. In 1733 the Frenchman du Fay learned of this, and cleared up one part of the subject by drawing our distinction between conductors and insulators (metals and electrics). Thus he gained the vital concept of electricity as capable of flowing, as a fluid. Then he went on to show that this fluid is of two kinds, like

* That is, insulators (see later).

Peregrinus' magnetic poles, like pairs repelling, unlike attracting, each other. The repulsion explained the confinement of electricity to the outer surface of hollow bodies, a fact noticed by Gray and by Hauksbee.

Kinnersley of Philadelphia made similar advances independently. With Franklin, he is one of the first of important American scientists. Franklin (1706–90) was the man who (chiefly 1747–52) identified lightning with sparks and brought in lightning conductors.

We cannot follow the twists of theory in the minds of these men. At that time electricity was not clearly distinguished from thermal, optical, and other " etheric " effects. Franklin replaced du Fay's two fluids by the excess or defect of one from an average. Then Aepinus (1724–1802), in 1759, and others, made it clear that the fluids do not themselves penetrate air or other insulators, across which they yet exercise force. Thus, as for gravitation earlier, the concept of action at a distance came in. With it came the Abbé Nollet's (1700–70) demand for a corresponding mathematical development, and Priestley's demand (1767) for instruments capable of measuring, as opposed to merely detecting, charge. The point is significant. Such quantitative development was coming to be regarded as a normal phase in the development of a branch of physics.

But the idea of action at a distance was by no means so easy to apply to this new case. Charge moves freely, yet not measurably, in at least some bodies. Mass does not, at least in any corresponding sense. This movability of charge by other charge had been studied, as " electrostatic induction," by Canton in 1753–4 and by Wilcke (1757). The phenomena seemed complex at first, but the appearance of charge in uncharged bodies on the approach of charged ones was eventually seen to be quite natural if there are two fluids which, normally mixed by their mutual attractions, are oppositely influenced by the new charge.

Franklin's one-fluid theory had the advantage of stressing that in all this the total quantity of electricity is unchanged. Some such conservation theorem was necessary before mathematical development could begin. Another necessity was some provisional narrowing of the problem. The old effluvium idea, now apt to be thrust into the background by the fluids, had had the advantage of keeping the medium in mind, but the disadvantage of gravely complicating the problem. Although the medium's influence was sensationally illustrated by the " Leyden " jar,* it was as well that Cavendish and Coulomb did not allow this to occupy all their attention in their work on the mathematical law of action at a distance.

We leave this latter to the next chapter, and turn to the physiology of the electrical effects given by the Leyden jar. To do this, we must anticipate on a few points the account of muscular and nervous physiology given at the end of this chapter.

Harvey, and a quarter of a century later (1677) Glisson, had taken " irritability " as the mark of living tissue. Haller (1708–77) went further and distinguished " irritability " and " sensibility." He

* Discovered, before Musschenbroek of Leyden (1746), by Kleist of Pomerania (1745).

showed that the latter is possessed only by tissues with nerves, and that all nerves showing it go to the brain. But he also showed that the "irritability," which is peculiar to muscle, could be called into play by stimuli other than those presumably given it by a nerve in action. It was a central advance in the mysterious region of nervous physiology. Here we are interested in it because among the other stimuli were the electric shocks of the Leyden jar and similar apparatus. The exhilarant action of these was soon being used by quacks and doctors (Schäffer 1752, Middlesex Hospital 1767), but analysis was much longer in coming, and when it came was soon swallowed up in the purely physical phenomena which it revealed.

In 1786 a worker in this borderline region, Galvani (1737–98), noticed that muscle-nerve preparations of frog's legs, hung from a copper hook, twitched when the hook was put on an iron support. The muscle was his chief interest, and he regarded it as *originating* the electricity which he took to be concerned in the action. In other words he took his discovery as one in physiology. Volta, on the other hand (1745–1827), held that the source of the electricity was the junction of the dissimilar metals, the muscle merely its detector. He concentrated on the action at the junction. Volta was an electrical experimenter of long standing, one who, with an inspiration like Priestley's, had long been striving for reproducible deflections and other forms of standardisation.

In 1792 he found that he could arrange metals in a series such that each was, in the terminology of static electricity, positive to all that followed it and negative to all that preceded it. By 1799 he had realised not only that the effect could be multiplied by using a "pile" of similar junctions, but that it could be *made to last* by moistening the junctions with acids. This was the decisive point, for, as we have seen, the idea of a momentary flow of charge had already been assumed.

Volta was right as to the source of the current, but the importance of the physiological connection—the hypersensitive detector it gave—should be noted. On the other hand, Walker (*Annals of Science*, Vol. 1, p. 66) has shown how, about then (e.g. 1787–97), purely physical methods of detection and measurement were advancing; and no doubt the discovery of the pile would soon have been made with these. Here, at all events, we shall abandon for the time the connection of nerves and electricity, too complex to get, at that period, beyond Soemmering's inconclusive analogy (1809) of nerves with electric telegraphs. The physics of electricity now joined with chemistry.

The first point which needed proof was that voltaic and static currents really were in essence the same. For there were contrasts. The former, for instance, would only give tiny sparks. In 1802 Erman showed that both give the same type of effect with an electroscope; and the view gained acceptance that the "potential," or level factor, was high in the static, the quantity passing high in the voltaic, experiments. In 1801, Wollaston compared their chemical effects. This brings us to a subject already not new.

Beccaria (1758) had got metals from their oxides and Priestley, inflammable gases from organic liquids, by applying discharges of frictional electricity. In 1789, Van Troostwijk and Deimann had similarly split up water into its component gases ; but electrostatic discharges are not easily disciplined to such uses. The voltaic currents were the ones which brought the subject to the front.

It is an instance of the speed already reached by scientific communication in some subjects, that in the year after Volta made his pile in Italy, Nicholson and Carlisle used it in England to make decisive discoveries on the decomposition of water, such as that the gases do not come off together, but at the anode and the cathode, and that their proportions are such as to re-form water.

Cruikshank (1745–1800) and, about 1801, Davy (1778–1829) began to electrolyse other salt solutions. In 1807, using a very powerful battery, Davy showed non-aqueous electrolysis, decomposing fused soda and potash to get two new metals, sodium and potassium. He also showed that pure water will not electrolyse, and extended, from water to electrolytes generally, the doctrine of the equivalence of the quantities liberated.* Ritter (1776–1810) pointed out that Volta's electrical series of metals is in the order of the metals' affinity for oxygen ; thereby opening a new realm for Lavoisier's oxygen theory.

In all this the physics of the current in itself was in the background, where we shall leave it until the next chapter. The purely chemical interest had many difficulties to meet. The actions in the pile and in the electrolyte had similarities yet baffling differences. Neither was usually simple. Why, if the current in the pile was essentially *due* to the chemical manifestations which undoubtedly *accompanied* it, did a potential difference, though not a current, exist between the metals without chemical action ? Why, too, did the products of electrolysis appear at different places ?

In 1805, Grotthuss (1785–1822) gave a theory of the subject. He supposed that on the application of a potential, the molecules of a dissolved substance arranged themselves in chains between the electrodes ; so that when, say, a hydrogen near the cathode was attracted to the latter and away from its oxygen, the latter took the hydrogen from the next molecule (on its opposite side) and so on right across until the unbalanced oxygen came off at the other electrode. This explained the equivalence of the quantities, and in fact was for so long accepted as an adequate explanation that we shall now leave electrolysis as such to Chap. XV.

Chemistry

The effect of these discoveries on Davy was to make him expound (1806) an electrical view of chemical affinity itself, which had hitherto been viewed rather as gravitational. We shall see later how Berzelius developed this view. Davy went on (1808) to another point. He made a discovery which traversed Lavoisier's oxygen theory of acids.

He showed that the strongest agents cannot elicit oxygen from either chlorine or hydrochloric acid, and in fact he asserted that chlorine is an element. The further suggestion was that if there is an

* Side actions apart.

element which all acids must contain, that element is hydrogen. It was a suggestion too novel to be easily accepted. Gay-Lussac did indeed come to admit that there *are* hydrogen acids. For he discovered (1813) the case of iodine and hydriodic acid. But the general oxygen view of acids, and indeed the general dominance of oxygen, was not displaced for a very long time yet.

With the coming of Berzelius (1779–1848),* however, a change did take place. By his Dualistic Theory he acknowledged that, of Lavoisier's two principles, the monistic must take the second place. Like Lavoisier, Berzelius was ardent for oxygen, and persisted that some acidity is due to it. But, with its new electrical meaning, Dualism swept into the dominant place in his thought. Chemistry had become two-complex.

Every inorganic compound, according to Dualism, consists of two electrically and chemically opposite parts. Acid and base, oxidising and reducing, metal and non-metal—all the opposing pairs—were viewed in the light of this fundamental dualism. Thus early was electricity built into chemistry. And yet the very closeness of the union, going as it did rather beyond the facts, promoted a reaction ; and we shall see that between the theory and its modern counterpart there was a period of scepticism.

Nor was theory the main preoccupation of Berzelius. He was concerned with everything which concerned chemistry. He made the subject autonomous, professional. From the twenties he was the recognised leader. His services deserved the place. He had ideas on colloids. He started the serious study of catalysis. He saw the capital importance of symbolism. He brought in the alphabetical abbreviations for the elements which made chemical formulae and equations as convenient and suggestive as their algebraic counterparts.

The main object of his prodigious activity was the determination of the atomic weights of the forty elements then known. In papers published 1810–18, he described the preparation, purification, and exact quantitative analysis, of about two thousand compounds. Where he encountered difficulties, he invented new techniques, standardising analytic methods and making many improvements in them. Yet he could not free his atomic weights from the uncertainty inherent in the rejection of Avogadro's hypothesis ; and he was constantly altering the ratios which he adopted of atomic to combining weights. For gases, he had nothing better than Gay-Lussac's view that all atoms occupy equal volumes. A standard for solids he found in a very different region, though one which had long fascinated scientists, that, namely, of crystals.

This is one of the subjects into the history of which we can scarcely enter. By "crystal" was for long meant only crystalline quartz ; while we have seen in Chap. VII that crystals were long thought to be within the province of biology. The existence of a variety of forms was occasionally recognised in the 16th century ; and in 1669, Steno announced the fundamental fact that in all variations of shape the

* Like Scheele, a Swede.

angles of any one crystalline substance are constant. In 1706 Guglielmini asserted that each salt has a form of its own, and in 1735 Linnaeus described about forty such forms.

The serious geometry of the subject began with Romé de L'Isle (1772 and 1783) and Haüy (1784). It matured between about 1810 and 1830. Guglielmini's view had by then been found inexact; but in 1819 Mitscherlich had noticed that similarity of formula and of crystalline form often goes together. This "law" enabled Berzelius to assign formulae to many salts (and so atomic weights to their elements) before Mitscherlich found that it had numerous exceptions.*

Another standard, to which Berzelius slowly came to give weight, was Dulong and Petit's discovery of 1819 that for most solid elements "specific heat × atomic weight" is a constant, about 6.4.† It was only a rough law with (then) no theoretic ground, but nothing exact was needed to decide between the small integers in question. Unfortunately at ordinary temperatures the law has marked exceptions, as Regnault showed. In particular, it is very wide of the mark for no less an element than carbon. Thus all Berzelius' standards failed him, and at the end of his life he saw his subject still in confusion.

The element carbon was very much to the fore after about 1820, for by then the centre of interest had begun to change to organic chemistry. We have noted that even in Scheele's time many organic compounds were known. But chemists were not easy about them. Berzelius, having analysed products of the body, took, as late as 1819, the (then not unreasonable) view that the Law of Fixed Proportions did not apply to them; so that they were hardly subjects for chemistry at all. Moreover, none of them had ever been prepared from their elements, or from compounds then regarded as originating outside the living creature. This was the crucial point; for the vitalistic view was in consequence taken that an absolute bar separated the two branches of chemistry, and that organic synthesis in the laboratory was impossible.

Analysis, however, remained—quantitative analysis, in fact, by Lavoisier's methods improved by Gay-Lussac,‡ Berzelius, and others. Scheele had already invented a method, still used, for isolating and purifying organic acids, that of decomposing their calcium salts with sulphuric acid. It was thus known that carbon, generally with hydrogen or oxygen or both, is the central constituent of all these compounds. There was, too, some knowledge of the distinctive properties of several *classes* of them: acids, alcohols, esters, ethers.

With atomic weights all insecure, formulae, even empirical ones, meant much less than they do now, a special handicap in organic chemistry. But they meant enough to stimulate constant improvements in gravimetric analysis. Balances were by then very accurate

* There were (e.g.) cases of compounds with more than one crystalline form.

† That is, the specific heat of each atom of all solids is the same, a view of the subject now very important.

‡ Chlorimetry (1824), alkalimetry (1828), silver and chlorine (1832). The estimation of iron by potassium permanganate belongs to a later period.

(see Graph II). Dumas' (1800–84) method of estimating nitrogen completed the circle of important elements which could be estimated (1830). Other great improvements were made by the German Liebig (1803–73). This apothecary's apprentice may be said to have started his country's pre-eminence in chemistry. His own great aim as a youth was to reach Gay-Lussac in Paris. Having achieved this in 1823, he was soon back in Germany and beginning his famous period at Giessen. His laboratory there nursed many great chemists, and results began to come in from the new mastery of analysis.

In 1823–4 Liebig and his friend Wöhler (1800–82) discovered that two very different compounds of silver, the fulminate and the cyanate,

GRAPH II

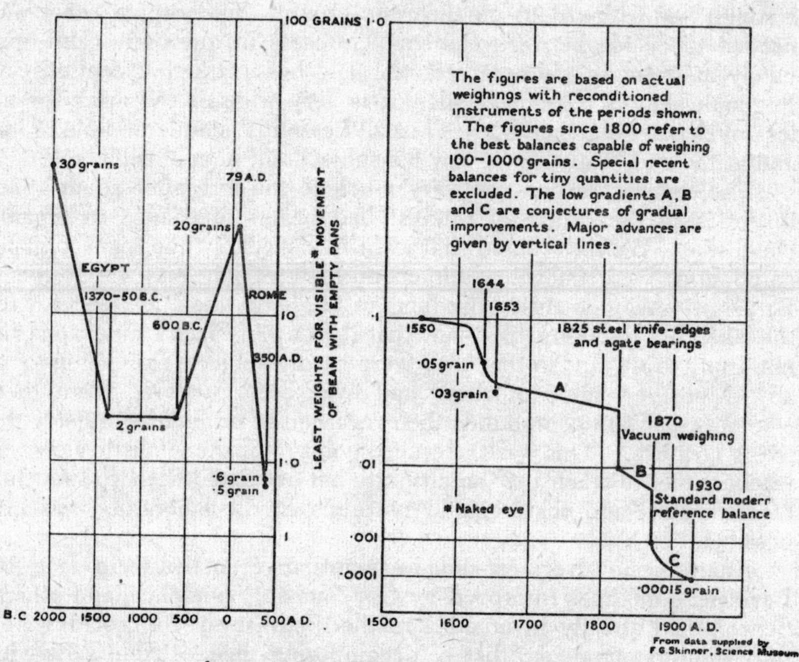

ACCURACY OF MEASUREMENT OF MASS.

have the same composition (AgCNO) by weight. We return to this subject of " isomerism " * later, but cyanates soon came to have an importance of a different kind. Because some not of living origin were known, they were reckoned " inorganic." It thus happened that when in 1828 Wöhler identified the compound obtained by treating potassium cyanate with ammonium sulphate and the " urea " obtained from urine (and so " organic "),† his work was greeted as an organic synthesis ; as a destruction of the vitalistic view ; as, in fact, a charter of liberty for organic chemists.

* A word coined by Berzelius for a second very important case of this kind 1830) : tartrates and racemates.

† Discovered in 1773 by Rouelle.

This was a great advance, but some of its results were equivocal. One was that Berzelius began pressing his Dualistic Theory into the organic realm, where its very inappropriateness stimulated a doubtful sort of ingenuity. The effort to divide every compound in two led to a new stress on the idea of "radicals," such as NH_4, SO_4, CN, which run through whole series of compounds. This idea, not unknown to Lavoisier, had been fixed by Gay-Lussac's work (1815) on CN. Our formulae for organic radicals like CHO, COOH, could not then have been written down, but the properties which they represented could be correlated.* This was done in relation to the ethers by Dumas and Boullay in 1828, but the formulation of the radical idea which concentrated the attention of the scientific world was that by Liebig and Wöhler in 1832.

It was an advance, but it was also a temptation. *Ad hoc* invention of radicals was so easy that the theory was strongly attacked by the older, empirical, chemists. But they could not check the theoretic ardour of the new men. Before 1840 there came another, subtler, theory of the same slightly too adaptable kind. This, the "type" theory, had been propounded by Laurent (1836) on the basis of Dumas' observation that chlorine and bromine can often replace hydrogen in radicals without much changing the latter's properties. In 1839 Liebig hit on the excellent instance of CCl_3COOH, derived from acetic acid CH_3COOH.

The notation here is ours, but much of the importance of this type of substitution came, later on, from its use in arriving at our present formulae. If, for instance (as here), there is a hydrogen unreplaceable by chlorine, the empirical formula CH_2O for acetic acid must at least be doubled or trebled. The doubled one $C_2H_4O_2$ agrees with the chlorine substitution. Other knowledge enables us to infer that a CH_3 group is involved ; leaving the acid group CO_2H to be identified with its fellow in countless other organic acids.

This theory, too, was fiercely frowned on.† For this there was a new reason. Unlike the radical theory, the type theory viewed compounds *as wholes*. In other words it was anti-dualist. Berzelius, now old, was stirred against it, as were many others. Under criticism, the theory lost its intoxicating quality ; but then so did dualism itself. It began to be realised that chemistry is neither one-complex, nor two-complex, but of many-fold complexity. The first symptom of this realisation was extreme discomfort. With the atomic weight table in

* The question whether "radicals" had to be capable of free existence was discussed at this time, and decided in the negative. For modern work showing that some are capable of a brief free life, see Chap. XXIV.

† With Gerhardt about 1850 it began to be usual to draw extreme analogies, such as that due to Williamson, which we should write

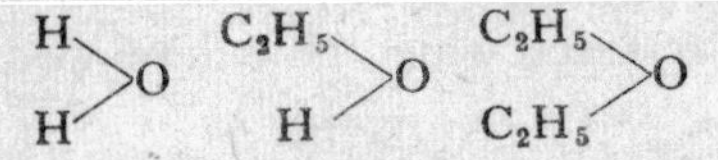

For such compounds may clearly be regarded as of the same type. If we can now safely draw such analogies, it is because of the greater safety of our empirical formulae, and the vastly greater mass of facts at hand to check the flights of fancy.

confusion, inorganic chemistry was in no better case. But we leave to Chap. XIV the story of how the troubles of the fifties led to a brilliant restoration in the sixties.

Physiological Chemistry

We turn to the relations of chemistry with physiology, by which we mean, discoveries in both subjects at once. But before doing so we will mention certain great figures in physiology pure and simple, whose work, scattered across our modern specialisms, we shall often mention in what follows, Magendie (1783–1855) and Bernard (1813–78) of France, J. Müller (1801–58) and Ludwig (1816–95) of Germany. The Frenchmen were typically positive and unspeculative, the Germans eager theorists, Müller vitalist, Ludwig anti-vitalist.

The two Germans, especially, were in the tradition of great teachers, so strong in physiology (see Chart III). Müller's pupils, Schwann, Henle, Helmholtz, Du Bois-Reymond, Kölliker, Virchow, account for most of the branches of analytic biology for fifty years to come. Ludwig had as many pupils, but Müller dealt with most branches himself; after him came specialisation. His great "Handbook of Physiology" (1834–40) made an epoch. Moreover his development foreshadowed that of physiology itself. By turning in later life to the microscope and to embryology, he showed what the science was to gain, as we shall see in Chap. XI, from a study of the very small and the coming-to-be.

In another way, Ludwig is as typical—namely, in his endless fertility in the invention of apparatus. This feature has been shared by many physiologists since his time, down to Keith Lucas in the recent past. Ludwig made many instruments automatically recording by (for instance) the addition of rotating drums. There is little that is intellectually remarkable about such ingenuity alone. What is remarkable is its union with a sensitive appreciation of facts of a totally different type.

In early physiological chemistry the name of Liebig is prominent. It was indeed as a pure chemist that Liebig became, in the thirties, as much a pontiff as Berzelius had been in the twenties. But towards the end of the decade he turned to physiology, to which he and his pupils applied, for the first time, specialised organic chemical methods. By his day the general animal cycle of O_2 and CO_2, with oxidation as the source of animal heat, was known; and Spallanzani (1729–99) had shown that it applies whether breathing is through the gills, or skin, or throat. As regards plants, the traditional theory was that the humus of the soil was their food. This view was largely undisturbed when in 1804 N.T. de Saussure (1767–1845) * showed that the carbon dioxide absorption (Ingenhousz, last chapter) accounts for nearly all their increase in dry weight during growth.

The currents of research which now developed were roughly concerned with:

1. the details of this cycle,
2. the similar cycle in nitrogen,

* Previous member of the family mentioned in Chapter VII.

3. the contrast of plants and animals in these cycles, leading up to,
4. the physiology of digestion.

1. In 1819 Magendie had shown that the nitrogenous "proteins" were necessary in animal nutrition, and in 1842 Liebig divided the chief organic constituents of both plant and animal tissues into fats, carbohydrates (such as sugar and cellulose) and proteins. In animal processes, the first two were ultimately oxidised to the expired and perspired carbon dioxide and water, while the last, which contained nitrogen, gave rise mainly to urea or uric acid.

These were obviously processes of breaking down, while in plants, as obviously, there must be processes of building up. As time went on, Dumas and Boussingault (1802–87) suggested that this was always true, so that plants were the necessary intermediaries between animals and the inorganic world. This is the subject of "3." But, meanwhile, some detail work on the CO_2,O_2 cycle had been accomplished. In 1817 the green pigment of plants had been isolated by Pelletier and Caventou and called "chlorophyll." In 1837 Dutrochet showed that the vital carbon dioxide assimilation only takes place in cells where this pigment is present. The optical pigmentary properties of chlorophyll seemed important, for synthesis only occurred in sunlight; but at that time the mere physics of the absorption of luminous energy was not fully understood. Finally, the decades about 1850 saw Boussingault, Sachs and others complete de Saussure's work by proving that plants derive only water and certain salts from the soil, the humus comprising not their food but their dead bodies. We return to this subject in Chap. XIX.

2. *The nitrogen cycle.* With Liebig, physiological chemistry passes another rubicon: it becomes capable of practical application. Much of his work had the object of compensating the poverty of the soils near Giessen by means of artificial manures. It was an effort to improve by chemical research the empirical technique of the agricultural revolution then spreading from Britain.

This may seem to have nothing to do with nitrogen, but in fact it was just through Liebig's neglect of nitrogen that his artificial manures (1845) were not very effective. And it was because his pupil, Gilbert, with Lawes at Rothamsted, made allowance for this element that greater success attended them. Even they, however, were only opening a vast field. Liebig himself knew in theory that plants need nitrogen and take it up by their roots as nitrate and ammonium salts.

In the fifties Boussingault improved our knowledge of this "nitrogen cycle" as another engineer, Ingenhousz, had that of the carbon-oxygen cycle. He showed (what was true so far as the plants which he used went) that if nitrates are present in the soil, plants can get all their nitrogen independently not only of the atmosphere but also of the organic matter in the soil. This was a great advance, but it did not explain the crucial puzzle in empirical agriculture—the usefulness of leguminous plants in the crop rotation. Liebig might easily have discovered a part of this explanation, but he did not do so, nor did anyone of his generation.

Liebig, in fact, was a man of strong limitations, and in particular he would never make that use of the microscope without which this particular action (and others like fermentation) could never be understood. His refusal, for instance, to believe that yeast was alive may be regarded as another instance of the frequent lack of relation between biologists' philosophic and scientific convictions. For Liebig remained a strong vitalist in philosophy.

3. *The parts of plants and animals in these cycles*. Meanwhile the general contrast of plants as synthetic, animals as analytic, had been attacked. Wöhler had shown (1824) that in at least one case an animal can make a more from a less, complex compound. Benzoic acid (C_6H_5COOH), administered by the mouth, is excreted as hippuric acid, more complex by a group containing nitrogen. For a dozen years after about 1845, Bernard had been experimenting on the action of the liver in a way which strongly reinforced Wöhler's work. In 1857 he finally showed that the liver synthesises the compound glycogen, in which form the body stores its sugar. Thus animals do sometimes build up.

Neither of these experiments, however, showed them building up from really simple or inorganic materials, and it remains broadly true that animals must be supplied with all three classes of food by plants, sometimes via other animals.

Bernard's work was the culmination of a series of advances in our knowledge of digestion. Its general interest arises from its being *experimental* rather than observational ; and experiment on digestive functions is obviously not easy. The first advances, indeed, came through an accident. Réaumur (1683–1757) had a curious pet, namely, a kite. This bird has the habit of vomiting whatever its stomach cannot attack, and by inducing his specimen to swallow experimental vessels, Réaumur was able (1752) to isolate gastric juice, and to show its solvent action on foods. Thus he destroyed the old view of digestion as fermentation.

Even then the next advance was long in coming, for it depended on advances in general chemistry. But in 1803 J. R. Young proved that an acid is normally one of the solvents in gastric juice, and in 1824 Prout showed that this acid is hydrochloric acid. In 1835 Müller's pupil Schwann found another solvent, "pepsin," in gastric juice. Spallanzani, also, established the preliminary digestive action of saliva.

But it remained for Bernard to show that gastric digestion is itself only a preliminary to pancreatic. He showed that the pancreas dissolves proteins not dealt with in the stomach, converts starch into sugar, and fatty foods into fatty acids and glycerine. Previous workers had only been able to put these ideas forward as suggestions.

To make them certainties Bernard carried experiment on living creatures a step further than anyone before, and in doing so incurred criticism. In 1822 the American army surgeon Beaumont had elicited many new facts (published 1833) about digestion from an accidental gastric fistula. Bernard produced these artificially in animals to watch the action of the organs concerned.

We now turn back to trace the history of another physiological subject, the nerves, and, in particular, the senses. These latter have a neurological, but also a subjective and a physical side. We have spoken of the Pythagorean association of number with music. The 17th century began the process of tracing this empirical correlation to a physical source, by associating pitch (and later colour) with the *frequency* of waves. In 1636 Mersenne determined the frequencies of the musical notes by combination of which all music is built up.

The same century began to analyse all the equally endless variety of colours down to combinations of three primary colours. Aristotle had tried to use only two, viewing all colours as mixtures of black and white. Newton was a worker on this subject; but for long after him the subject was obscured by failure to realise the necessity of distinguishing the combination of pigments and that of beams of light. We turn for the present to the nerves, on which, it will be recalled, Haller had made fundamental advances in the middle of the 18th century.

Ever since Descartes' theory of the automatism of animals, the idea of unconscious, "reflex," responses had been growing familiar. So had the practice of experimenting on animals by removing parts of their nervous systems (for instance, Bohn, 1686, Duverney, 1697, decerebrate frogs). A beginning had been made by Winslöw (1732) and others with the classification of nerves, while in 1751 Whytt showed that the spinal cord was necessary for the reflexes which he examined. It had been known since Vesalius that the nervous system contains grey and white matter, the former located in the centre of the spinal cord and on the surface of the brain.

But the usual attempt to relate function and structure failed for the nervous system, and even into the early 19th century effective differentiation of parts or localisation of function was denied. In fact, Galen's "animal spirits" were still referred to; although the idea that their entry from the nerves blows out and shortens the muscles ought not to have survived the old demonstration (Glisson 1677, better Swammerdam, published 1738) that contracting muscles do not increase in volume. About 1800, Gall was making dissections which showed that there *is* some localisation of function; but he was discredited by an assistant who carried the doctrine to ridiculous lengths ("Phrenology").

In 1811 Sir Charles Bell (1774–1842) discovered that the posterior spinal nerves could be cut without convulsing the back, while the merest touch on the front ones had this effect.* In 1822, Magendie, by experiments on puppies, conclusively showed that the posterior and anterior nerves are very different, and that their difference applies also to their branches. The posterior are the sensory, the anterior the motor, nerves.

With others, Bell began, by refined dissections of the brains of animals, to make localisation of function scientifically respectable. The

* It was a day of private medical teaching; and Bell's school of anatomy in Great Windmill St., London, was a well-known centre.

Frenchman Legallois showed in 1811 that breathing is stopped by injury to a specific part of the lower brain. Flourens, by excisions done on pigeons (1822–4), showed that the "cerebellum" (at the base of the brain) coordinates the parts of a complex movement, while the upper "cerebrum" supplies the factors commonly called purposeful. In 1833, and later, Marshall Hall (1790–1857) localised in the spine a number of reflexes (chiefly of cold-blooded animals), thereby fixing with precision the concept of reflex action.

Later, it has been shown that a decerebrate frog will jump out of boiling water, but will die of starvation though in reach of food. The further removal of the spinal cord—not easy to effect without fatal results—causes complete lack of response in parts with nerves leading thereto, and very few activities remain intact. In 1861 Broca (1824–80) fixed on the area of the cortex concerned with speech; and this mapping was carried in the seventies, and later, into considerable detail. But by this time the cell theory had come in, with new kinds of neurological information and technique. We therefore leave the subject until Chap. XVII.

We return to the question of the senses, which, of course, is a strong instance of the physics-physiology connection. Indeed, the reviver of the wave theory of light, the London specialist Thomas Young (1773–1829), began from the physiological end. In 1792 he published his first paper on the action of the eye. Clearing up the problem of pigments and of coloured light, he gave in 1801 his theory that colour vision is due to the existence of three classes of elements on the retina, each sensitive to one only of the three primaries, which he gave as red, green, and violet (lights).* There the matter rested for half a century, while certain results on sensation in general began to be reached.

John Hunter and others had given isolated instances of a peculiar fact about the sensory nerves which it was one of the great Müller's main achievements to establish as a general truth. This "doctrine of specific nervous energies" (1826) was that the sensation which we get from a particular nerve-ending is the same in kind whatever the stimulus. Thus, light, pressure, and electricity applied to the eye, all yield luminous sensations. It came to be realised that, for instance, the nerves of thermal and pressure sensation are separate, and that the specificity is true of individual fibres.

Another law as general in aim was stated by E. H. Weber in 1846, in a great work on touch and thermal sensations. This was, that the least noticeable increase dW in a sensation-causing influence W is given by $\frac{dW}{W}$=a constant, dW and W being supposed measured by some purely physical means. Later work has shown that this relation only shows the general trend of the facts, not their detail. Weber did further work with the general aim of establishing quantitative

* At about the same time Dalton was investigating colour blindness. Another prominent name here was Goethe's. Goethe had a fine scorn of Newton and revived Aristotle's theory. We cannot pursue this particular theory of his, but some general features of his point of view are discussed later.

" measures " of sensations. He also noted that the threshold of nerve injury is the threshold of pain.

The great man who visibly bodies forth the union of physics, physiology, and indeed of psychology and aesthetics, as regards the senses, is the Prussian army surgeon Helmholtz (1821–94), one of the greatest of scientists, great equally in pure physics, in pure physiology, and in their junction. We refer in the next chapter to his work in physics. In physiology proper he was, like Ludwig, fertile in the invention of apparatus, as witness the myograph (1850–52) by which he measured the velocity of nervous impulses ten years after his teacher Müller had declared this to be impossible. By inventing the ophthalmoscope * and the phakoscope he revolutionised the work of the oculist.

He revived and strengthened Young's theory of vision. We defer this interesting subject to Chap. XIX, but his " Physiological Optics " (1856–7) and " Sensations of Tone " (1863) are among the few full-length books which, in the new age of brief, swiftly-superseded papers, retain their place as works to be read.

Helmholtz was an accomplished musician, as became a German in a great age of German music ; and he realised that the subtle, yet highly definite, development of musical taste is a mine of evidence for the physiology of hearing. In particular he gave a partial reason for the simple integer relations of consonant notes, and of the problem of the pleasantness of harmony. He found that only sources giving a number of well-developed harmonics † are musically pleasant (doubtless because the ear, viewed as a vibrator, also has its harmonics). He pointed out that, mathematically, the consonant intervals are those which give coincidences among these harmonic overtones.

He developed one of the only two serious theories of hearing, the " resonance " theory of which hints had been given by Duverney (*c.* 1680) and Cotugno (1760). This views the fibres of the basilar membrane (or conceivably some other parts of the cochlea) as a series of resonators each of which excites a different afferent nerve. Here we find a hint of the reason why music is organised round discontinuous, not continuous, changes of pitch : the fibres, though very numerous, do form a discrete, not a continuous, series of frequencies.

In 1865 Rinne suggested the alternative theory, that all or many of these resonators must in general be excited by every sound, and that the analysis must be performed in the brain itself. This theory seems to require the (undifferentiated) afferent nerve to be capable of registering far more stimuli per second than any other nerve is known to do, while giving little explanation of the complexity of the inner ear. It is, therefore, in less favour. But no theory yet explains all the facts.

* 1851 ; anticipated by Babbage 1848.

† Mersenne, and later (1677) Noble and Pigott, had noticed that strings vibrate not only as wholes but in halves, etc. The general history of acoustics is ignored in this book, but Mersenne determined the velocity of sound, and Gassendi proved it constant when pitch or intensity were varied.

CHAPTER X

MATHEMATICAL PHYSICS

THE events to be told in this chapter form one of the characteristic scientific movements of the 19th century. They are an example of close and fruitful interrelation.

The wave theory of light, revived by Young and others soon after 1800, led (up to about 1840) to a series of mechanical " models " for the ether, none of which proved quite satisfactory. At the same time, Poisson and others, inheritors of the experiments of Priestley and Cavendish and of the mathematics of Laplace and Lagrange, were building up the field theory in electricity which was to replace these mechanical models later on.

Meanwhile, 1820 onwards, Oersted, Ampère and the rest discovered the laws of force between currents and magnets, and a little later Ohm, Faraday, and Henry established the ideas of resistance and induction. It remained to connect their empirical laws with field theory.

The discovery of thermoelectricity in 1820 had not only given Ohm an E.M.F. steady enough for the discovery of his " law," but had made possible experiments identifying radiant heat and light as but forms of radiant energy. Then came energy doctrine in general, first (soon after 1800) as to heat and work in gases, then (especially after 1840) as to chemical, electrical, luminous, and other " powers of nature." Joule, Kelvin, Helmholtz, Clausius, in kinetic theory, related heat and gas properties to the Laplace-Lagrange mechanics. The mechanical energy of the 18th century grew into the richer concept of energy in general, governed, as to its availability for conversion, by a new set of conservation and minimum laws. The applicability of these laws where the underlying mechanism was unknown gave crucial help in developing the empirical laws of electricity into a general electromagnetic field theory. With Maxwell, this was found to account brilliantly, also, for many properties of *light*. Thus the earlier mechanical ether-models were superseded.

Meanwhile the spectroscope, a development of the radiant energy line, was analysing light into discontinuous spectra which prepared new difficulties for Maxwell's theory. But by that time a new type of experimental work, intimately connected, as usual, with chemistry and physiology, demands a new chapter.

We treat this story in the following order : wave theory ; empirical electricity and magnetism ; heat, gases, and energy ; radiant heat ; electromagnetic theory.

WAVE THEORY

There are certain things, vortices and waves among them, which exert a perennial fascination and are found arising as key ideas in every

PLATE X

HELMHOLTZ. Physicist, physiologist. *From a pastel by Von Lenbach, 1894. By kind permission of the artist's widow.*

PLATE XI

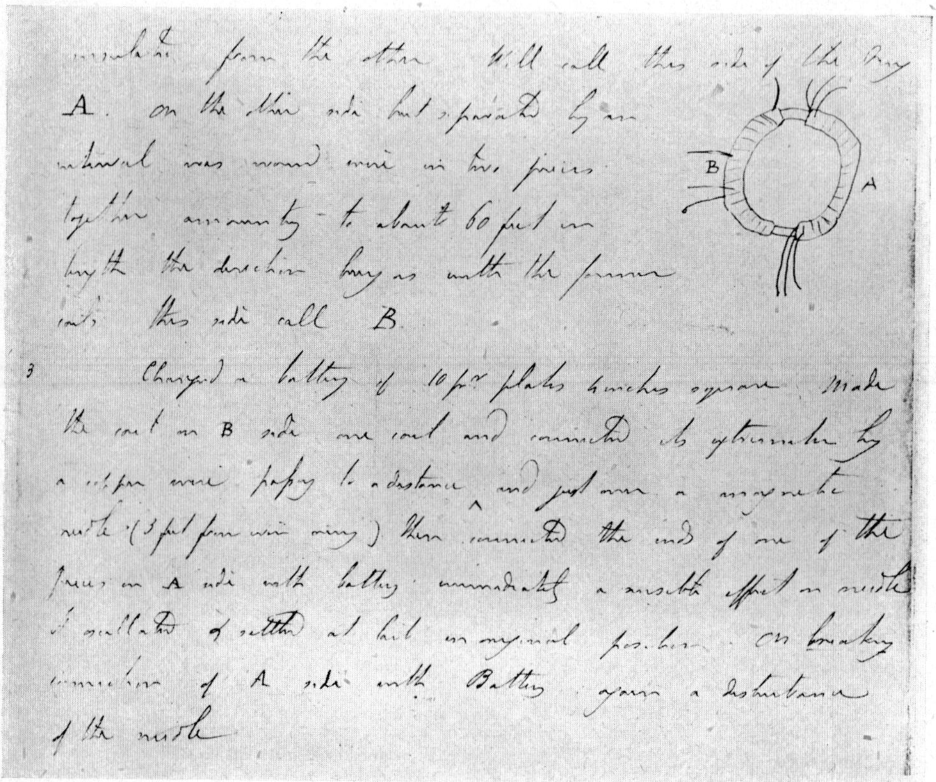

insulated from the other. Will call this side of the ring
A. On the other side but separated by an
interval was wound wire in two pieces
together amounting to about 60 feet in
length the direction being as with the former
coils this side call B.

3 Charged a battery of 10 pr. plates 4 inches square. Made
the coil on B side one coil and connected its extremities by
a copper wire passing to a distance and just over a magnetic
needle (3 feet from iron ring) then connected the ends of one of the
pieces on A side with battery immediately a sensible effect on needle
It oscillated & settled at last in original position. On breaking
connection of A side with Battery again a disturbance
of the needle

B

A

FACSIMILE III. The sheet of FARADAY'S diary recording his first successful experiment in electromagnetic induction. *By kind permission of the Royal Institution.*

century. The Cartesian vortices of the 17th century are revived in the 18th in Bernoulli's luminiferous ether of fine vortices, and in the 19th in Helmholtz's decisive work in hydraulics and in Kelvin's knotted-vortex-ring theory of the chemical elements.* But, apart from Bernoulli and Euler and Franklin, Newton's authority kept the ether so far out of 18th-century optics that no decisive experiment was provoked.

Further, optics had for long no novelties as stimulating as the lens systems which had provoked the advances of the 17th century. Towards 1800, however, the mineralogy of Werner and the chemistry of Lavoisier were giving currency to hosts of examples of another natural object of age-old fascination, the crystal. It is true that these had been optically fruitful in the 17th century, but in the early 19th almost the whole of optics revolved around them. In chemistry, as we have seen, their influence, with isomorphism and atomic weights, was only less great. Thomas Young, however, received his impulse from physiology, passing to pure physics only towards 1800.

Young reflected that a corpuscular theory, unlike a wave theory, offered no natural explanation of the fact that intense light travels no faster than weak light. He saw a resemblance between the periodicity of sound beats and of Newton's rings ; and, taking up a suggestion of Newton about the interference of tides, he used it (1801) to explain the colours of these rings. He had hitherto held Newton's explanation of diffraction ; but in 1803 he, for the first time, explained that too by interference.

He and Wollaston proceeded to verify Huygens' wave-surface construction for crystals. But this roused the French corpuscular school, and its pontiff Laplace, bringing to bear the full resources of his mathematics, produced (1808) a corpuscular explanation. This point should be noted, for a hundred years or so later men were again glad to recall that the corpuscular theory had had successes to its credit. For the moment, however, the opposite was true, and corpuscles were rapidly approaching their eclipse. The particular point which overturned them was polarisation.

At first, indeed, polarisation was a difficulty on either theory ; for Young, like earlier workers, had supposed light to consist of longitudinal (and so non-polarisable) waves. In 1808, Malus found polarisation in light reflected from transparent surfaces ; and in the same year Arago (1786–1853) showed that there is an equal amount in the refracted beam. Biot (1774–1862) then began to study the beautiful coloured figures got by examining crystals with polarised light. Young's attribution of these to interference led Fresnel and Arago in 1816 to find out experimentally when two polarised beams can interfere, with the conclusion that they can only do so so far as the planes of polarisation have a component in common.

But the nature of the polarisation itself remained a mystery until,

* And often the vortex-ideas prove unprofitable ! Those of Descartes did. Bernoulli's did. So did Kelvin's. Tait found that the numbers of knots of different orders of "knottiness," 1 . 2 . 3 . 10 . 27 . 41 . 123 . ., bore no special relation to the chemical elements.

in 1817, Young wrote to Arago the suggestion of transverse waves, which are evidently in their nature polarised. To Fresnel the difficulty with these was that they demanded something like an elastic *solid* ether, through which, it seemed, matter could hardly move as easily as it does. But he now saw that he had been premature in his rejection. In a brilliant series of papers, 1818–21, in which he had to create the mathematics of transverse waves as he went along, he showed that Young's new theory accounted for all the facts.

He had, however, to make assumptions about his ether which not only differentiated it from ordinary elastic solids but suggested alternatives. We now enter on the period during which these alternatives were examined in detail by a series of gifted mathematicians. We cannot follow the process very far. Outstanding was the question whether it is the density or the elasticity of the ether which alters in refracting and doubly-refracting, bodies. In 1828 the great French mathematician Cauchy (1789–1857) reduced to a minimum the assumptions needed to build any theory; and in the same year Poisson (1781–1840) raised a difficulty for all subsequent workers when he showed that a longitudinal wave would accompany the transverse one.

Irish, Scotch, English, and German schools entered a mathematical region hitherto almost exclusively French in suggesting ways out of the various difficulties. MacCullagh in Dublin, Neumann in Germany, Green, the founder of the Cambridge school, all put forward theories. These, by 1837, had exhausted the possibilities of ordinary "elastic" models; yet all were in one way or another unsatisfactory.

In 1839, Cauchy examined a contractile ether (much later revived by Kelvin) and MacCullagh, one resistant only to rotary, not to ordinary, strains. But these were physically so odd that interest migrated for a time from these ambitious schemes to narrower subjects. Only one point, of great importance later, must be noted. The great Dublin mathematician Hamilton (1805–65) put the equations of optical systems in a form in which they had formal analogies with certain equations which he had given for general mechanics (see Chap. XIII).

Among the narrower subjects were dispersion and metallic reflection. By the middle thirties it was realised that the absorption of light must be accounted for by some kind of friction or viscosity in opaque substances. Viewing matter primarily as loading the ether, Cauchy obtained (1835) a definite formula for dispersion, that is, for refractive index in terms of wave-length. This was founded on the supposition that the mutual distance of the molecules is small compared with the wave-length; but unfortunately it failed to agree with observation in just the region—the infra-red—where this assumption should have been soundest. In fact one entire factor was ignored in Cauchy's work, and the subject needed a contribution, yet to be noted, from spectroscopy and radiant heat.

Another difficulty, at once faced by Fresnel, will be kept to a later chapter, since it was not solved for a century. This was, the effect of motion on the optical properties of bodies, and, generally, the difficulty that an ether destroys the basic assumption that all motion is relative.

A related point, however, must be noticed. This was, that for the first time it had become possible to measure the velocity of light. Arago suggested methods of doing this in 1838. Others were implicit in Wheatstone's rotating mirror method (1834) for measuring the (exceedingly short) duration of electric sparks.

The crux was that the apparatus would be costly, and it was 1849 before there came a man, Fizeau, with private means to build it. He used a rotating-cog-wheel method. Foucault in 1850 and 1862, used Wheatstone's idea of a rotating mirror, and thus obtained values agreeing with those of Bradley in 1728. More, he was able to show that the velocity in water is less than that in air in the ratio of the refractive index. This seemed to the Victorians to dispose of the ancient question between the wave and corpuscular theories, wholly in favour of the former.

Electricity and Magnetism, mainly Empirical

Franklin, reasoning from the already current observation that charge is concentrated on the outer surface of bodies, had suggested that there may be no force within them. Priestley (1767) found that this was so. He argued from it to an inverse square law between charges, on the ground that in the case of gravitation a spherical shell of matter exerts no force on internal points. In such work, Nollet's demand for a mathematical electricity was beginning to be satisfied. In the seventies Cavendish used a closer form of the same argument, but his work was not published until far into the next century, and the proof accepted by the world was that made by Coulomb (1738–1806) with a torsion balance.

In 1750 the Englishman Michell had used this instrument (in which the known torsion of a thread balances an unknown force acting at the ends of a bar) to reach an inverse square law between magnetic poles. This, with Cavendish's unpublished discovery (before 1781) of the similar law for charge, is another English case of work done before its time.* Coulomb (1777) reinvented the torsion balance and with it verified (1785) the law both for poles and for charges. This gave the French mathematical school one of the axioms needed for mathematical development ; but *physical* insight was still the limiting factor, since there was a difficulty, which did not arise over gravitation, as to the relation of charge and matter. It was 1812 before Poisson developed a sound mathematical theory of electrostatics.

That for magnetism was made harder still by the need of a physical supposition to explain why magnetised matter has both the opposite poles upon it ; but in 1824 Poisson laid the foundations here also. He used an idea of Coulomb's that each molecule is a small magnet complete with two poles, so that a magnetised body resembles in structure the solution between the plates in Grotthuss's theory of electrolysis.

Poisson made much use of the functions developed by Lagrange, Legendre, Laplace, and others for gravitation and mechanics generally ; and in particular of one which Green (1828) called the " potential," the

* See Chapter V for the application of Michell's instrument to gravity.

charge or pole-strength being substituted for mass in the definition given in Chap. VI.

Poisson and his contemporaries went as far as was possible with the "action at a distance" point of view, but in practice neither charges nor magnets can be handled without using solid materials to transmit or bar their effects. This at once raises the old, complicating, question of the "medium"; and a new point of view was necessary before this could be taken into account.

By 1824, too, another factor had entered: voltaic currents (Chap. IX) had been shown to exert mechanical force on magnets, and vice-versa. Very much earlier—early in the 18th century—it had been thought that lightning produces magnetic effects; and in 1807 Oersted of Copenhagen (1777–1851) began a long series of experiments to clear the subject up. In 1820 he showed that a current makes a magnet in no other field set at right angles tc it. The news went all over Europe with extraordinary speed, and before the end of that year the brilliant French group, Biot, Savart (1791–1841), and Ampère (1775–1836) had discovered the laws of force not only between a magnet and a steady current, but also (Ampère) between two of the latter.

A current element of length ds, was found to exert on a magnetic pole at distance r a force $\propto \frac{ids}{r}$, current being i. The force could be doubled, that is, by doubling the batteries, or by doubling the wire into a coil. Ampère, not naturally neat with his hands, developed the greatest ingenuity in testing this formula and the similar one for two currents. By bending the wires used, he showed (for instance) that there is no force-component *along* a current, only at right angles to it.

In 1820, Davy and Arago independently had shown that a current can magnetise steel; and by 1825 powerful electromagnets were being made. In that year Ampère summed up the subject in a great memoir introducing the idea that magnetism consists in resistance-less molecular currents. He showed that identical effects are produced by a circuit of any shape and by an indefinitely thin magnetic shell of the same shape.* This at once made it possible to measure current and magnetism in strictly mechanical terms; though it was 1832 before Gauss defined units for magnetism and 1846 before Weber did the same for current. By 1821, Schweigger and others, using these discoveries, had made the first "galvanometers." Nobili's astatic instrument (described 1825) was a great advance in sensitiveness. But these appliances were at first of small quantitative value. For instance, the distinction of throw and steady deflection was not understood.

In fact, in the twenties quantitative experiment was still beset with difficulties. Until the Daniel cell (1836) and the Grove cell (1839) voltaic sources did not give constant currents for any length of time. It was Seebeck's currents at heated junctions of dissimilar metals (1822) which gave Ohm the steadiest E.M.Fs. which he could find for his

* Given a suitable current in the circuit, and normal magnetisation in the shell.

work on current and resistance. In the twenties, this "thermo-electricity" had the further importance of showing that current could be produced without chemical intervention; while in the thirties, as we have said, it was crucial in enabling Nobili (1784–1835) to make a detector sensitive enough to measure radiant heat.

Others before Ohm had made studies of resistance. Beccaria in 1753 had shown that wide water tubes resist the static discharge less than narrow ones. Cavendish (1775) had made some surprisingly accurate comparisons of the resistances of water, salt solutions, and iron wires to these discharges. Davy in 1821 had shown that the same figures hold for voltaic electricity and had examined the effect of shape and size of conductor. But none of the quantities in Ohm's now familiar $C=\frac{E}{R}$ could then have been precisely defined, and exact analysis was badly needed.

Ohm was inspired by Fourier's work (1822) on the flow of heat through conductors to make (1826 and 1827) a similar analysis for the flow of electricity, defining "current" like quantity of heat, "E.M.F." like temperature. But his contemporaries were not impressed by the analogy, and in fact it is for his experiments that he is remembered. He compared the fall of potential along identical sections of wire in different parts of one circuit, along identical lengths of wire with cross-sections different in shape, and also in area. He compared "piles" or "batteries" (and also resistances) in series and in parallel, and in sum concluded that all the effects fell under the familiar law.

Ohm did not identify his E with electrostatic potential, but the connection of static and current electricity was in many minds. Early in the thirties it led Faraday (1791–1867), Davy's protégé, to open a new field by looking for the current analogue of electrostatic induction. After some search for effects near wires bearing constant currents Faraday realised that the matter was rather one of non-constant currents, which had hitherto not been investigated. He found, in fact (1831), that currents are produced in neighbouring conductors whenever a current changes in magnitude or its conductor moves. Henry in America had noticed in 1829 (published 1832) a particular case of Faraday's discovery which the latter himself did not see until 1834, namely the *self*-induction of a current at make or break. He may very well, also, have known of induction in the full sense.

The decisive peculiarity of Faraday's mind now began to emerge, the fact that he thought in pictures. In particular, he brought back the old ideas of lines of force and of a medium, the ether, as against the mathematics of the French school, which was not concerned with how forces were transmitted so long as their laws were known. The living idea of continuity, which Europe had brought back into Greek mathematics and evolution had brought into the formalism of 18th-century natural history, was thus brought into 19th-century physics.

Faraday conceived all the ether filled with as many unit lines, or rather (as we should say) tubes, of force as there were units of magnetism concerned. Each tube was trying to expand sideways and contract endways. In fact, the lines of force were lines of elastic strain in the ether.*

In 1832 Faraday showed that it is the electromotive force, not the current, induced which is directly related to the number of unit tubes cut by the conductor through motion or change of inducing current. By 1833 he had satisfied himself that electrostatic discharges could always be treated, so far as effects went, as transient currents. But he avoided any decision on one idea originally bound up with the identity of " frictional " and " voltaic " electricity, the idea that currents consist of moving charge or charges. Thus, when later he extended " lines of force " to the electrostatic field also, he did not make use of the conception, later so fruitful, that the magnetic field consists of moving lines of electrostatic force, that is, of *changes* of strain in the ether.†

In all this, the medium was still the ether—still a vacuum.‡ If we call conducting solutions media, we may say that Faraday now began to consider other media in starting his great work on electrolysis (deferred for convenience to Chap. XV). At all events, he went on (1837) to consider the effects of different insulators in Leyden jars. Cavendish, to express these differences, had used a quantity " specific inductive capacity " (k), proportional to the charge on the plates for a given potential difference between them. To Faraday, a " k " greater than unity could only mean that the lines of force were k times more crowded in the dielectric than in air or a vacuum. As to why this crowding should take place, Faraday advanced a theory which we shall briefly discuss in connection with Maxwell. Similarly, magnetic lines of force were crowded into an iron magnet, dispersed in the rest of their circuit, in the proportion of another coefficient, μ, expressing the magnetic capacity of iron compared to that of a vacuum.

But for long it seemed that magnetism was unlike charge in occurring only in iron and one or two other substances. Then (1845) Faraday discovered that some substances, in a field, acquire a magnetism opposite to that of iron, " diamagnetism." He and Arago showed that most substances can be (feebly) magnetised in one way or the other. So the position was regularised, especially when Weber pointed out that diamagnetism, on Ampère's hypothesis of molecular currents, is merely a case of induction.

On the basis of these conceptions, numerical specifications of fields in dielectrics and magnetics, as well as in free space, became possible in absolute measure. But before developing this we must turn to other subjects.

* The magnetic lines were closed curves, " poles " being merely a convenience of thought.

† Inventors seized on induction as a means of turning motion into current. By 1850 Jacobi had realised that a machine to do this could also do the opposite. By 1856 Werner Siemens began to have practical success with these " dynamos " and " motors." At many points he seems to have been anticipated by the Hungarian Jedlik. By the sixties, with the coming of telegraphy, many inventors were at work. Though this belongs to the history of technology rather than of science, it had many scientific repercussions ; perhaps especially in making very *powerful* currents, and costly apparatus, available.

‡ Or air, since for this purpose no difference was found.

Heat, Gases, and Energy

We have seen that Black and other practical workers in heat had used the concept of a material heat in preference to the kinetic view. Its constancy in quantity was then only a case of conservation of matter generally. But with the Lavoisierian hardening of caloric into a definite theory came also a hardening of the kinetic theory, in terms of which the constancy was a case of another already known conservation law : that of kinetic energy. Heating by friction, for instance, was on this view a conversion of large-scale motion into invisible, atomic, motion. It was of course realised that it is only under certain conditions that kinetic energy does remain constant, but it was long before this became the dominant consideration. At first, the kinetic theorists had to disprove the calorists' account of friction : that it diminished the " heat capacity " of a substance by squeezing out some of the caloric.

The argument (1798) of the American Count Rumford—that friction (for instance, in boring cannon) can produce an *indefinite amount* of heat—was inconclusive ; but in 1799 Davy claimed to have shown that the heat capacity of the water produced by rubbing together two pieces of ice was *greater* than that of the ice.* Many experiments began to show the intimate connection of mechanical energy and heat. Dalton noted that sudden compression heats a gas. Laplace reasoned that therefore the compressions in a sound wave would not proceed according to Boyle's law, as Newton had supposed in his calculation of the velocity of sound. The erroneousness of Newton's result had long been a puzzle. The heat in the compression, having no time to escape, would increase the pressure, and so velocity, in a calculable ratio (" adiabatic expansion "), that of the specific heats at constant pressure and at constant volume. Clément and Désormes (1819) found this ratio experimentally. It gave the required correction of Newton's value.

There was, moreover, a growing accuracy of gas technique, as we have seen in the last chapter. Dalton and Gay-Lussac, in 1802, had given $\frac{1}{267}$ as the constant in what we now call Charles' law. Later Regnault altered this to $\frac{1}{273}$, implying that the volumes of gases would become zero at $-273°$ C.

Other developments, however, and in particular a wider view of energy in general, united to a deeper knowledge of mechanics, were needed before these matters could be cleared up. In 1807 Gay-Lussac had done one typical experiment, one, namely, designed to show that when a gas expands (for instance, into an *empty* vessel) *without* doing work, there is no net change of temperature. In 1819, Dulong showed that the heat evolved in compression varied as the work done. But many gifted workers continued to take a material view of heat until the middle of the century, and the conclusive developments of kinetic theory were deferred for thirty years.

* Whether he can really have done this has lately been doubted by Prof. Andrade.

Another aspect of heat—its transference or motion—had begun to give important results.

Newton had investigated the rate of cooling of bodies, and had concluded that if they are at T degrees above their surroundings, their temperature drops at a rate proportional to T. But thermometers were then very imperfect, and it was Scheele's thorough investigation (1777) which first drew a clear distinction * between convection and radiation. The third type of transfer—conduction—can only be noticed here as leading to Fourier's great work of 1822, inspiration to later mathematical physicists. We shall confine ourselves henceforth to radiation.

Pictet (1752–1825), like others as early as 1684, had observed that "cold" could be radiated no less than heat—for instance from ice, via concave mirrors, to a thermometer. Prévost (1751–1839) suggested that, rather, all bodies are always radiating *heat* ; only some radiate less than they receive. This "theory of exchanges" (1791) contains one of the fundamental ideas of science : that of dynamic equilibrium. In 1800, W. Herschel (1738–1822) found heating rays in the solar spectrum beyond the red, and in 1801 Ritter found rays beyond the violet with the photographic action discovered by Scheele. It was thus revealed that the visible spectrum is but a part of a more extended thing with properties continuous except as to visibility through a much greater range of refractive indices and so (on the now reviving wave theory) of wave-lengths. Leslie (1777–1832), Forbes (1809–68), and others, began trying to prove that radiant heat has all the ordinary properties of light : not only straight line propagation and ordinary reflection (already admitted), but refraction, diffraction, interference, and polarisation. But the effects are often small and hard to distinguish from those of convection, so that it was not until Melloni and Nobili improved the thermo-couple in the thirties that the proof was really given.

Motion had been linked to heat, now heat to another form of motion—wave-motion or light. The kinetic view, in fact, positively called for something which, sure enough, was standing out, as the thirties ended, in the minds of those who could survey the whole field of experimental science. This was, the convertibility of the "powers of nature." Chemical change had been converted into current and back, current into magnetism and back, heat into electricity and back, electric current and chemical action into heat. Many cases had heat as the final form ; but one very striking instance, the steam engine, had not. As early as 1824 this had inspired the young French engineer Carnot (1796–1832) to take up a harder side of the subject which we treat later.

But was the conversion quantitative ? A bold suggestion that it was had been made long before, in physiology, when Lavoisier and Laplace had tried to equate the heat evolved by a candle and by a living creature in "burning" the same amount of carbon and hydrogen. At this time there would have been widespread repudiation of the view that such laws could apply to living matter, but we shall see that the crucial ideas emerged in physiology as soon as in pure physics.

* It must be remembered, however, that Scheele was a phlogistonian.

In 1829 Roget applied energy ideas to the voltaic cell, and in 1830 Carnot wrote some unpublished notes containing most of the essentials of the mechanical theory of heat. In 1839 Séguin tried inconclusively to compare the heat put into a steam engine and the work done by it. But these were not the influential advances. About the same time Joule (1818–89) * began a long series of experiments on the heating effects of current (which he found proportional to C^2R), on those of the chemical actions producing them, and on their equivalence. In 1843 he went on to experiment on the heat produced by given quantities of mechanical work ; but the scientific world viewed him coldly.

In 1842, from the physiological end, the German doctor Mayer (1814–87) announced the equivalence of heat and work, and in 1845 added a numerical element to his statement. Taking Gay-Lussac's experiment of 1807 as showing that a gas in expanding does no work against internal attractions but only against external pressure, he gave a value for the heat-equivalent of work. At the same time, like Colding in 1843, he gave a statement, not free from metaphysics, of the whole doctrine of the Conservation of Energy.

But even accurate statement was less important than proof and amplitude of exact detail ; and for this the world waited for the far greater genius of Helmholtz in 1847.† Helmholtz also approached from the physiological side, but unlike others he was as strong in mathematics and physics. The essence of his approach, and of that of Joule, W. Thomson (later Lord Kelvin) and Clausius (1822–88) about the same time, was a many-sided appreciation of the energy-conservation equations of the French classical mechanics, and of the fact that if the whole universe consisted of atoms connected by central forces, then the conservation and convertibility of energy followed simply as propositions in mechanics. At the same time (1847) Thomson "discovered" Carnot's work, already mentioned ; though unfortunately in a "caloric" form due to Clapeyron (1832). Clapeyron made serious mistakes through not realising, with Carnot, that heat is lost as work is gained. Carnot's work was in what we should now call thermodynamics. This subject and the kinetic theory of heat and of matter entered exact consideration simultaneously.

Bernoulli had shown (1738) that the motion of a swarm of particles could account for gas pressure and for its rise with temperature. Le Sage and Prévost had worked at the subject. Since 1845 a manuscript on it by Waterston had lain neglected at the Royal Society. Joule, in 1848, calculated the velocity of hydrogen molecules needed to account for the observed pressure.

Carnot's work was concerned with the fact that, though energy might never be lost, it might be rendered *unavailable*. Thus, the heat in the steam going to the condenser of an engine is no longer available for producing motion. It was typical of the two nations that while England had had steam engines for decades without taking up this theoretical point, full-size examples had no sooner arrived in France

* Like Black and others already noted, Joule was connected with brewing.

† Even he was at first accepted only by the physiologist Müller and by the mathematician Jacobi.

TRANSCRIPTION OF FACSIMILE, WHICH APPEARS OPPOSITE,
of sheets from
SADI CARNOT'S NOTE-BOOK,
anticipating the First Law of Thermodynamics.

La chaleur n'est autre chose que la puissance motrice, ou plutôt que le mouvement qui a changé de forme. C'est un mouvement dans les particules des corps. Partout où il y a destruction de puissance motrice, il y a, en même temps, production de chaleur en quantité précisément proportionnelle à la quantité de puissance motrice détruite. Réciproquement, partout où il y a destruction de chaleur il y a production de puissance motrice.

On peut donc poser en thèse générale que la puissance motrice est en quantité invariable dans la nature, qu'elle n'est jamais, à proprement parler, ni produite, ni détruite. A la vérité, elle change de forme, c'est-à-dire qu'elle produit tantôt un genre de mouvement, tantôt un autre ; mais elle n'est jamais anéantie.

D'après quelques idées que je me suis formées sur la théorie de la chaleur, la production d'une unité de puissance motrice nécessite la destruction de 2,70 unités de chaleur.

Une machine qui produirait 20 unités de puissance motrice par kilogramme de charbon devrait anéantir $\frac{20 \times 2{\cdot}70}{7000}$ de la chaleur développée par la combustion ; $\frac{20 \times 2{\cdot}7}{7000} = \frac{8}{1000}$ environ, c'est-à-dire moins de $\frac{1}{100}$.

PLATE XII

La chaleur n'est autre chose que la puissance
motrice ou plutôt que le mouvement qui a changé de forme. C'est un mouvement dans les particules des corps.
partout où il y a destruction de p.m. il y a
en même temps production de chaleur en quantité
précisément proportionelle à la q.té de p.m. détruite
Réciproquement partout où il y a destruction
de chaleur il y a production de p.m.

on peut donc poser en thèse générale
que la p.m. est en quantité invariable dans
la nature qu'elle n'est jamais à proprement
parler ni produite, ni détruite, ~~qu'elle~~ à
la vérité elle change de forme c.à.d. qu'elle
produit tantôt un genre de mouvement, tantôt
un autre ~~mais qu'elle existe~~ toujours mais
elle n'est jamais anéantie.

D'après quelques idées que je me suis formées
sur la théorie de la chaleur, la production
d'une unité de puissance motrice nécessite la
destruction de 2,70 unités de chaleur.

Une machine qui produirait 20 unités de
p.m. par kilog. de charbon devrait anéantir
$\frac{20 \cdot 2,70}{7000}$ de la chaleur développée par la
combustion; $\frac{20 \cdot 2,7}{7000} = \frac{8}{1000}$ environ c.à.d. moins de $\frac{1}{100}$

FACSIMILE IV. Sheets from CARNOT's notebook, with transcription facing.
By kind permission of Messrs. Gauthier-Villars, Paris, and of the Institut de France.

FACSIMILE V. A 13th-century statement of heat as motion, with translation facing. *Reproduced, by kind permission, from an MS. in the possession of the Jewish Theological Seminary, New York. Translation by Miss R. S. Mark, reproduced by her kind permission and that of the Editor of "Isis."*

PARTIAL TRANSLATION OF EXTRACT OPPOSITE

from a work of

LEVI BEN ABRAHAM (1246–1315)

". . . the heat of the stars and especially the sun, although as we have said they have no inherent quality and are neither hot nor cold, the cause of their warmth is either the motion or the light for the reason that motion produces heat as is evident in many cases; for instance it is apparent in the case of the arrow that is sent forth and the lead of which melts [during its flight]. And thus it is with the heat-giving property of the light when its rays are reflected. So much more this swift motion that sends its rays to the earth like an arrow travelling the enormous distance in no time with attachments that are reflected back continually and uninterruptedly through the air [produces heat] and this cause is stronger and more specific and therefore the sun and stars give heat more than other portions of the firmament."

than Carnot produced an almost Euclidean piece of reasoning upon them.

In fact, it is no bad summing up of Carnot's work to say that, as the Greeks gave us the abstract ideas (point, line, etc.) with which to think of space, and the 17th century those (mass, acceleration, etc.) with which to think of mechanics, so Carnot gave us those needed in thinking of heat engines. In each case the ideas are so pervasive that we use them even to state that they never apply exactly to visible objects.

Carnot's " unit of thought " was the well-known perfectly frictionless, perfectly insulated engine, which gains and loses all its heat at two standard temperatures T and *t*, and imparts motion to nothing except the crankshaft ; in particular, not to the particles of the steam. It is therefore " reversible," that is, capable, on reversal, of transferring all the heat back from sink or condenser to source. The expansions and contractions in it are all either isothermal or adiabatic,* and we can reason only about a complete cycle of its operations, that is, one which returns the working substance to its original state in *every* respect.

With such an engine it can be shown to follow that the work done per unit of heat transferred (" efficiency ") is independent of all details, such as the nature of the working substance, and is in fact simply equal to $\frac{T-t}{T}$; otherwise we can get an unlimited amount of work from it without recourse to the source. Now it will be recalled that as early as the 16th century it was the mark of a scientist to give up the attempt to secure perpetual motion. But it will also be recalled that the subject was not understood. For instance, the " perpetual motion " of a barometer thread, was taken as a case in point. It was prime movers which first made the question vital, and it was the triumph of the movement now described to have cleared it up.

It did this by recognising that there are *two* principles involved, not one : that is, by making the subject, like chemistry, two-complex instead of one-complex. The " First Law of Thermodynamics " was simply that already discussed, the constancy of the actual amount of energy in the universe. The " Second Law " was now precisely stated as the impossibility of getting an unlimited amount of heat or work out of a Carnot engine (and, *a fortiori*, out of any other less efficient engine). Clausius (1850) and Thomson (1851) gave equivalent statements of the law ; though Clapeyron's error clung to Thomson until Joule weaned him from it.

Thomson had been much concerned at the dependence of " temperature " on the properties of a particular gas or liquid ; and it was because he saw in Carnot's work a method of defining an " absolute " (that is, a work) scale (1848) that he welcomed it. To give efficiency not unity $\frac{T-t}{T}$, T must be finite. Thus the suggestion, implicit in Charles' law, of an absolute zero at about $-273°$ C. was confirmed.

Actually to realise an absolute scale, it was necessary to repeat, more exactly, Gay-Lussac's work of 1807 on the question whether there is in

* That is, involving no loss or gain of heat.

fact any energy used or freed * by any particular gas in expanding without doing external work. Any such energy must be reckoned as a correction in using that particular gas to construct an exact gas thermometer. From 1852 to 62 Joule and Thomson tested this point in their porous plug experiments. The correction in question, tiny in ordinary circumstances, becomes large enough at low temperatures to be used in the liquefaction of gases.

Clausius, meanwhile, was applying the new theories to change of state and to vapour pressure, and was founding physical chemistry. In 1854 he re-stated the Second Law in terms of the quantity " entropy," characteristic of an adiabatic as temperature is of an isothermal. For a reversible change in which heat dQ is transformed at absolute temperature T, the entropy is given by $\frac{dQ}{T}$. The law then states that entropy always increases in any but a reversible change. In 1857 he first stated the kinetic theory in full mathematical form, but was unable to allow for the evident fact that the velocities of the molecules must vary widely and constantly. That is, his treatment was not truly statistical.

Now, as we shall see later, the statistical concept was one of the great central concepts of the 19th century. Its mathematics had been given (see Chap. VI) about fifty years earlier.† In particular, the normal error law had been enunciated. In 1860 Maxwell applied this law to the velocities of molecules of gases, and was able to raise confidence in the theory by predicting the unexpected but true result that viscosity is independent of pressure. The theory was also in agreement with the contemporary revival of Avogadro's hypothesis. Maxwell went on to found with Boltzmann what we now call equipartition theory, but we leave this development to Chap. XXI, and go on to the advances in thermal radiation and spectroscopy, which added further provinces both to optics and to thermodynamics.

A Scotsman, Melvil, in 1752 saw that metals do not give continuous, but line, spectra. The observation, however, led to no result, even when Wollaston in 1802 first noticed the dark lines crossing the solar spectrum. Fraunhofer (1787–1826) rediscovered these in 1814, and gave the first full examination. These discoveries were due to using spectra focused by a telescope. Unfocused spectra do not show the lines. In 1823, J. Herschel had suggested that the lines which Melvil had seen might be used in chemical analysis to detect metals ; and many

* E.g. owing to attraction or repulsion of the molecules.

† The theory of probability is one of the vital subjects to which we cannot give the place it deserves. Plana (1812) was the first to use the equivalent of a coefficient of correlation ; but neither he nor Bravais (1846) were concerned with the degree of relation between the *independent* variables. This essential feature of the modern subject of correlation was first introduced in the seventies by Galton (see *Isis*, Vol. 10, p. 466). From Tshebysheff (1821–94) onwards right down to the present day, some of the most important workers in probability have been Russian. Doob (*Math. Statistics*, Vol. 6, September, 1935) states that, at bottom, it is identical with the theory of measure in " sets of points " (see Chap. XIII). Statistics are, in different ways, central both in quantum theory and in experimental biology.

spectra were mapped. In 1849 Foucault identified some of these, in position, with some of Fraunhofer's *dark* lines. Stokes (1819–1903), of the Cambridge school, suggested in lectures, before 1850, that this must be a resonance phenomenon, substances absorbing what they emitted because both processes were due to the vibration of their atoms with definite periods.

Meanwhile Tyndall (1820–93), Magnus (1802–70), and Balfour Stewart (1828–87), in trying to disentangle the relations of the adsorptive, reflective, and emissive powers of bodies for *heat* radiation, were forced, with Kirchhoff (1824–87), to recognise that they must treat each wave-length separately to get simple laws on Prévost's lines.

They began the process, later completed by Boltzmann and others, of making radiation amenable to thermodynamic reasoning. They found it necessary to postulate a new unit of thought to add to Carnot's, that of a uniform temperature enclosure. By reasoning from such an enclosure, they were able to show that, whatever mosaic of substances made up its walls, the absorption and emission of every part separately must be equal in both quality and quantity, and so equal to that of a perfect radiator or absorber—a perfectly "black body," with no reflective power. Thus they reached the concept of a full or black body radiation, characteristic of each particular temperature and independent, like the absolute scale, of any particular substance.

On the basis of this, Kirchhoff and Balfour Stewart were able in 1859 to clear the subject up. They arrived independently at the resonance theory of Stokes. They showed that the dark lines in the solar spectrum, by then nearly all identifiable with those of terrestrial substances, indicated the presence of those substances in the outer layers of the sun. With this result, astrophysics came into existence.

From 1855, Kirchhoff, Bunsen, and Roscoe had been developing the technique of spectroscopic chemical analysis, a method which led to the discovery of several fresh elements. The effect in physics of the resonance theory was, as we shall see, the enunciation by Maxwell and by Sellmeier of a better theory of dispersion and metallic reflection than that given by Cauchy.

Electromagnetic Theory

Faraday's stresses in a medium did not appeal to the continental mathematical school,* but in the forties, whatever view was taken, the subject challenged theoretical unification. In 1834 Lenz had given the abstract principle that all induction effects are in a direction to oppose the cause producing them ; and in 1845 Neumann, using this principle, formulated axioms to cover the three cases, static, current, magnetic. In 1846, Weber (1804–91) gave mathematical form to Fechner's theory of current as two simultaneous opposite streams of opposite charges. Riemann also briefly discussed an ether-model capable of resisting both compressive and rotary stresses. These theories held "in solution" alike the irreducible difficulties of the subject and the resources available against them. But they were far

* Gauss, never of the French school, was an exception here.

from simple. Thus, Weber had to suppose his opposite streams connected by forces dependent not only on distance but also on relative velocity.

Of more practical value, though of less scope, were the theories of Helmholtz and of W. Thomson (1847 and 1851). These men used their new energy doctrines, of which Lenz's law was seen to be only a particular case.

Neumann had used Ohm's concept of potential. It was not until 1849 that Kirchhoff identified this with the potential of electrostatics, expressed in different units. When he gave the conversion factor, he turned continental minds in a direction to which those of England had already turned. For the factor had the dimensions of velocity, and the magnitude (3×10^{10} cm./sec.) of the velocity of light.

Herschel had suggested to Faraday that magnetism and electricity might, as disturbances in the same ether, have their effect on light. In 1845 Faraday, acting on this, discovered the rotation of the plane of polarisation in a magnetic field. The thread of mechanical ether models began to be taken up again. W. Thomson compared electrostatic fields first with thermal (1842) * then with elastic (1846) " fields," which, like light, have the significant feature of finite velocities of propagation. Maxwell (1855), a new arrival on the scene, gave a mathematical comparison of Faraday's tubes of forces with tubes of flow in a liquid.

Now in 1856 Thomson had formed the impression that magnetism has a " rotary character " and in 1858, Helmholtz, in hydrodynamic researches on vortices (see Chap. XXIV), had compared magnetic field with fluid velocity, and the accompanying electric current with vortex filaments. In 1861–2, Maxwell, on this impulse, supposed magnetic lines of force to be the (closed) axes of vortex rings.

If space is to be filled by such rings, contiguous ones will have opposite velocities where they touch ; and it was in getting over this difficulty that Maxwell exhibited a resolution lacking in his predecessors. Between adjacent rings he supposed particles of electricity to lie, which by rotating in the sense opposite to the rings, preserved continuity of motion.† Such electric particles, then, must be available throughout the field.

To understand how well this worked in, we must recur to Faraday's view of dielectrics. The crowding of electrostatic lines in these had been linked by him with the supposition that in dielectrics, unlike conductors, the intermingled opposite charges are bound together ; so that an applied field experiences an elastic resistance when trying to separate or orient them. In fact, it separates them only a small distance, the resulting condition being one of " polarisation." In reaching it there is of course a transient " current."

Maxwell now extended this theory to suppose that *everywhere*, even in the empty ether, an electrostatic field causes a similar " displacement current " among the electric particles already described. These, in fact, constituted a new kind of electricity and a new kind of current.

* Like Chasles, 1837.

† A crux also in hydrodynamical vortices (*loc. cit.*).

The sum of the two kinds of electricity was constant. When both kinds were taken into account, every current of the ordinary sort—every change of displacement—became a closed one.

It might have been thought that, with so particulate a view of ether and charge, Maxwell would have gone straight on to modern concepts of electrons and even of photons. What happened was different. Having found that his concepts led him to new equations of vastly greater scope than any reached before, he let the scaffolding of models drop into the background.

The scope of the new equations was indeed sensational. They gave, as the models enable us to anticipate, a finite velocity for the propagation of the linked electric and magnetic disturbances. And this velocity turns out to be that of light. In fact, these disturbances *were* light; and their equations at once gave many of the familiar properties. They got rid of the old questions of the longitudinal wave and of the plane of vibration in polarised light. Maxwell, moreover, was able to give (1864) successful theories of the action of metals and of crystals on light and (1869) a better dispersion formula than that of Cauchy.

But his work was not at first well received.* The idea of a second kind of current, making all currents (even, for instance, in the charging of a condenser) closed ones, was repugnant. Also, many of the theory's best predictions were only verified much later. Hertz found his electromagnetic waves only in 1887. It was 1899 before Lebedew decisively verified the predicted value of that pressure of light on matter which the theory required. Moreover, Maxwell himself tended in practice to revert to simplified forms of his theory. He was apt to view matter as only ether with a different μ and k. This barred him from attacking many urgent problems.

When acceptance came, it was acceptance of the equations with over-simplification of the model. The simple model: static strain in the ether as static field, changing, rotating strain as magnetic field, was mathematically very useful, but in the end it obscured fundamental difficulties. Some of these were being faced by the mechanical ether-models the imagining of which went busily on, with men like Kelvin, for the rest of the century.

They were being faced, but they were not being solved. The very kind of solution that could be hoped for had to be examined before that was possible.

* A different, and less suggestive, theory was given L. Lorenz (1829–91) of Copenhagen in 1867.

THE NINETEENTH CENTURY

A man born about 1800, wanting a serious view of the whole of science, could no longer be a dilettante. The activities of science had become multifarious and specialised, its literature voluminous. That literature, too, was more difficult as well as more copious. In particular, much of it demanded a deep grasp of mathematics.

Nor would the Grand Tour of Europe long have sufficed to keep him informed. Before 1850 such a vital medical advance as anaesthesia had been made in America. Before 1900, Japan had made contributions. White men, with a new consciousness of power and of a mission, were spreading all over the world, recovering the Sanskrit classics, interpreting (new sphere for " cryptography " !) the Egyptian hieroglyphics, recovering ancient units of measurement, acquiring a new sense of the breadth and length and living continuity of human history, preparing for the idea of the continuity of the whole of life.

Our observer, any time up to 1850, would probably have made his headquarters in Paris. The French revolution had opened the floodgates of French genius. From the Government schools—the Polytechnique, the Ponts et Chaussées, the Génie—rushed brilliant spirits like Cauchy, Fresnel, and Carnot. It was a time of men, such as Galois of France, Abel of Norway, and the English poets Keats and Shelley, who were snuffed out incredibly young in a very passion of creation. This was one side of the Romantic Movement, in which Germany, especially, was engulfed, in her case to the detriment of science. One German study, indeed, that of history, was becoming a scientific industry, smoothing out conventional periods, destroying partisan blacks and whites. But German science as a whole was long in throwing off the sweeping theories of Hegel, Goethe,* and Naturphilosophie (see next chapter). When it did so, the reaction was so violent as to delay acceptance of true scientific generalisations of comparable scope, such as evolution and the conservation of energy.

Before considering England, we recall that in the 19th century, the agricultural and industrial " revolutions " were spreading in every direction, carrying with them England's ways of thinking. Industrial invention and scientific theory grew much closer than in the 18th century. The science of electricity brought forth the corresponding industry ; the industry of steam brought forth the science of thermodynamics. If the words have definite meanings at all, both these types of case have constantly occurred. But the position is not simple. For example, many " pure " researchers have almost a mechanic's view of their apparatus.

* Goethe's point of view had the same attractiveness, and ultimately the same root, as that which, in early sections of this book, we have called " philosophy." Science, he felt, should serve culture, not culture, science. Science should explain what we notice (such as coloured shadows with white light, and the other points in his attack on Newton) not notice only what it can explain (which is often quite uninteresting from every other point of view). Curiously enough, this is also the " practical man's " objection to science. Unfortunately, no man progresses from the harder to the easier ; and science, if it is to advance, is obliged to insist on its own criterion of what is interesting, and of what is an explanation.

But then a mechanic's view is not entirely a mechanical one ; his way with a machine may resemble that of a groom with an animal : instinct, not reason, may rule.

It was by its insistence on mechanical models in science and life that England sustained its part in the 19th century. No country was so mechanised, no country was so statistically-minded, nor had such mechanical, utilitarian, ideas of goodness. We shall see detailed applications of these ideas later, in scattered places. But it is convenient to treat a few general points here.

Statistics and economics were typical of the time. Taking up hints made in the 17th century, many human studies, archaeology, history of religion, history generally, sociology, came to be called " sciences," and to boast that they would show Man that he, no less than matter, was bound by inexorable laws. Many of those who made these boasts would have been profoundly disappointed with the century's harvest : results have been plentiful, but they have borne little resemblance to the laws of physics.

There were, of course, undoubted scientific applications of statistics and probability. The science of " vital statistics " was founded by the Belgian Quetelet (1796–1874) who (1835 onwards) showed that the normal error law applies not only to physical quantities such as heights, but also to some mental qualities. From 1840 onwards, W. Farr applied these methods to the rise and fall of epidemics.

Probability may be viewed as a means of retaining continuity in discontinuous situations, and this was a view of it which appealed to the Victorians. For while discontinuity, creationism, dominated popular religion, continuity dominated intellectual life through much of the century, continuous field theory in physics, continuity in mathematics, evolution in biology.

The " classical " political economists, like Adam Smith, grew up in a world more interested in the profits of the unrestricted struggle for existence than in the fate of its victims. One Englishman, indeed, the clergyman Malthus, brought the latter into prominence by his calculation (1798) that human beings would increase geometrically until there was only just food for all, and that any additional children would then die. But it was a cold-blooded calculation,* and when the theory gave Darwin the idea of natural selection, it was not so much the cruelty of this process as its gradualness which aroused popular clamour. Creationism felt itself assailed.

As we shall see in later chapters, both biology and physics swung half a century later over into discontinuity, and religion the opposite way. With the religious struggle we are not concerned, but it influenced the biologists themselves, who, as in former controversies, made contributions in detail which seem strangely opposed to, or rather independent of, their views on life as a whole. Both sides took their doctrines in a narrower sense than would now be possible ; and it was this which made their argument so fierce.

Another controversy typical of the Victorian age had this strange

* As well as erroneous in several ways which cannot be considered here.

peculiarity that its relevance to the detail of biology is illusory, that both sides seem concerned to show that their doctrine could explain any possible facts, and so predict none. This was the old controversy of mechanism, vitalism, and animism. This is often associated with the evolutionary question, but it was different not only in its chief technical points d'appui (e.g. metabolism, experimental embryology) but also in tendency. For evolution was not anti-vital : a machine does not struggle for existence. It is the very negativism of Darwin which shows this, the fact that he did *not* try to explain how variations arise, only how they are cut off. Life itself, burgeoning into every possible corner, was taken for granted.

On the other hand, life is a sterile category in regions of experiment such as physiology ; and the instinct of experimentalists has always been against it. Mechanism, here, has much more vitality than vitalism—suggests more experiments—always provided that the word is taken in the broadest sense reached by the physics of the time. The Victorian objectors to it did not see all its philosophy. Vitalism makes life one force among many in the arena. Mechanism, even when it has no such intention, has the effect of removing life to a position where it is beyond competition, to the position of the experimenter himself. The true criticism on it is possibly that at present it makes the experimenter so very abstract a creature. But such a criticism was beyond the mental horizon of the Victorians.

Our onlooker, starting from Paris, would certainly have left it for England when he heard of Darwin and Maxwell. But it is probable that he would have ended his life in Germany.

CHAPTER XI

EVOLUTION AND THE MICROSCOPE

EVEN apart from physiology in the narrower sense, already sketched up to about 1850, Darwinism was by no means the only great biological movement of the 19th century. In Chap. VII we roughly divided biology into two streams, the synthetic, understanding things by viewing them as parts of wider and wider wholes, and the analytic, understanding them by cutting them into smaller and smaller parts. Darwin represents the former, but he must not blind us to the latter, which may be illustrated by new work with the microscope, leading up to the cell theory and to Pasteur's micro-organisms. Each of course reinforced the other, bacteria lending a new piquancy to Darwin's concept of the struggle and interrelation of life.

In this chapter we pass chronologically from cell theory, embryology, and morphology to evolution and the distribution and classification of living forms. We end with Pasteur and Lister.

CELL THEORY

It is in plants that cells are easiest to see, in plants that they were seen in the 17th century,* by botanists that, throughout its formative period, the cell theory was developed. But *as* a theory, a theory of the cell as a universal " brick " of which all tissues were built, it needed a stream of influence from a study developed primarily in connection with animals, and indeed in connection with medicine, that of tissue structure. Stahl had seen the importance of this latter (" histology "), and the universal brick idea had had an early attraction. Haller, observing how widespread are fibres in the structure of living things, went so far as to say that the fibre is for the physiologist what the line is for the geometer. In 1781 Fontana spoke similarly of " tortuous cylinders." But the idea was long in germinating.

Bichat (1771–1802) took a great step. He analysed the body into twenty-one different types of elementary tissue, each with its own type of sensibility or irritability. These tissues had for him a partially separate life, and the organism was the result of their conflict as well as of their unity.

But these men had two weaknesses. In the first place, even by Bichat's day there was none of that sense of things as *growths* without which analysis of so complex a matter as tissue structure cannot succeed. In the second place, the old non-achromatic microscopes had shot their bolt, and these men were no microscopists.

Then about 1812, Amici in Italy began making advances in the

* Especially Grew and Malpighi, both 1671.

microscope, and in 1830 Lister's father in England achieved the crucial advance of constructing an achromatic microscope objective.* Results soon followed. We shall see later on to what use Amici put his new instrument. But he not only used it, he still further improved it. In fact, about 1840, he made the first (water) immersion lens. For a long time, however, few of these latter were in existence, and most of the advances of this chapter were made without them.

Leeuwenhoek had seen in the cell, itself small, a thing smaller still; and about 1831 R. Brown (1773–1858), a botanist of whom we shall hear again, realised that these "nuclei" are nearly always present. From 1830–9, Valentin, a pupil of Purkinje's, did much work on them and on the other cell-inclusions destined to cause so much perplexity. The German botanist Schleiden (1804–81) also sensed the importance of the nucleus. Schleiden took up cell theory in reaction from the arid, systematic atmosphere of his time. A new feeling was abroad.

Various writers like Sprengel had by then brought the *name* in; while Dutrochet spoke (1824) of both animals and plants as built up of masses of globular cells. Dutrochet, like Brown, was much concerned (e.g. 1826) with what we should now call osmotic inflow and outflow through membranes, as bearing on the cell and on vital activity in general. This was a great advance,† but the cell movement was started rather by Schleiden's work of 1838, applying it throughout the plant kingdom.

Purkinje was a great improver of microscopic technique, and in 1835 his microscope revealed the cellular structure of the skins of animals. But in general Purkinje rejected cellular in favour of "granular" structure. In 1839 Schwann (1810–82), a pupil of Müller's, announced the general theory of the cell as the basis of all life, by the various combinations of which all plant and animal tissues were built up. Like Bichat's units, Schwann's new ones each had their own partly separate life, and were constantly building up and breaking down (metabolism) and exchanging substances with the non-cellular fluids of the body. It was a bold theory, but almost everything remained to be done.

Schleiden and Schwann and their school foreshadowed two particular applications of the cell-idea, later of capital importance. The first was that of microscopic creatures ‡ as responsible for putrefaction, fermentation, and (Henle, 1840) disease. The second was that of cell-division as, sometimes at least, responsible for both growth and reproduction. Schwann was the first to view eggs as essentially cells. In 1837 Sars and Siebold showed cell division in invertebrates.

Schleiden had thought that new cells arise by budding from the nuclei of old ones, and in 1840, Von Mohl, studying cell-division, did not exclude the possibility that new cells might arise within old ones. Remak in 1841 denied this, as did Nägeli in work 1842–6.

The subject, it began to appear, was far from simple. Von Mohl

* Chevalier (1820) had also worked on compound objectives.

† In 1833 another writer, Raspail, spoke of the cell as the laboratory of life.

‡ The idea of unicellular organisms (Siebold, 1845, "protozoa") as against multicellular ones began to have importance in systematics.

in 1846 distinguished the "protoplasm" from the sap of vegetable cells; and Nägeli showed the nitrogenous nature of this substance as compared with the constituents of the cell wall. This latter was an early stumbling block, its importance being overstressed. By 1850–53 it began to be seen that protoplasm is very similar in plant and animal cells, but it was 1861 before Schültze, and also Brücke, gave the cell its classic definition as a lump of nucleated protoplasm, with the wall as secondary or even absent.

Meanwhile a new and most significant province had been conquered for the cell theory. Virchow (1821–95) had made a detailed examination of the histology of *diseased* tissues (1858). But Schültze was the man who, gathering together all these threads, gave cell doctrine and protoplasm (1863) their full dominion over histology, embryology, and protozoology. He did so in a different sense from the original workers. The early tendency had been, as we have seen, to elevate the cell into the position of a universal unit of thought, like Carnot's cycle in contemporary thermodynamics, or interchangeable parts in the new mechanised industry. Such a tendency had been exhibited when Schwann had reduced Bichat's twenty-one kinds of tissues to five kinds of cell. But by Schültze's time, the fifties and sixties, this hope had had to be abandoned: the microscope had revealed too much mutability and variety. The cell was indeed omnipresent, but it was not self-explanatory. It was a beginning, not an end. Only in the nucleus was it still hoped that a universal unit might be discovered. We follow the fortunes of this hope in Chap. XVI, where it will be seen probed by new advances in microtechnique.

Morphology and Embryology

Merely to envisage, cinematographically, the fluent lines of the growth of a creature from its tiny egg, is to see why morphology and embryology are associated and why their early stages were a happy hunting ground for literary scientists like Goethe with a fine sweeping eye for analogies. There grew up in the last quarter of the 18th century a distinctive line of thought, still detectable in modern morphological reasoning, and reaching, very early, a tempting degree of success. With Goethe, it brought together plant and animal biology, long divorced to their disadvantage. It viewed all parts of a plant (except the stem) as modified leaves, all parts of the skull (Goethe) or even all bones whatever (Oken) as modified vertebrae, the jaws of insects (Savigny) as modified limbs.

There is obviously some truth in such views, which were, like cell-theory, a very natural reaction from the 18th-century atmosphere. Müller and von Baer, whose work we soon approach, were deeply influenced by them.

Unfortunately, apart from wildness in detail, these views went beyond mere analogy to something which was only the old assertion of "philosophy," as opposed to science, in a new disguise. All living creatures were built on one plan (Bonnet); each great group had its own "type" (Goethe), of which individual species were but imperfect "realisations." Such views might have been harmless, might even

PLATE XIV

LINKS OF SCIENCE AND ART III. GOETHE. Literature and biology. *From a print in the British Museum.*

PLATE XV

Von Baer. Biologist.

have grown into evolution, had the types been viewed as relative to the environment, or to other factors. But this was just what they were not. They were eternal, laid up in heaven, secure alike from refutation and from the usefulness of provoking inquiry. That essential relativity which marks science, no less in biology than in physics, was absent from this " Naturphilosophie " ; and the first third of the 19th century saw discredit gradually overtake it. The more embryological study suffered, no less than in former centuries, from an equally " philosophical " atmosphere. The old controversy of preformationism and epigenesis still went on. It was indeed against the former that Wolff had shown how, in the growth of the plant, petals and leaves are initially indistinguishable, a view with obvious relations to one of those mentioned above.

Coming now to the definitive line of attack, we find it in Wolff's germ-layers, already noted, in the embryonic chick. They were rescued from obscurity by another Halle man, Meckel, in 1812. In 1797 Cruikshank made the discovery of the ovum in mammals, but the announcement which moved the scientific world was that of von Baer (1792–1876) in 1827. From ova in the Fallopian tube, von Baer was led back to the unfertilised ova in the much larger follicle of De Graaf, formerly regarded as itself the ovum. He was also led forward to the germ-layers which Wolff had seen.

Introducing now a comparative element, he showed that, in a variety of species, these germ-layers led to the same organs ; and yet they are transient and never seen in the adult. He distinguished four of them. Remak in 1845 reduced these four to three. From the outermost of these grew the skin and nerves, from the middle the muscles and skeleton, while from the middle also grew (in vertebrate embryos) the " notochord " which was one of von Baer's great discoveries. It was another of the transient structures which embryology began to demonstrate, a rod of tissue never seen in adults except in some fishes. This was valuable for systematics because invertebrates do not possess it, and valuable also, when evolution came on the scene, as evidence of a common ancestry of all vertebrates.

Of similar bearing was Rathke's extension (1829) of an old observation of Malpighi's (1673). Rathke described the vessels which, in the chick, and also in mammalian embryos, arise from the aorta and encircle the gullet. They disappear in later life, but Rathke's point was that these show a fish-like stage in birds and mammals ; for in fish they persist through life, and are connected with the gills.

J. Hunter (1790), Meckel (1811), and Serres had suggested that the embryos of higher animals resemble adult lower animals. Von Baer improved this by saying rather that the further we go back towards the embryo, the more the species resemble one another. He also said (as did Aristotle) that characters distinguishing two species are later in growth than the common ones.

After evolution came, this " biogenetic law " was unconsciously transformed, especially by Haeckel, into the doctrine that the individual recapitulates the history of his race.* Von Baer never held it in this

* For the later history, see Chap. XVII.

sense, for, true to his "naturphilosophisch" background, he connected the common characters of species with the Eternal Types at the back of them; and the coming of Darwinism left him anti-evolutionist. In fact, while embryology contributed its quota to evolution, the actual impulse to Darwin came from a different source.

The next great advances in embryology sprang from its union with the cell theory (1861) at the hands of Kölliker's school, and these we defer to Chap. XVII.

EVOLUTION

Buffon (1707–88), reacting against the classifiers, had moved in the direction of the mutability of species.* It must be remembered that some of the theories which the human race has always held link us by implication with the animals. The Aristotelian orthodoxy had always stressed an animal, and even a vegetable, side to man. There was another old idea which could be twisted to a new meaning, the teleological idea that every part of an organism must be perfect for its purpose. Buffon argued that since some organs seem no longer to have any use, *times must have changed*. This was a vitally important notion. One side of it was the idea of degeneracy.

Meanwhile, over on this side, there was beginning that extraordinary exhibition of hereditary aptitude and concentration which gave us the theory of evolution. Erasmus Darwin (1731–1802), the grandfather of Charles Darwin, was among the first to recognise the scientific interest of the immemorial farming practice of artificial selection and crossing. An allied recognition was noticed in Chap. VII. Such practices were immemorial, but they were especially in the air in the England of the agricultural revolution. With Erasmus Darwin the idea of the genetic continuity of species enters, but does not achieve general currency.

Like nearly all early workers, he assumed that environmental influences, pre-natal, climatic, nutritive, etc., were transmitted to the offspring.† The credit or otherwise for this assertion usually goes to the Frenchman Lamarck (1744–1829) whose systematic work we have already mentioned. Unfortunately this systematic work was too speculative to give him authority, and Lamarck, with Geoffroy de St. Hilaire, came to be associated in the minds of biologists with the discreditable school of Naturphilosophie. This was unfortunate, for, though Lamarck still held his "ladder of nature" in too rigid, simple, and geometric a sense, yet he was an evolutionist so far as concerned the assertion of a series of forms *growing out of one another in time under environmental influence*. But the 18th century had not yet lost its hold, and this idea of continuity was still eccentric and stiff in the joints. It lacked the *historical*, relative, tone.

But with palaeontology (coming especially from France), embry-

* In general atmosphere, the late 18th century was much less favourable to the mutability of species than the late 17th, as the reader may have gathered from Chapter VII.

† This was not always held to be incompatible with Special Creation. See *Isis*, Vol. 25, 1936, pp. 286–7.

ology and the biogenetic law (coming especially from Germany), and the struggle for existence (arising especially in England) the stage was set. It is not fanciful to say that insularity did the rest. It is not only that no one was as likely as the insular aristocratic Englishman to achieve that detachment from each of these viewpoints necessary in order to fuse them into one. It is not only that an island's maritime tradition prompted those explorations of which Charles Darwin's in the "Beagle" (1831–5) was but one. It was that the island floras and faunas examined on these expeditions were what gave Darwin his clue.

Darwin (1809–82) had been advised at Cambridge to correlate botany with Lyell's then new geology; and he noticed that the more remote in geological time was the junction of island and mainland, the more different were their forms of life. The facts of difference were too striking to be ignored, but Darwin by no means rushed incautiously into print about them. He was content to accumulate evidence and to gain authority as a systematist.

In 1852 H. Spencer began to put forward his philosophy of evolution as a process of progressive differentiation from simple to complex: and at last evolution began to be in the air in scientific circles. Russell Wallace (1823–1913) had had the idea of natural selection almost as soon as Darwin, and the two men were coupled in the first publication of the idea (1858). In 1859 came the "Origin of Species."

Its essential argument looked simple: no two specimens of a species are alike; no two will survive their environment equally well; hence (assuming that variations are inherited *) cumulative changes in the race will always be going on. It looked simple but it was not; and the book turned out to be a watershed from which views flowed in all directions. Not only was there the well-known lay sensation; there was technical division also.

In Germany, Haeckel (1834–1919) became rashly enthusiastic. F. Müller was in favour. Von Baer was in opposition. In English-speaking countries, Hooker, Huxley, Lyell supported Darwin. Owen, Agassiz opposed him. But even granted acceptance, which soon became general among scientists, there was the question of meaning. On this the possible attitudes were rapidly defined and taken up by individuals, all except those vital ones of Mendel which, later, were destined to give natural selection a new lease of life.

Thus, Nägeli, the one link with Mendel, favoured one possible view, "orthogenesis"—that, whatever the environment, an internal momentum or directing force inherent in a creature's constitution ordains the lines along which it can evolve. In France, once an initial disfavour was overcome, the movement took a Lamarckian turn. In Germany, Kölliker pointed out how few are even reasonably continuous series of fossils. Like Huxley and Nägeli, he concluded, in fact, that variation is not always gradual, but occurs by jumps. Upon these various points, Darwin did not always feel called on to decide.†

* An assumption not made explicit by Darwin (see later).

† He was copious in detailed suggestions. Among particular forms of natural selection, he stressed "sexual selection" of the fittest individuals as mates.

Darwin did realise that the tendency of sexual reproduction would be to cross variations back to the mean, and so to swamp them. For he believed in tiny, gradual variations. This was the point at which Mendel could have helped him, over, for instance, the awkward point that *incipient* organs would generally be a liability, not an asset. We refer again to the views on heredity then current.

Moreover, Darwin's work made unescapable a number of exceedingly important technical distinctions, of which the preceding doctrines, such as Naturphilosophie, had been innocent, or which they had slurred. Such was the distinction between analogy and homology: between resemblance or correspondence due to similarity of function, and that due to descent, even when function has grown dissimilar. The recognition of this distinction by E. Geoffroy St.-Hilaire and R. Owen was a vast gain in morphology, giving the subject *two* fundamental categories instead of one. We shall see in Chap. XVI a further extension brought into the idea of homology by recent work in genetics.

Nowadays, we feel that Darwin's doctrine should have been split into two, evolution, which, in some shape or other, is not doubtful, and natural selection, which merely shifts back one place the real question—the cause of the variations. Certainly Darwin did not so split it. His was the preliminary task of insisting on variation as a fact; and, after all, it is the very essence of science that its theories never do more than shift questions one place back, than open them and keep them open. Darwin opened a whole new world. The stimulus was terrific.

It was terrific, but it was not in the direction imagined at the time, Men's minds were full of the great new laws of physics; but natural selection turned out to be not a law but a litany—sung over the graves of those who were not fit. Neither Darwin nor anyone else at the time saw fully that "fitness," taken in a broad enough sense to cover all the cases, merely *means* survival. Probably these men's ideas of fitness came from the human case. It was the rough fitness of the doctor, who contrasts well and ill and does not need to consider that what is fitness in the parasite would be unfitness in its host. This perception transforms the subject into ecology and physiology—into the classification of limiting factors and the enumeration of the interrelations of life. Darwin was responsible for countless ecological perceptions, from pollination to parasitism; but ecology as a separate discipline was still in the future, and one of the evil effects of Darwinism was to divert attention from physiology. The immediate stimulus was along two or three lines—man and heredity, and the geography, geology, and classification of plants and animals. We will begin with the first of these.

From the beginning the evolutionists had sought to show man himself as only one unit in their kingdom, subject to the same laws as the "lower animals." We must eschew the history of anthropology, but from the thirties there had emerged evidence of man, and of tools made by him, at unexpectedly early geological epochs. In 1856 the Neanderthal skull had been discovered, appreciably nearer to the ape than anything found up to then.

Huxley and Haeckel and Galton (1822–1911) pushed the matter on. Galton was a cousin of Darwin's, and heir to several strains of extreme ability. In him aristocracy's pride of blood yields up its unexpected contribution to science. From about 1865 he took up Quetelet's statistical methods and applied them to human and other heredity, especially intellectual. We cannot consider the eugenic advocacy to which this led him (*c.* 1883 onwards), but it reversed one strong Victorian trend on the immemorial question of heredity versus environment, scornfully urging the predominance of the former. The resulting analysis of genetical ideas joined hands with several other current tendencies. One of these, the crucial one from cell theory and from Mendel, we treat in Chap. XVI.

DISTRIBUTION

Amid all this turmoil the bulkiest work of biology, collection and classification, went on, often in a spirit very little altered by the new atmosphere. But even here the volume increased, aided, of course, by Europe's new economic imperialism. From an early date men fired with enthusiasm went forth to fill in in detail, by new work in the field, the great and enthralling scheme of the descent of one species from another, of the succession of flora and fauna down geological time. The rocks were searched with a new zest; the minerals themselves were almost felt to evolve. The oceans were explored on a far costlier scale than before.

As some picture of the oceans is indeed necessary to complete the new picture of life, we must pause a moment upon the new science of oceanography. We have seen Darwin's association with a voyage of discovery. From the forties onwards, countries the world over began to send out vessels. J. Müller, in Germany, the Scandinavian brothers Sars, and others, began to realise fully, for the first time, the teeming variety of microscopic and other life in the ocean. Maury, a U.S. naval officer (1806–73), concentrated on the physical side. His work on winds and currents made possible such a shortening of ocean routes that the finance of oceanography was henceforth assured. Dohrn founded his marine zoology station at Naples in 1872.* The British "Challenger" expedition (1872–6) dealt with the Atlantic, the U.S. "Tuscarora" expedition, with the Pacific. The poles were attacked, the depths were plumbed.

In each case, the work done was only a beginning; but the harvest was enough to send men to almost mystical lengths, as when the organic remains in the deep oozes were supposed ("Bathybius," 1868) to be the earliest form of life, as yet undifferentiated into individuals. Less mystical and no less suggestive was the idea (Quinton, 1897) that the salt solution of the "internal environment" has the composition of the primeval ocean where life must have taken its rise. This idea has lately been re-examined by Macallum (1926) and Pantin (1931).

Henson (1835–1924) gave the name "plankton" to the floating

* Where F. M. Balfour worked (see Chap. XVII).

life of the ocean, and stressed the ecological interest of this great swarm of migrant and circulating life which drifts with the ocean currents. German and American observers (1853) described the calcium (" globigerina ") ooze of protozoan skeletons on the ocean floor. The " Challenger " expedition discovered the red clay of the Pacific deeps and the ooze of the radiolarian skeletons.

As a result of all this, some seas, and especially the North Sea, are now fairly well known. The migrations and ecology of fish will be referred to again in Chap. XIX. Sea creatures, living in relatively simple and unchanging environment, have proved to be the most convenient for many physiological and embryological investigations, and thus the subject has a further importance.

We return to the general question of the distribution of species. As early as Buffon it had been noticed that surprisingly sharp distinctions of flora and fauna may occur at seemingly slight barriers. This is sometimes clearly due to past conditions. Indeed, since the present distribution is not a static or finished thing, the subject is only one aspect of general geology and palaeontology ; but it began separately. Its first literary monument concerned the distribution of plants, Humboldt being the author (1845–7). His work, however, was ignored until, in 1872, Grisebach drew attention to it. By then Sclater (1858) and Wallace (e.g. 1876) had long been at work, and had shown that the regions are simpler and more definite for animals than for plants. Each of them reckoned six main faunal regions, Palaearctic, Nearctic (now often combined as Holarctic) Ethiopian, Oriental, Australian, Neotropical. When they tried to go into greater detail, they met great complication. For instance, very old groups, like invertebrates and reptiles, gave different regions from later ones.

In serious plant geography early dates were 1872 and 1878. A later classification was that of Thiselton-Dyer in 1911, into North-Temperate, Tropical, and South-Temperate. We refer again to plant distribution in Chap. XIX.

Turning now to past geological ages, early palaeontological successes had been most sensational among *animal* remains. Palaeobotany had its pioneers (for instance, R. Brown, 1851) ; but its definitive workers were Williamson (1858 onwards, appreciated only in the eighties) and especially Scott. We note only one result of their studies, the demonstration that at several distinct periods sharp and extremely profound changes have taken place in the whole complexion of life on earth. The need was felt of correspondingly profound physical changes to account for them. Such changes were found in the ice-ages with which we have (Chap. VII) coupled the name of Agassiz. He himself had been converted by the suggestions (1834–7) of Venetz and Charpentier ; while it was his work (1840) with Buckland on evidences of glaciation in England which put the matter in its most interesting light. Since his day three ice ages in all, each followed by lesser oscillations, have been demonstrated.* Since the last one is (relatively) recent, it linked up with evidence of climatic changes in

* Several others are discussed.

historic times and of their effects on human history. This point cannot be treated here, nor can the geophysical events behind climatic changes (see, however, Chap. XXIV).

The first evolutionists had rarely been wholly free from relics of the 18th-century feeling that some regular, invariable, series of phases would prove to be traceable in evolution. Their evolutionary tree was a specimen of topiary art. But the endless details of geography and palaeontology dispelled all such feelings. Complex has not always followed simple, nor have the "highest" forms of one great group given rise to the "lowest" forms in the "next higher" group. Most main divisions go back in parallel very much farther than the early evolutionists anticipated, getting closer and closer but rarely affording hope of demonstrable junctions. We signalise this more chastened mood by ending our section on synthetic biology with an account of the humbler, more technical, subject of classification.

Systematics and Related Topics

The necessity of *some* classification for field work, combined with the difficulties of erecting an evolutionary scheme,* have tended to keep working classifications somewhat backward, and this in spite of repeated international efforts to rationalise the subject. The field is too large for complete syntheses to be numerous. Economic and other considerations have concentrated attention on certain groups, such as fungi and other pests, and given the subject a scrappy appearance. As in geology, too, the "University Zone" in the North Temperate region has received undue attention. Beauty may concentrate attention on a group, as with the diatoms.† Fortunately, these happen to be among the most important constituents of the plankton; while algae in general have the importance of being the most primitive plants known.

As an instance of the economic bias we may cite the case of insects before concentrating our attention mainly on plants; though the enormous size of the insect group would in any case have drawn attention to it. Brauer in 1885 made fundamental the distinction of wingless and winged (the latter including forms which have *lost* their wings).‡ In 1899 Sharp divided wings into those developed outside the body and those which remain internal until pupation. Another important distinction is that of complete and incomplete metamorphosis. The former, the most highly evolved, is seen, in evolutionary light, as an extraordinary tour de force of variation, whereby the insect is enabled to gain the advantage of both unspecialised and highly specialised structure. One of the few fairly safe ecological generalisations is that the primitive, unspecialised creatures, though rarely dominant, keep their humble niches for very long periods.

* In any case, since parenthood is only one factor in life, there is no reason why an evolutionary classification should have been the most convenient for all purposes.

† E.g. O. F. Muller (1773); Agardh (1785–1859).

‡ Linnaeus had recognised "aptera," but with him these had included crustacea and other types not now regarded as insects at all.

Dominance is the lot of those at the other extreme, that of specialisation. Such specialisation, being irreversible, is nearly always temporary.

We have mentioned the conservatism of classifiers, who, indeed, are usually only stirred to decisive reforms by discoveries, in physiology or embryology, which make these unavoidable. We have noted (Chap. VII) that the Linnaean system itself, though mainly motived by the whole spirit of a century, was correlated with the physiological question of the sexuality of flowers, while de Candolle's system was affected by new work on the vascular bundles. Animal classification, apart from its broad outlines, is so completely formed on discoveries in other subjects that our few notes on it are given piecemeal under those subjects, such as the notochord, the gill arches, and metamorphosis in crustaceans. But one such discovery concerned plants and animals equally, especially those lower groups where de Candolle had failed. This was the alternation of generations. Chamisso introduced this phrase in 1819 to describe the complex life cycle of certain tunicate worms. In 1842 Steenstrup drew general attention to the fact that jelly fish and certain worms have offspring which are not in the least like the parent, but which, for their part, have offspring of the original type. The same thing was then discovered in plants. Certain macroscopic features of the sexual process in flowers—pollination and its accompaniments—had been worked out by Sprengel (1750–1816), following up Camerarius and Koelreuter (1761). But the corresponding processes in non-flowering plants remained a mystery. Then came the discovery among them, by the German Hofmeister, of this same alternation of generations (1851). Henceforth all these groups were seen to have a common life cycle. It was another effect of the new microscopes.

In all cases, and in the lower animals, the essential point was that one generation was sexual, producing ova and sperms which unite, the other (thus produced) asexual, producing spores. Hofmeister showed that in primitive forms the relation of the generations is apt to be indefinite, only growing definite and invariable for higher ones.* In many algae the generations are indistinguishable but for their contrasting modes of reproduction. In mosses the asexual generation is parasitic on the other ; in ferns the asexual one is independent and is the larger. Now, however, comes the point. Hofmeister was able to show that the same two generations are present in the conifers and in the flowering plants, but the contrast of sizes is enormously greater than in the ferns, the sexual generation being so completely parasitic as to be normally regarded as an " organ " of the asexual one. For the final extension—to the higher animals—we must await the interpretation in terms of cell theory in Chap. XVI.

The systematic consequence was the recognition that the old distinction of flowering and non-flowering plants was only a rough convenience. The discovery also constituted a new common element between animals and plants, especially among the lower forms. In

* A tendency of evolution.

this connection it may be noted that Haeckel (1866) set up a third group, " protista," to cover those forms where the obvious distinctions broke down. He also drew a more enduring distinction between Siebold's unicellular animals and " metazoa " or multicellular ones.*

By the sixties plant classification with an evolutionary background began in earnest. It is too large a subject for full treatment here, but it has been dominated by two rather divergent schools, those following Bentham and Hooker (1862–83) and those following Engler and Prandtl (1887–1909). On one point, indeed, the formers' views are no longer held. They placed the gymnosperms between mono- and dicotyledons. Since Eichler (1883) it has been agreed to group the two latter together as angiosperms, much later in time than the gymnosperms, and derived from them.† It is the manner of this derivation which divides the schools, Engler holding flowers to have been derived from a unisexual, the others from a bisexual, gymnosperm strobilus (flower). This results in a different sense of proportion between the chief characters used in classifying.

In 1882 Sachs took advantage of the new microscopic facilities to be noted in the next section to make a much better classification of the lower plants ; but we must leave this aside. A recent English work in the Bentham and Hooker tradition is that of Hutchinson (1926 and 1934) in which Ray's old distinction of tree-like and herbaceous habit is revived in a more scientific form.

Of great future interest is the pioneer work of Mez of Königsberg in deserting morphological criteria altogether in favour of one from chemical physiology, serum diagnosis. In this the degree of relationship of two plant species is measured by the capacity of the protein of a species to produce a precipitate in the serum of (for instance) a rabbit inoculated with the protein of another species. Fortunately, the tree of relationships thus produced corresponds fairly well with the morphological one. The word " tree " is perhaps no longer appropriate, for very recent work has thrown doubt on the implied assumption that the relationships never form a network. Plant species may be produced by hybridisation. Recent ecological and genetical work is suggesting another idea : that many species are polyploid (see Chap. XVI) forms of others, ecologically distinct. Thus at the time of writing a new era is foreshadowed in this subject.

Pasteur and Lister

C. Singer has suggested that the best Greek medical schools may have rejected the idea of infection in disease just because the " barbarians " held it.‡ At all events the " barbarian " peoples of the Bible gave the Middle Ages what the Greek doctors did not. This idea of infection seems thus to be (perhaps like that of the physical) a primary one which we did not derive from the " high " philosophy of

* A distinction with no similar significance for plants.

† Not, of course, from present ones.

‡ Primitive peoples hold very much this idea in every sphere. Even the domestic dog appears to believe that he can impart motion to his master infectively by fervent imitation of the actions of going for a run.

Greece, but from "lower" cultural stages via the religions of the Middle Ages.

We have noted that an early European assertion of atomism was Fracastoro's (1484–1553) living "atoms" of infection. Fermentation and the leavening of bread were already associated with fever in the minds of men like Boyle because of such common properties as rise in temperature; and at the same period fermentation and digestion were coupled. A further element in this complex of ideas was spontaneous generation. Fermentative processes, which had some of the characters of life yet seemed to start from nothing, began to be taken as cases in point. J. T. Needham (1713–81) thought that he had proved spontaneous generation. Spallanzani showed (1765) that these processes could be indefinitely delayed by boiling, but his proof was inconclusive. The limiting factor was the available microscopes. As early as 1687 Leeuwenhoek had seen bacteria. In 1773 some attempt was even made to classify them and the other minute creatures observable in infusions of various kinds. But the lenses of the time were not adequate.

In 1836 Schwann showed that putrefaction is due to living bodies which, like the effect itself, are destroyed by boiling. But the vitalist Liebig held, oppositely, the not unnatural view that putrefaction is a form not of life, but of death—of chemical breakdown when the vital force is withdrawn. About 1837 both Schwann and Cagniard de Latour proved that yeast is made up of minute plants: the influence of the improved microscopes was being felt. In 1836, Bassi had shown that one of the silk-worm diseases is transmitted by a minute fungus. In 1840, Henle, a pupil of Müller and teacher of Koch, set forth in detail the doctrine of infection by micro-organisms. But he could not prove it, nor use it therapeutically. It remained heretical. "Spontaneous generation" still prevailed, any organisms observed being supposed to arise out of nothing, as effects, not causes, of the changes associated with them.

The mind of Pasteur (1822–95) * seems to have followed, despite its great power, the windings of a curiously specific interest. Perhaps his father's tannery induced a fixation on the group of subjects traditionally associated, as we have seen, with fermentation. Even in his earliest stroke (see Chap. XIV) the tartrates in question had a fermentative origin: they formed the deposits in wine-vats known to the Egyptians. He went on to study yeast and to show the micro-organisms in sour milk. The subject became a storm centre. He planned decisive experiments to show that in media initially sterile of bacteria none of the controverted processes will occur.

By this time several lines of research were converging on a most important realisation. Cleanliness was becoming a note of the best medical practice; though scornful dissentients were not wanting. By 1843 Oliver Wendell Holmes in America had been strongly urging it in maternity cases. By 1846 Semmelweis in Vienna had achieved

* Perhaps like that of his contemporaries, Mendel (1822–84) and especially Bernard (1813–78), of peasant or yeoman tradition all three.

great reductions in puerperal mortality by preventing students from coming straight from the dissecting room to the maternity wards. But orthodoxy remained unconvinced.

From 1853, when the Crimean War drew painful attention to the point, Lister (1827–1912) began to consider the causes of surgical sepsis; and soon the stage was set. In 1861 Pasteur at last carried general conviction against spontaneous generation. He boiled meat broth in a flask with a very long thin neck until no bacteria were left. This was shown by the fact that he could now keep the broth in the flask for an indefinite period without change setting in, the narrow neck admitting nothing. Then he broke off the neck and in a few hours the liquid showed micro-organisms, and the meat was in full decay. That the air carried such organisms he proved by twice filtering it through sterile filters and showing that with the first filter, but not the second, he could set up putrefaction.

In 1865 Lister heard of these experiments and turned to the discovery of means of killing micro-organisms more suitable than prolonged boiling for use upon living tissues. Independently of Lemaire (1860) he hit on carbolic acid; and from his definite statement in 1867 the concept of sterilisation has spread through the medical world.* If it was in one aspect a negative one—that of merely keeping life free from enemy forms of life—it was, like the other negative conceptions already discussed, also an assertion that life itself, *vis medicatrix naturae*, is the one positive thing, and needs only to be left to itself.

The discoveries of Pasteur and his followers gave positive content to the idea of "enemy forms of life." Following his practical bent, Pasteur attacked (1862–5) the terrible disease which in a dozen years had nearly destroyed the silk-worm industry of France. About 1863–6 he went on to the equally terrible "phylloxera" of the French vineyards. Both proved very complex and difficult. His next tasks were anthrax and chicken cholera.

Bacteria now began to be regarded as a subject for the systematist. Cohn (1828–98) began work here in 1872; but the complex life-cycles of bacteria, and their apparent extremes of variability, have made their classification difficult. One primary division is that of aerobic, or oxygen breathing, and anaerobic. Bacteria are usually placed among plants, near the fungi, but they lack sexual processes. They also lack chlorophyll, but the extent to which they possess other means of synthesising their body substances from inorganic material forms, at present, one basis for their classification, as also (Lwow) for that of protozoa. Disease-causing species are the ones which, having lost this power, die unless living hosts provide nutrition. Knight has lately (1936) suggested that this classification follows the lines of evolution. But we must return to a much earlier stage and notice the great improvement in microscopic technique which made possible these advances and many others in connection with cells.

Work of this kind was for long done with a few instruments for teasing out and dissecting tissue. Valentin, with Purkinje, improved

* Asepsis has, however, replaced anti-sepsis—prevention, rather than cure, of bacterial invasion.

the razor by using two close parallel blades. About 1866, W. His made his sliding microtome, which was improved in the decade following. Automatic machines began with Threlfall's made in 1883. These demanded rigid embedding of the specimen in substances like paraffin wax. Soft tissues had for some time been hardened with alcohol or chromium salts.

Under a powerful microscope ordinary pigmentation usually fails, and as early as 1770 efforts at staining had therefore been made (Hill, wood-structure). In 1847 Gerlach began experiments on this subject, and in 1854 he made the crucial advance of showing that carmine will selectively stain certain parts of cells, the nucleus especially.* Virchow and others seized upon the discovery. Indigo, logwood, and finally (1862) the newly discovered aniline dyes, were found to give similar effects. About 1873 Golgi began to use silver salts to bring up cell-boundaries. By then a wonderful selectivity of staining had been achieved. In 1878 came the much improved Abbé microscopes, in 1886 the compensating ocular, substage illumination (1888) and the oil-immersion lens.

Koch (1843–1910), who made bacteriology a professional study, is remembered for further advances in technique. He showed how to cultivate bacteria outside the body, in a transparent medium to make microscopy and photography possible. One of the greatest difficulties of early workers was to separate the various organisms which they usually found together in nature, that is, to get pure cultures. By 1881 Koch had his method of diluting the cultures until the different organisms are separable.

In 1876, dealing with anthrax, he drew attention to the importance of the spores, so much more resistant then the bacilli themselves. In 1878 he pursued surgical sepsis into detail. In 1882 he discovered the tubercle bacillus, and eight years later prematurely announced a cure. In 1883 he discovered the organisms of cholera. But we cannot follow all the diseases which he, Pasteur, and their followers attacked. In many cases research only showed complexity beneath complexity and left solution far ahead. The same applied to diseases caused not by minute plants but by minute animals, protozoa and others.

The general question of immunology plunges us back into those 18th-century days when, as noted elsewhere, the practice of inoculation came in from the East. Long before Jenner (1749–1823), it had been an unlettered belief in his own Gloucestershire, and in parts of the Continent, that persons who had had cow-pox did not get smallpox. And as soon as Jennerism came to the front, cow-pox vaccinations were claimed from Dorset (1774–89) and from Holstein (1791).

But it was Jenner who, inspired by Hunter, first *tested* these local beliefs (about 1778 onwards). He made his first vaccination in 1796, and in 1798 published an account of twenty-three cases. The idea was quickly taken up (amid strong opposition) in England, on the Continent, and in the United States. By 1800, thousands had been vaccinated.

* Other workers share Gerlach's credit over carmine. See *Isis*, Vol. 22, 1934–5, p. 404.

Fifty years later, however, there was still no branch of science to which the then universal practice could be referred for explanation; and it was 1880 before accident presented Pasteur with a case in which lapse of time had destroyed the virulence of a cholera culture and endowed it with the power of immunising subjects against active cultures. He had been investigating vaccination and the immunity conferred in some diseases by a former attack. He now proceeded (1881) to find a vaccine against anthrax. His triumph against hydrophobia was in 1885.

Here was another field opened, another which, unfortunately, has proved of extreme complexity. In 1884 Metschnikoff discovered "phagocytes" in the body, which take up invading organisms and render them harmless. The serum of animals which have recovered from a disease was found often to contain an "anti-toxin" which confers immunity on a human being. The German Behring (1854–1917) found this to be the case (1890) for tetanus and diphtheria. In 1892 the mosaic disease of tobacco, and in 1897 foot and mouth disease of cattle, were traced to "viruses," entities apparently capable of propagation, and so "alive," yet so much smaller than bacteria as to suggest that they contained at most a few hundred protein molecules. They passed through filters, eluded ordinary microscopy, and—a great hindrance to their investigation—could not be cultivated outside the body.

Each of these new principles gave means of control over a few more diseases; but no one, nor all together, covered the whole field. Of late, the tendency has been to utilise the fact that, whether living or dead, all these factors act in and through the *chemistry* of the internal environment. The subject thus becomes one province of general biochemistry.

The broadest significance of bacteria and protozoa would be missed if disease were regarded as their only widespread effect. In Chap. XIX we sketch some of the other life-relations in which they play a vital part, especially those in the soil.

CHAPTER XII

NINETEENTH-CENTURY MATHEMATICS. I

In the 19th century mathematics grows quite beyond elementary exposition. All that is possible is an account of the century's basic ideas and of the continuity of purpose which led to them. For it is this continuity in certain highly abstract, almost philosophical, purposes which distinguishes the mathematics of the 19th century. Thus, there is the purpose of rigour, beginning with Cauchy's work in correction of the classical French school, a purpose to be compared with the new impulse to *precision* in every sphere, from the precision tools brought in by Whitworth (e.g. 1833) to the precision diagnosis brought in by Laënnec's stethoscope. A particular expression of it was the separation of analysis and geometry, too easily used since Descartes' time to bolster up each other's weaknesses : Von Staudt's geometry without algebra, Weierstrass' algebra without geometric intuition. The union of the two, however, retained its fecundity, as with Riemann and in transformation-theory.

Again, there was the purpose of generality : in Cauchy's introduction of the imaginary (of the most general type of quantity obeying certain rules), in Hamilton's extension of this (continuous quantity), and Kummer's (number-theory) to several units. This led to algebras with fundamental axioms different from those of the ordinary theory, and so to the theory of the most general types of algebra and logic. Correlative with this was the development of the most general kinds of geometry : *n*-dimensional, non-Euclidean. Another exhibition of the philosophical tendency was Kronecker's refusal, on *a priori* grounds, to leave the domain of rational numbers.

There was in this a new capacity for abstraction, a power to formulate entities, such as " groups," cutting across the subject boundaries of previous generations. Algebraic number-fields and non-intuitive spaces form other instances. The idea of a " field " may perhaps here be said to be affecting mathematics as it affected contemporary physics.

In all this, the 18th-century sense of form did not die, but was greatly enriched in some, though not all, branches of mathematics. It is perhaps legitimate to add that to a certain extent " emancipation from 18th-century formalism " was simply admission of defeat. When mathematicians cannot solve a problem—such as quintics, elliptic integrals, and many types of differential equations—they surround it with a thicket of ascertainments about it which gradually assume an independent vitality and put on airs of generality to forget their slightly shameful ancestry ! For us, however, this ancestry serves as a useful link in understanding ; and it is in the wake of these " insolubles " that we enter the 19th century. Our path lies through real and complex function-theory to algebraic numbers and so to theories actually outside the rules of arithmetic. Transformations, invariants, matrices, groups, and the like lead over into geometry. Noting the systematisation of certain stages of geometry by Klein and of analysis by Poincaré, we touch on applied mathematics and conclude with the opening of a new domain of difficulty in point-set theory.

Paris was the centre of mathematical teaching until about 1850, but the greatest mathematician of all was rather outside its influence. In the great German Gauss (1777–1855) most of the above tendencies and subjects can be found budding strongly, but his initial isolation grew with time, and the very ubiquity of his genius makes it hard to identify him with any one trend and to use his name as a keynote in the inevitable split into subjects. Impatient of formal exposition, though insistent on rigour, he often concealed not only his methods but his results; so that he was in many cases not the discoverer who occasioned the later advances in the various subjects.

It is easier to understand the development by following the central orthodox French school from the great figure of Lagrange to that of Cauchy (1789–1857). We have spoken of the 18th-century failure with quintics and elliptic integrals, of the perplexity over trigonometrical series, and of the consequent nervousness over infinity in general with which the century closed. This was well shown in Lagrange's effort (1787) to confine himself to functions expressible near x (say up to $x+h$) by a convergent Taylor's series:

$$f(x+h)=f(x)+hf'(x)+\frac{h^2}{2!}f''(x)+\ \ldots\ .$$

and to found these, at least, on a thoroughly sound basis.* But he failed: Taylor's theorem itself was too insecurely founded.

Then in 1807 Fourier came out with a better form of D. Bernoulli's assertion that any arbitrary function can be represented by infinite trigonometrical series, an assertion which, attacked by Lagrange, prompted Cauchy to the decisive advance of subjecting the whole question of the meaning of functionality to the closest analysis which it had yet had. The idea of continuity in particular was transformed. For the 18th century, with its obsession with formulae, a continuous curve was one with the same equation throughout. Cauchy, adapting a re-definition due (1787) to Arbogast, made continuity mean roughly "without break." Now this has a most important consequence in the calculus. For if, at a certain value of x, y jumps from a to b, we cannot differentiate at that point; but, provided we take proper precautions, *we can evidently still integrate* over a range containing that point. Hence this work of Fourier in 1807 prompted Cauchy, in 1814, to go back on the 18th-century habit of making the differential coefficient fundamental, and to make the definite integral basic instead. For the formal 18th century, the crux had been that (formally difficult) integration can be derived by inversion from (easy) differentiation. But from

* Landen (1758) had proposed something similar.

Cauchy onwards the crux was the older one (see Chap. VI) that a sum is simpler than a quotient.

We spoke of proper precautions. Cauchy reverted also to the careful old definition of an integral as the limit of a sum, and, here and in infinite series, founded all his reasoning on the idea of limits. In his " Cours d'Analyse " (1821) he began (like Gauss in 1812) to give precise criteria for when " the limit of the sum " of a series is finite and definite—when it converges. He dealt especially with Taylor's theorem. But he did something else, in a different but connected field, where the use of definite integrals was again crucial. To explain this we must briefly retrace our steps across the centuries.

Cardan had verified that the mysterious " imaginary " roots of equations do really satisfy them; but the serious admission of such roots must have seemed hazardous. For one thing, if we are allowed imaginary roots, can we not invent an unlimited number for any equation? Girard saw (1629) that this is not so; that a quadratic still has only two roots; and that, granted roots of the type $a+b\sqrt{-1}$, it *always* has two roots—the first hint of a capitally important fact. Wallis (1673) pointed out that the difficulty of conceiving imaginaries would equally have excluded negative numbers until the idea of directions on a line was hit on, and reached the point where it became obvious that to get a similar picture of a complex number we must depart from a single line or dimension.

The men of the 18th century realised the importance of the existence theorem that every algebraic equation has a root, whether we can find it or not; though it was not until 1801 that Gauss gave a proof of it which satisfies all but perhaps the latest critics. At all events, no one doubted the connection of the imaginary with algebraic equations (see next section). But, in spite of the formal convenience of its use, there was no such clarity about its relation to function-theory in general. For instance, Legendre (1752–1833) had been working since 1786 on the elliptical integral, but the connection of this with $\sqrt{-1}$ (henceforth " i ") was not realised. Fagnano (1682–1766) had pointed out the resemblance of the functional relation involved to that of the (periodic) circular functions, and by 1800 Gauss had realised that the inverse, the elliptic function, has actually two periods, whose importance for our purpose here was that *one or both of these periods was imaginary or complex*.

About the same time Gauss obtained that graphical representation of complex quantity on a *plane* instead of a line which was to prove so fruitful. But he did not publish his results. Nor did the similar vector representation of Wessel (1797) or the " diagram " of Argand (1806) gain, for the time, any notice. Thus, when Cauchy began examining what differentiation and integration *mean* when the variable is not necessarily real, he was without this priceless aid to imagination.

Such examination at once leads to a question of meaning. By a function of z (z not necessarily real and, say $=x+yi$) Cauchy could not merely mean any function, not necessarily real, of x and y, such as $x-yi$. The function must be a function of $x+yi$ as a whole. Even so, however, and considering only continuous cases, such a function was very different from one of a real variable in this fundamental respect: in passing from one value of z to a neighbouring one, for say integration or differentiation, he had not just one "path" at his disposal but an infinity ("the whole plane," from the graphical point of view not possessed by Cauchy). This posed a question: was the result of the operation independent of the path chosen?

Was there, for instance, a unique differential coefficient at any point? It was found that there was, so far as concerned most ordinary functions of "$x+yi$-as-a-whole": though they had points at which there was not. Cauchy concentrated attention on such "analytic" functions. They obeyed the equations first considered by D'Alembert:

$$\frac{du}{dx}=\frac{dv}{dy}, \quad \frac{du}{dy}=-\frac{dv}{dx}, \text{ (where } u+vi \text{ is the function in question).}$$

These equations were rediscovered by Cauchy.

Then in 1814 Cauchy showed that for such functions (definite) integration is also independent (as we should say) of the path chosen.* This theorem links up at once with his assignment of the fundamental position to definite integration, and it was from it that there sprang the huge structure of the theory of functions of a complex variable. We shall not give the exact conditions under which the theorem is true, but we may note that Cauchy made them needlessly narrow.

While the theory of functionality in general was thus growing, Legendre was grinding out results on elliptic integrals. But he was missing the crucial simplification. If we write one such integral in the equivalent form $\int\frac{dx}{\sqrt{(1-x^2)(1-e^2x^2)}}$, it seems (delusively!) as if it should not have needed the genius of the young Norwegian Abel (1802–29) and of the young German Jew Jacobi † (1804–51) to grasp the analogy with $\int\frac{dx}{\sqrt{1-x^2}}$, which$=$arcsin x. Now we find it easier to deal with sin x, which is one-valued, than with arcsin x, which is infinitely many-valued; so the analogy suggests the crucial step of re-defining elliptic *functions* "sn x," etc., by inverting the above relation, just as we might *define* sin x by inverting the relation arcsin $x=\int\frac{dx}{\sqrt{1-x^2}}$. The analogy works: for instance, Jacobi in 1829 gave addition-theorems for sn x, etc., similar in form to those for sin x, though more complex. These men rediscovered (about 1825) Gauss's property of double periodicity; and this suggested yet another definition of elliptic functions as simply functions with two periods (whose ratio is not purely real).

Of this, the obvious generalisation was: why only *two* periods? Abel, in fact, went on to show that n-ply periodic functions can be constructed, and that with their aid all algebraic functions can be integrated. But we must leave this subject, which became immensely technical.

* Result known to Gauss in 1811, but not published. Cauchy made the result public first, but his paper of 1814 was not itself published until 1827.

† We cannot discuss priorities between these two.

MAP IV

BIRTHPLACES OF SCIENTISTS

Please consult Explanation on p. 10 before studying the Map

The Baltic region since 1500

Number on Map	Name	Birthplace *
1	Abel	Findoe
2	Aepinus	Rostock †
3	Arrhenius	Upsala
4	Baer, von	Esthonia
5	Bartholinus	Copenhagen
6	Bergman	Upsala
7	Berzelius	Linköping
8	Bjerknes	Christiania
9	Bjerrum	Copenhagen
10	Bohr	Copenhagen
11	Brønsted	Varde
12	Buch, von	Stolpe
13	Cantor	Leningrad
14	Clausius	Köslin
15	Copernicus	Thorn (Torun)
16	Euler	Leningrad (worked; as did D. & N. Bernoulli)
17	Fahrenheit	Dantzig †
18	Gadolin	Worked in Sweden
19	Grassmann	Stettin
20	Grotthuss	Lithuania
21	Guldberg	Christiania
22	Hertzsprung	Copenhagen
23	Hilbert	Königsberg †
24	Jung	Lübeck †
25	Kirchhoff	Königsberg †
26	Klingenstierna	Upsala
27	Krafft	Leningrad
28	Kunckel	Rendsburg
29	Lenz	Dorpat
30	Linnaeus	S. Sweden
31	Lorenz	Helsingör
32	Oersted	Rudkjöbing
33	Ostwald	Riga
34	Pallas	Leningrad (worked; born Berlin)
35	Rathke	Königsberg †
36	Raunkaier	Denmark
37	Richmann	Pernau
38	Römer	Aarhus
39	Scheele	Stralsund †
40	Seebeck	Reval
41	Severinus	Ribe, Jutland
42	Sommerfeld	Königsberg †
43	Steno	Copenhagen
44	Svedberg	Upsala
45	Thomsen	Copenhagen
46	Tycho Brahe	S. Sweden
47	Virchow	Pomerania
48	Waage	Flekkefjord
49	Warming	Denmark
50	Weyl	Holstein
51	Wiechert	Tilsit
52	Wien	E. Prussia
53	Wilcke	Strengnäs
54	Wilhelmy	Stargard
55	Winslöw	Denmark

* In a few cases, work-places, or the nearest larger place, have been substituted. Numbers within a circle on the map indicate that only the region, not the town or village, is given.

† Hanse towns. Marked because this commercial eminence, past or present, was one factor in the clustering of scientists in this region.

MAP IV

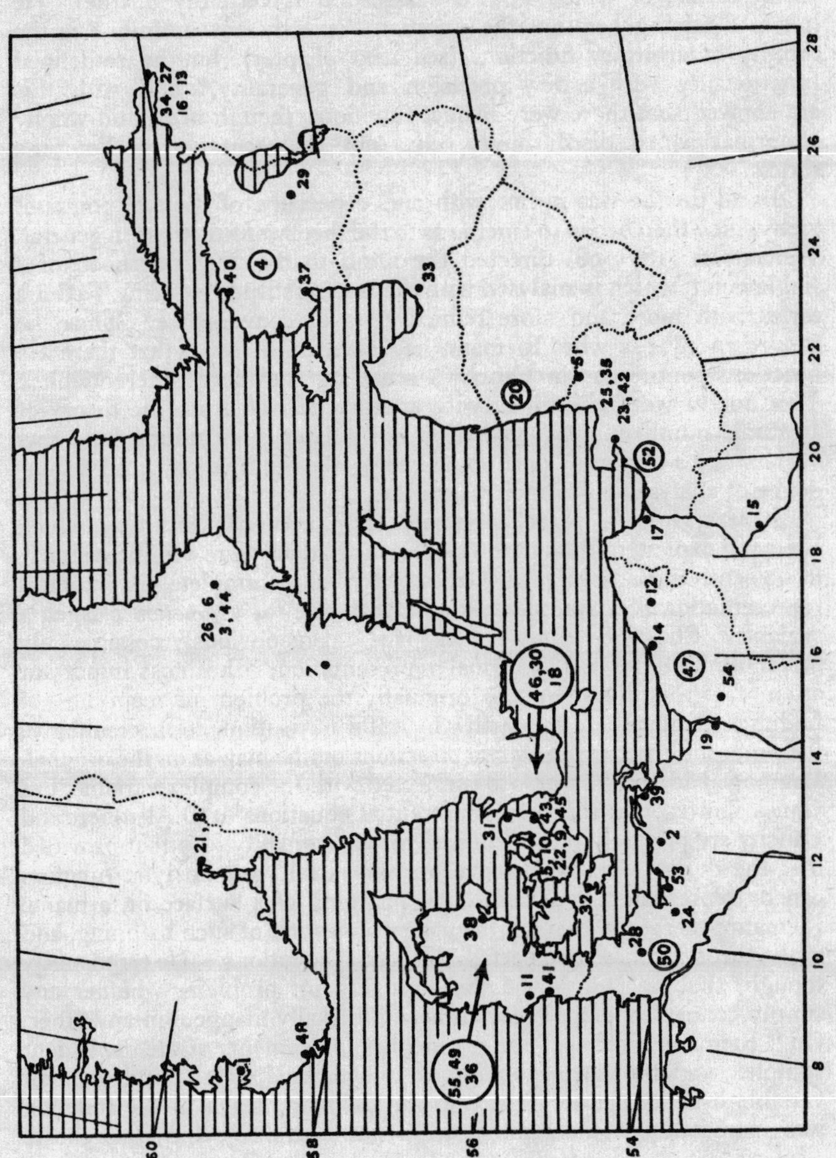

60
58
56
54
8
10
12
14
16
18
20
22
24
26
28
34, 27
16, 13
29
4
40
37
33
20
51
25, 38
23, 42
52
15
17
26, 6
3, 44
12
14
7
46, 30
18
47
54
19
39
31
43
45
2
21, 8
53
24
38
50
28
11
41
4A
55, 49
36

Meanwhile, Dirichlet had been examining in detail Fourier's assertion that a finite stretch of any arbitrary function could be represented by an infinite trigonometrical series. He pointed out (1837) the unsoundness of the original argument from the infinity of adjustable constants available, which would apply equally to a simple infinite power series, of which Fourier's assertion is certainly untrue. He left to a later generation the question " exactly how infinite " is the infinity of arbitrary functions (see next chapter), but he re-defined functionality with a new precision and generality,* and with this aid showed that there were in fact functions, though of a kind wholly abnormal as to discontinuity, etc., not representable by Fourier's series.

In all this he was in line with one, especially, of the two points of view which then began to emerge as to the theory of functions in general. Weierstrass (1815–98) directed the mind to the neighbourhood of a single point, which is analysed with the last exactitude by using Taylor's series with more and more refined tests of convergence. Thus, he proved in 1871–2 what Riemann had asserted in 1861, that there are functions continuous in Cauchy's sense but nowhere differentiable.† This line of work chimed in with another derived from the theory of algebraic numbers. We shall recur to it later, only noting here that Weierstrass's aim was " arithmetisation," the elimination of all dependence of analysis on geometric intuition.

It was, in one sense, precisely this geometric intuition which Riemann exploited, though the two lines of attack were complementary, in no sense rivals. In 1824, Gauss had tried to complete his graphical representation of a complex variable with one for a *function* of such a variable. Short of 4-dimensional space, this is obviously possible only under limitations. " Conformal representation," the most important attempt in this direction, was originally the problem of mapping—of finding a plane map of the earth which (for navigation, etc.), should give the same angle between any two directions on the map as on the original. Lagrange had seen that this is connected with the complex variable, but Gauss showed that the above differential equations of D'Alembert and Cauchy are precisely those required for conformality ; so that provided that the derivatives given above do not vanish, any analytic function can be represented by a conformal mapping of a surface on a plane. Riemann, therefore, took up the general question of such mapping, and of the theory of these two partial differential equations. He erroneously thought that he had solved the very difficult problem, whether any simply-connected plane area can be conformally mapped on any other. On " Riemann's surface " he mapped the movements of the dependent complex variable corresponding to movements of the independent variable over its plane. This surface had many leaves if the function was many-valued (the number need not be finite), and, though we

* His definition, that to each value of x there corresponds a definite one of y, *whether we know how to pass from x to y or not*, had been verbally anticipated by Lacroix. It consummated the movement away from the 18th-century view of a function as an expression in x and y.

† His proof lay in constructing such a function.

cannot pursue the subject, we note that the tracing of the intricate paths of his variables over these leaves led Riemann to the earliest serious interest in a kind of geometry then germinating (" Topology." See later).

By this time it had, of course, been realised that imaginaries are no more imaginary than any other mathematical quantities, and the procedure of regarding the complex variable as a whole was familiar. Hamilton (1805–65), of Dublin, had pointed out (1837) that $\sqrt{-a}$ is meaningless when it stands alone, and that only in the " domain " of ordered pairs of quantities does it acquire meaning. It was being realised that the complex number is simply the most general kind of quantity to which the operations of arithmetic and their inverses apply, or can lead ; and that it is " real " quantity which, on this view, is the abstraction. We shall now go on to see that the idea of a closed " group " of operations, leading to nothing outside themselves, had other fields of use.

One of these was that other " insoluble," the quintic equation. Euler and Tschirnhausen in the 18th century had shown how to derive from any algebraic equation another, the " resolvent," which gave its roots. This reduced to a uniform procedure the methods hitherto successful in solving equations, and it was therefore suspicious that it would not work above the quartic ; for in spite of improvements by Lagrange the resolvent for the quintic was of *higher* degree (6th) than the original equation, and therefore useless for solving it. It therefore occurred to several workers (e.g. Ruffini, 1799) that it might be possible to prove the quintic insoluble. They failed, not having sufficiently considered what they meant by insoluble. For instance, roots could always be found roughly by trial and error.

Then in 1812 and 1815 Cauchy did an interesting thing. He proved that if we permute among themselves in every possible way the letters (say n in number) in any rational expression, the number of different values so obtained cannot be less than the greatest prime in n, unless it is 2 or 1. From this Abel (1802–29) was able (1824) to prove the quintic insoluble by radicals—that is, by algebraic functions of the coefficients. For he was able to show that any such functions must be *rational* in the roots themselves, and at most of the fifth degree, and so fall under Cauchy's theorem. Yet to furnish solutions, the permuting of these roots must give five and only five different values. He was then able to exclude Cauchy's alternatives piecemeal. His proof by no means gained acceptance, and in fact contained oversights. One contribution of Galois (1811–32) lay in fixing the conception of the group of all the possible substitutions, implicit in the above. Using this, he not only gave an explicit statement of the necessary and sufficient conditions that an algebraic equation shall be algebraically soluble, but he put on its way towards systematic development the whole notion of a group taken in abstraction from particular cases (such as substitutions) to which it applies. Towards the forties several workers, for instance Cauchy himself, were approaching this point of view. We shall find instances later.

It is possible to approach algebraic equations in a different way, by considering the types of number which can be solutions of them. For it began to be clear (Liouville 1844–51) that there are numbers not of this sort. At all events, the "algebraic number field," as it came to be called, was discontinuous and fell within an extended theory of numbers. We must therefore briefly note the advances which had been made by Gauss in this subject.

Beginning as a boy, in the age-old fashion, with experiments on numbers, Gauss was the first to systematise the theory and to carry it to a stage where general methods can really take hold.* He classified its problems and its methods, and standardised its notation. Thus, we can consider the question, is a number divisible by another? Here Gauss brought in the notation of "congruences." If the difference of two numbers, a and b, is divisible by c, a and b are "congruent with respect to c" ($a \equiv b$, mod c) and each is a "residue" of the other. This amounted to a new algorithm (see Chap. IV).

Other questions included that of finding the integral roots of indeterminate equations (e.g. single equations in two unknowns) and of finding whether a number can be expressed in a given "form," for instance the binary quadratic form $am^2+bmn+cn^2$.

The "gem" of the subject was the theorem of quadratic reciprocity. This was stated by Euler and imperfectly proved by Legendre (1785). The first sound proof was that published by Gauss in 1801. This theorem will not be given here, for, like so many in this stage of the theory of numbers, its importance cannot be made plain in a short compass. But the search for a corresponding theorem for biquadratics led Gauss (1831) to the conception of complex *integers*, which may fitly launch us on the subject of algebraic numbers in general, of which they are an instance.

Gauss had been interested in dividing the circumference of the circle into any given number of equal parts, which leads to the "cyclotomic equations":

$$x^n - 1 = 0$$

When $n=4$, i enters; and it was here that complex integers came in. Dirichlet, studying them, saw their awkwardness. For instance, primes of the form $4p+1$ (i.e. 5, 13, 17, . . .), being also of the form a^2+b^2, i.e. $(a+bi)(a-bi)$, are no longer primes if this more extensive field of factors is allowed.

In the "field" of rational numbers, i is by no means unique (as it is with continuous quantity) in constituting a new "unit" not itself a member of the field. For no surd is a member of the field: $a+b\sqrt{m}$ is as much a 2-unit number as $a+bi$. But it is $a+bi\sqrt{m}$ which gives us the worst difficulties, when m assumes certain values, say 5. For if we are allowed to use as numbers (and so as factors) any number of the form $a+bi\sqrt{5}$ (a and b integers) the number twenty-one, for instance, no longer has the fundamental property of being *uniquely factorable into primes* (of the extended sort), being, indeed, factorable in three

* Numerical experiment is still the only way into some of its regions, but nowadays this usually needs calculating machines. Gauss's book on number theory was written in 1797 but only published in 1801.

ways. Kummer found a similar difficulty with $x^n - 1 = 0$ when $n = 23$, and to get over it he invented "ideal numbers." In the field built on these, ordinary primes may be factorable, but all factorisation is again unique. Dedekind's different approach to "ideals" put them in a more general light, belonging to a later phase of thought but mentioned here for convenience. Essentially he re-defined divisibility: a is divisible by b if the both-way infinite set:

$$\ldots -3a, \quad -2a, \quad -a, \quad 0, \quad a, \quad 2a, \quad 3a \ldots$$

is wholly included in the similar but more "fine-grained" one:

$$\ldots -3b, \quad -2b, \quad -b, \quad 0, \quad b, \quad 2b, \quad 3b \ldots$$

This definition clearly agrees with the ordinary one for ordinary numbers, but we cannot show how this apparent complication simplified the situation outlined above. Our object has only been to stress the new kinds of generality entering mathematics. We now turn to instances of this in the region of continuous quantity.

We have so far mentioned the name of no single mathematician in these islands except Hamilton; and it was after 1800 before we awoke from our "Newtonian sleep." While Cauchy and Gauss were leaving Lagrange and Laplace behind, Woodhouse was just expounding them at Cambridge; and although he, likewise, criticised their want of rigour, it was long before continental rigorists had much following in England. But in a brilliant group of undergraduates, Peacock, Babbage, W. Herschel's son John, the insular virtue of originality began to reappear, in Ireland as well as in England.

Hamilton had (see next chapter) done much work in analytical dynamics, and was anxious to develop means of representing vectors in three dimensions as neat and suggestive as that of Argand and Gauss for those in two. We have already noted his abolition of the pseudo-mystery of i by viewing $a+bi$ as simply an ordered number-couple. In this language, what he wanted was ordered number-triples: in the other language he wanted several units, i, j, k, instead of merely i. But when he tried ("Quaternions," 1843) he was brought up against the perplexing fact that (somewhat as with number fields) not all the ordinary rules of algebra can be simultaneously true. Like Grassmann (1809–77) in Germany, and Boole in England (1815–64) about the same time, he began, in effect, to consider an "algebra" in which the rules $a+b=b+a$, $ab=ba$, $a(bc)=(ab)c$ are not all obeyed. If only the second was disobeyed and $ab=-ba$, Hamilton and Grassmann independently found that they had the algebra of the "vectors" in question, which were just then assuming such enormous importance in physics—fluid flow, elasticity, electric field, current flow, etc. Hamilton's notation seemed eccentric to the continental school, and that of the German Grassmann is nearer to our modern usage in vector analysis. This point should remind us that, over the calculus also, the German Leibniz had a better notation than the insular Newton.

We indicate in the next chapter how this issued in tensors and matrices, and conclude here by noting that it was in some ways Boole whose development of these ideas went furthest. Much work has been done on algebras when, say, only $a(bc)=(ab)c$ is retained—together with some linear multiplication table such as $ab=c$ (Pierce, " linear associative algebras," 1870, etc.). As an instance of another, we may consider an " algebra " of areas like a 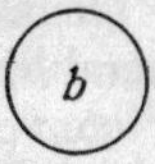 b defining cases thus a-b ab b-a $a+b$ so that while $a+b=b+a$ and $ab=ba$, $a-b \neq -(b-a)$. Such ingenuities helped the English school, De Morgan, Jevons, Clifford to free themselves from narrow associations of ideas, and to formulate one of the very few possible views of the nature of mathematics as a whole, the view of it as a " game " with meaningless symbols. We shall see in the next chapter how this developed. When it did so it was with substantial help from another related notion, which Boole developed from Leibniz. This was the view of logic itself as an algebra—as a game, and a very simple one, with symbols. Thus were some, though not all, the advantages of an algorithm brought into logic and even into philosophy.

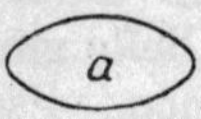

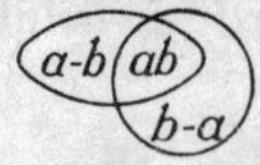

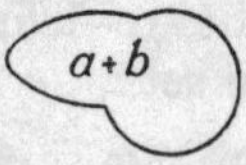

The subjects, transformations, invariants, matrices, groups, which we now approach, form a link between analysis, which we leave behind, and geometry. Long before the 19th century it had often been convenient, in Cartesian geometry, to move the origin and tilt the axes. Provided that the axes remain straight lines, all such changes to a new variable x, y, can be expressed by $x=ax_1+by_1+c$, $y=dx_1+ey_1+f \ldots$, but these relations might also be taken to mean that, with unchanged axes, we considered a new point (x, y) related in a particular way to the old point. Giving (x, y) a series of values, we should get a series of values of (x_1, y_1)—a series of points into which we had " transformed " the first set of points. If we took expressions for x in terms of x_1 of higher degree than the first, we should get, in the first point of view, *curvilinear* coordinates (Gauss) instead of straight ones. In the second point of view we should get not one but more than one x for each x_1 (etc.), but not more than one x_1 for each x; so that the reciprocity of the transformation would be lost. Hence most work has been done on linear transformations.

These perceptions suggest a number of problems. One is this: the equation of a given curve is altered by a change of axes, so that the equation is not a pure expression of the nature of the curve but is also in part an expression of the nature of the coordinates. This is inconvenient if we are interested only in the curve itself or only in the coordinates. Another is, what happens when we perform one transformation after another?

As to the first, some singularities of a curve (cusps or double points) are intrinsic to it, while others (maxima or minima) are clearly related to the axes. Hence, either the differential equations for the former sort of singularity, or the solution of these equations in terms of the coefficients in the curve's own equation, must contain expressions which are " invariant " when the axes are changed. That this was in fact the case became clear both as to differentials and as to coefficients in 1841,* when Jacobi stressed that this property is possessed by certain differential determinants arising in the theory of curves, and Boole hit upon an algebraic invariant. What he found was this. If we subject the binary quadratic form: $ax^2+2hxy+by^2$ to the above linear transformation, changing it into $a_1x_1^2+2h_1x_1y_1+b_1y_1^2$, the expression $ab-h^2$ (which occurs constantly in the geometry of the conic) is equal to $a_1b_1-h_1^2$ except for a factor dependent only on the coefficients in the transformation; $ab-h^2$ is thus an "invariant" of the form. Others were soon being found. In 1844 Eisenstein found the invariants of a binary cubic and quartic. These began to be so complicated that systematic methods of finding them and abbreviated methods of writing them were desired. In 1845 Cayley of Cambridge began to supply the first demand, while the symbolic notation brought in by Clebsch and Aronhold (1858, 61, 63) covered both. We cannot show the working of this in an example, but the symbols have, in common with *i*, a significant peculiarity: they can be operated on according to the ordinary rules for real numbers, yet any single one taken alone is meaningless. It only acquires meaning in certain combinations with others, or with itself. In 1851 Sylvester began his many decades of work on algebraic invariants; but we must eschew this highly elaborate side of the subject. Of greater note for us is the question, raised by a mistake in one of Cayley's papers, whether for binary quintics, and higher degrees, there are an infinite number of these "concomitants." It was known that most of them were combinations of a few simple ones, themselves not expressible as combinations ("irreducible system"); but was this irreducible system itself finite in number? In a series of papers, 1868, 1871, etc., Gordan showed that it is finite for all binary *n*-ics, *n* finite; and in 1890 Hilbert proved it so for any finite number of *m*-ary *n*-ics. The full bearings of this theorem are still not understood.

As we shall see, invariants have proved to be the strongest weapon in dealing with non-Euclidean and general geometries. They have been vital in relativity, which is essentially concerned with changes of observers, that is, of coordinates.

We now turn to the effects of successive transformations. In 1858, Cayley, examining this point in relation to substitutions of the type:

$$x_1=\frac{ax+b}{cx+d},\ x_2=\frac{a_1x_1+b_1}{c_1x_1+d_1}$$

found it convenient to isolate the "matrix" $\begin{Vmatrix} a & b \\ c & d \end{Vmatrix}$ and to regard the question of successive substitutions as a particular case of the theory of the combination of such matrices. If by the "product" of two matrices we mean the single one which would have given the same result, it is clear that, as with Hamilton's *i* and *j*, multiplication is not commutative.

These matrix ideas were developed from other roots by other writers, and have lately found vital applications in quantum theory. For us here, such arrays, like determinants, are an instance of the use of the geometric instinct for form in regions where geometry in that sense does not *seem* to be in question.

* J. Gregory had some idea of invariance by 1667.

CHAPTER XIII

NINETEENTH-CENTURY MATHEMATICS. II

But the effects of successive linear transformations may also be regarded as a question for the theory of groups; for such transformations form a group. There are other very simple geometrical examples of groups. One is rotations: any number of turns of a plane through a right angle about one of three mutually perpendicular axes always gives the same result as one of the single turns. This has immediate applications in crystallography. We have seen that algebraic equations form another field for groups. By the end of the century it was being found possible to work out a number of properties of closed groups of operations as such, without reference to such particular applications as these. There were finite and infinite groups, according as the number of distinct operations possible was finite or not, continuous and discontinuous groups according as the operations formed a continuum or not, groups in which the order of operations was immaterial, and so on. We shall see that, in the hands of Klein, groups became a very convenient means of classifying many kinds of geometry according to their basic ideas.

It was in this matter of the basic ideas of geometry that some of the most characteristic work of the 19th century was done. We shall treat it under two heads: first, non-Euclidean geometry, or the demonstration that self-consistent geometries could be constructed on axioms and postulates contradicting those of Euclid; second, geometries constructed by simply dropping some of Euclid's postulates, or by taking as spatial units not the point but the line. With these are involved the effort, correlative to the arithmetisation of analysis, to show that geometry could stand on its own feet, without any algebraic assistance.

The Greeks realised that the parallel axiom is a pure assumption, and also seem to have discussed the possibility of a fourth dimension of space. For Ptolemy is credited with trying to prove both the axiom, and the impossibility of four dimensions. Certain Arabic writers about 1200, Wallis (1660), Saccheri (1733), Gauss and others about 1792, all tried to prove the axiom, a line of efforts which should be compared with those to solve the quintic, since at about the same time men turned in each to the alternative of proving such proof impossible.

Göttingen gave birth to non-Euclidean geometry. Gauss and F. Bolyai (1775–1856) discussed the subject as students there, and later (e.g. 1804 and 1808) corresponded about it. In 1808 Gauss indicated that any geometry denying the parallel axiom must have a particular length (for instance, a "curvature of space") associated with it: so that here again—and about contemporaneously with dualism in chemistry—we have an instance of human thought advancing in complexity unit by unit—first considering no-constant space, then one-constant spaces.

In 1826 the Russian Lobatschevsky (1793–1856), apparently also influenced by Göttingen, gave lectures on a geometry constructed without the parallel axiom; and in 1832 F. Bolyai's son (1802–60) published his similar work, written perhaps in 1825. These men wrote later papers, but the subject did not gain general discussion; and indeed their point of view was a limited one. Using only geometric methods, and considering only two dimensions, they were concerned to keep valid as many other Euclidean axioms as possible. They proved that, for transferability of figures to be retained when a "plane" is distorted into a "curved" space, the product of the principal "curvatures" of the new space must be constant. If the product is negative, space may remain infinite; otherwise it must be finite; so that about a hundred years before Einstein, finite space began to be considered in theory. In the positive case, no straight line can be drawn parallel to a given one (as on a sphere); in the negative case, a pencil of them, of constant angle, is possible. Difficulties of defining straight line and parallelism were ignored at the time. But the existence of this negative case was only realised, again at Göttingen, by the man who made the subject respectable. This was Riemann in 1854; and he realised it because he used analytical methods, which he applied at once to the three-dimensional case.

Riemann envisaged the most general possible triply-extended manifold which is Euclidean when small enough parts are taken. Transferability was also retained. But he substituted for the Pythagorean relation $ds^2=dx^2+dy^2$ the more general one $ds^2=adx^2+2hdxdy+bdy^2$.

It thus became clear that with the new unit of complexity came the possibility of more than one viewpoint: the subject could be treated by supposing that the measuring rods used vary from place to place in a Euclidean space—that bodies are transferable only according to a definite law—and so, that the nature of a geometry can be defined by the "metric" it uses. We shall notice other cases of this trend of thought.

Riemann made it clear that the question: Is our space Euclidean? is one for purely empirical decision. After him came Cayley, whose different point of view we shall notice later. Einstein's handling of non-Euclidean geometries in general relativity was made possible by a further development of vector and invariant ideas. The use of these "tensors" developed by Ricci and Levi-Civita (1887, but especially since 1900) may be illustrated by the fact that, just as not every new set of coefficients in $ax^2+2hxy+by^2$ gives a different conic, so not every new set of such in Riemann's "quadratic differential form," $adx^2+2hdxdy+bdy^2$ gives a different space. We thus need certain invariants. Just as we can describe a force by the components of a vector, so we can describe a non-Euclidean space by viewing a, h, b, etc. (for the three-dimensional case, further coefficients are needed) as components of a higher kind of vector—a tensor, in Einstein's case a tensor named after Riemann and Christoffel (1861 and 1869) who had some notion of tensors. Other tensors include the shears in a solid. We can only state here that a tensor's fundamental property is that, if it is zero in one coordinate system, it is so in all others (the systems being subject to certain limitations). An equation in tensors true in one system of coordinates thus has the vital property of being true in all others. Another practical advantage which tensors bring is the extraordinary neatness of notation possible for entities possessing the tensor property. On the other hand, this highly condensed way of displaying certain formal characters of a situation is often unsuited for immediate calculation. It is a much more recent development than our present subject; but before returning from recent times we may note that the question of parallelism, put in the background by Riemann's approach, has lately recurred. With some non-Euclidean spaces, and definitions of parallelism, the ultimate direction of a vector after moving "parallel" to itself from one place to another varies with the path taken. The question thus becomes very complex.

Given Cartesian methods, it is easier to handle *n*-dimensional than non-Euclidean geometry. As we have noted, D'Alembert and Lagrange viewed mechanics as a four-dimensional geometry. Cayley (1843) and Grassmann (1844) were among the earliest to bring it to the fore.

Many *n*-dimensional operations belong essentially to plane geometry.* But each additional dimension adds great complications in that detailed development which is excluded from this book; 3-space brings in torsion, 4-space, half-parallelism. Even in Euclidean 3-space the theory of surfaces (2-spreads) is very complicated.

We now turn to geometries which ignored, rather than reversed, some of Euclid's demands. Our best clue is to follow the fortunes of projective geometry (see Chap. VI.), or rather, of what we should now call by this name; though the whole point of the story is that the clarity and self-consciousness implied in the name were long in coming.

Frezier in 1738, but especially Monge (1746–1818) from 1768, revived what were essentially Desargues' methods, for the purpose of specifying and describing solid bodies on plane diagrams.† In the ordinary "plan" and "elevations" of (say) a machine, the vertices of projection are infinitely distant and the planes at right angles. But with conical pencils we could get the three of them on one plane. In 1822, with Poncelet (1788–1867), the revival went beyond the 17th century original as regards the Principle of Continuity, which had referred only to the infinite and not to that other "mysterious" idea, the imaginary. Already, Monge had examined whether, if he got results by using a subsidiary quantity (say p), his results became erroneous when for a part of their range p became imaginary. He found that they did not; and Poncelet made much use of these "imaginary elements" of space. Imaginary points and points at infinity extended the idea of space as complex variables were extending that of function.

* On the other hand, some apparently two-dimensional properties, especially in projective geometry, are three-dimensional in that they cannot be proved without the use of a third dimension.

† The connection was with military engineering.

From Poncelet's time we may roughly distinguish two streams of tendency: Steiner (1796–1863), Plücker (1801–68), Gergonne (1771–1859), interested chiefly in that old feature of projectivity, its miraculous manipulative fertility; Cayley and Von Staudt (1798–1867) interested to know what these peculiar new devices and points of view ultimately *meant*.

Gergonne about 1825, Plücker and Steiner later, grasped a peculiar fact about a line and a point in plane geometry: that each needs the same number (two) of coordinates to fix it, and that there is a correspondence or duality between them, so that for every theorem about points we could get one about lines, and *vice versa*. Again, the number of tangents to a given curve from a general point (its "class") is in duality with the number of points on it on a general straight line (its "order"), and, reasoning analytically on these lines, Plücker (1839) found relations among the singularities which a curve can have, such as double points or double tangents. A cubic can have one double point, a quartic, three; but they need not do so. Riemann found (1857) that, in connection both with Abel's functions and with transformations, a curve's "deficiency" from the maximum number (now called "genus") was crucial. Thus several branches of mathematics were brought together, and the classification of curves, with which we began in Chap. VI, was carried a stage further.

Plücker (1865), going a step further with duality, began to consider 3-space as made up not of points but of lines ("line geometry"). This fact is mentioned here as showing that men were slowly realising that it is entirely a matter of convenience what ultimate elements we use; and that geometry can be viewed as a matter, not of space as such, but of figures in space. The final step, hardly yet completed, is that of realising that the figures themselves are but cases of logical "structures," to the logic of which their spatial origin is irrelevant. But perhaps the word "structure" should remind us that we can eliminate spatial imagination more easily from our results than from our methods.

Meanwhile, in philosophic Germany, Von Staudt began on foundations. From Desargues to Steiner, projective workers had used the concept of length in their proofs whenever it was convenient. They had had no sense of incongruity. But Von Staudt proposed, not only to free geometry of analytic entanglements—of x and y—but to free it of what they usually represented, namely length and its transference (as in proofs of congruent triangles).

Questions of what we can do when only certain constructions are allowed were not new. Indeed Leonardo (though hardly with the 19th-century objective) had considered what figures can be made with only one opening of the compass. The Dane Mohr in 1672, then, about a century later, the Italian Mascheroni, examined what was possible with compasses only, and thus came near to the point in showing that a straight edge, let alone one with fixed graduations, is needless for most of the common geometrical constructions. Finally (1847 and onwards) Von Staudt succeeded in showing that if we eliminate all idea of measure and length, but not of straightness, we still can build up the whole of projective geometry ; so that the latter really does occupy a more general position than Euclid's.

In 1859 Cayley succeeded in giving a *definition* of length on its principles, thus completing the projective triumph by making length not a base but a mere result. He used certain invariants of points A, B, C on a line cutting a quadric (e.g. a sphere or an ellipsoid) at say P and Q, to define distance in terms PA, PB, QA, QB, in a way which retained the fundamental property AB+BC=AC. This definition cannot be explained in a short compass, and is not given here. The number measuring AB according to it becomes infinite as either end approaches the " absolute " quadric, and imaginary beyond ; so that, as Klein saw, non-Euclidean geometries became particular cases of a projective geometry with a proper choice of meaning for distance—a reinforcement of the notion of geometry as concerned not with space but with figures in space. About 1860–70 there was a positive deification of projective geometry.

Then Klein showed that certain of Von Staudt's proofs could not stand without a further analysis of the distribution of points in the continuum ; and recourse had to be had to subtler, more general, geometries still. Long before this, indeed, Gauss had dreamt of something more general. The Greeks and, for instance, Leibniz and Euler had dealt occasionally with a peculiar class of spatial problems, such as mazes, in which, so long as the lines do not break or cross, their bending or stretching is immaterial. Möbius (1790–1868) and Listing (1806–82) had partially systematised these " topological " problems. Gauss in connection with electric currents, Tait in connection with vortex atoms, had tried to solve the very difficult case of knots—of closed curves (in 3-space) not deformable into a circle. Riemann's many-leaved surfaces linked the subject (1857) with the theory of functions of a complex variable. Von Staudt's omission was not supplied at the time by workers in this precise field, but by Cantor (see later). But as, lately, the two fields have converged, it is convenient to notice them together.

In 1872 the " Erlangen Programme " of Klein (1849–1925) summed up the progress of geometry up to that time. Klein, who had gathered ideas both in France and in Germany, pointed out that most of the geometries by then suggested, line or point, metrical or projective, Euclidean or non-Euclidean, can be completely described by a certain group—that of the operations they permit ; being concerned with properties invariant under that particular group of operations. For Euclidean metrical geometry, the group is that of translations and rotations ; for Cayley's geometry, it is the group leaving a certain quadric unchanged ; for topology, it is the group of one-to-one continuous transformations. Other groups include inversions and reflections. This apotheosis of the theory of groups in the seventies had the great value of making it clear that operations and operands are correlative and that each has, or *is*, only a certain domain of validity. The precise definition of such domains is in one sense the central problem of mathematics.

Not long after Klein systematised large regions of geometry in this way, Poincaré similarly cleared up a vast region of analysis. Hermite in 1858 had correlated our two insoluble problems of 1800 by showing that the quintic could be solved with elliptic functions. These latter could also carry another, related, problem one stage further than algebraic functions. This was the "uniformisation" of curves, putting $f(x, y)$ in the form $x=\phi(t)$, $y=\psi(t)$. For curves of genus 0, this can be done by algebraic, for those of genus one, by elliptic, functions. Now in a plane representing the complex variable, the trigonometrical functions repeat themselves along the real, the hyperbolic, along the imaginary, axis, the elliptic, at corners of equal parallelograms covering the whole plane. To generalise this, Poincaré considered certain groups of transformations (studied by the German Fuchs, e.g. 1880) which make a discontinuous selection from the points of the plane. He found (1881 and onwards) that if he could construct functions unchanged by the general linear transformation $z=\dfrac{az_1+b}{cz_1+d}$, subjected to this discontinuity condition, they would solve * all algebraic equations and uniformise algebraic curves. Clearly, if the group concerned were "continuous"—did not make a discontinuous selection—the needed subtle analogy with ordinary periodicity would vanish.

These "automorphic" functions repeat themselves at corresponding points of a certain network of curvilinear polygons. They will also solve all linear differential equations with algebraic coefficients.† In (e.g.) 1873 Lie showed that *continuous* groups have somewhat similar reference to *partial* differential equations. It remains to be added that Poincaré succeeded in actually constructing such functions; but that over the whole field he left many "loose ends" to be tied up by later workers.

Differential equations form a convenient bridge to applied mathematics. Both are too technical for proper treatment here, but a few notes on applied mathematics are necessary for the elucidation of contemporary physics. Emancipated from the 18th-century point of view, workers in differential equations began to make advances which often enabled them to answer particular questions about equations which they could not solve. The calculus of variations continued up to Jacobi (1836) to be kept in close contact with differential equations; while minimum action principles continued, up to Hamilton, to be the basis of the most generalised work in mechanics. Lagrange's "variations" had kept the energy constant, and had given him differential equations of the second order. Hamilton (1834), by varying the energy and keeping the time constant, got linear equations, a great advance, though one partly anticipated by Poisson (1809), Pfaff (1814–5) and Cauchy (1819).

Hamilton applied his methods to optics. Thanks to the fact that waves had not in his day quite banished corpuscles, he treated "systems

* In the formal sense. For numerical purposes, other means were already well-developed.

† Fuchs had previously made some advance beyond the case of constant coefficients, up to then almost the only one generally soluble.

of rays " in a very general dynamical manner. In this way he made it clear that there are, on any theory, profound formal analogies between light and matter. This realisation has lately had important effects in physics.

Both Lagrange's and Hamilton's minds, however, were pinned down to the actual *paths* of points of the system, being concerned with what happened to the " action " when these were varied. But after the 1830's, it will be remembered, pure mathematics became much concerned with transformations; and, in tune with this, dynamics came later on to view the change-over of an assembly of points from their initial to their final positions as a particular type of *transformation* (one which left curves previously in contact still in contact), the question of their actual paths being, so far, ignored.

Boltzmann (1844–1906), the generaliser of certain results of Maxwell in thermodynamics, was much concerned with assemblies of particles, using for this purpose both the dynamics which we have been discussing and the theory of probability, statistics, etc., developed by Gauss and others about 1800. In the middle of the century Clausius, discussing the kinetic theory of heat, had pointed out that, for molecules made up of several atoms, the mutual vibrations and rotations must take toll of any (heat) energy supplied; and that this would explain the different ratios of specific heats of different gases.* In 1860 Maxwell showed that, in equilibrium, these modes of motion or " degrees of freedom " must in fact each take an equal toll. Boltzmann showed that whatever n independent numbers (generalised coordinates) we use to specify the state of a system, the averages of each of the fractions of the total energy expressed in terms of them must approach equality as the system approaches equilibrium. This " Equipartition of Energy " was in advance of its time, and we leave it here to the final chapters (esp. XXI) on physics. We may note, however, that " statistical mechanics " makes use of elementary n-dimensional geometry; 3-space enables us to represent the position of a particle, 6-space its three components of velocity as well. Hence the path of one point in $6n$ space suffices to enable us to envisage the paths of n particles as the path of one representative point, and so to treat all the components of a gas as one unit.

Poincaré used this method in his new investigations of (for instance) the problem of three bodies, that old " insoluble " † of gravitational mechanics; but the features of n-space which he needed were anything but elementary. On the one hand he needed (e.g. 1895) the topology of n-space, as Riemann had needed that of two, in order to clear up the mazes or knots described by his representative point; and on the other hand his rather intuitive methods raised questions which assimilated

* It will be remembered that Avogadro's hypothesis was revived about 1860. The idea of different numbers of atoms per molecule was thus " in the air."

† In the light of the " transformation " view of mechanics, the difficulty with the problem of n bodies is seen to lie in its needing transformations of 3- or higher-dimensional regions, which are immensely more complex than those of two.

the whole subject to that of point-sets or aggregates to which we now proceed. But, apart from Poincaré's personal equation, it was inevitable that the ultimate analysis, both of statistical and of individual cases, should lead to this subject. For questions whether a certain path will ultimately return upon itself (be periodic) or whether, if slightly disturbed, it will eventually leave a definite neighbourhood of its original motion, are clearly questions very similar to those of Fourier.* Now these, above all others, had led Weierstrass to " arithmetise " analysis : to eschew geometric aid and to avoid delusive " formal " solutions which conceal particular cases where the operations they enjoin become meaningless or endless.

Weierstrass had been led by his ever-finer analysis to flood the subject with such things as undifferentiable continuous functions, which aroused the wrath of the Jew, L. Kronecker (1823–91), a worker in algebraic numbers to whom only the positive integers appeared a sound basis for mathematics. Like Boole and the English school, who regarded mathematics as a game with meaningless symbols, Kronecker was thereby giving very unphilosophical expression to one possible philosophy of the subject.

Another worker in algebraic numbers, Dedekind, joined hands with Weierstrass in developing a method for dealing with the position. It lay in working out an analytically usable definition of irrational numbers, by extending that of Eudoxus. Dedekind's " cut " (1858, published 1872) divided the *rational* numbers into a L(eft) and R(ight) class, each with at least one member, such that every L$<$every R. A rational number makes a cut in which it is either the greatest of the L's or the least of the R's ; which both exist. An irrational is then defined as one for which neither exists, but which (Weierstrass) can be defined, and reached in practice, by a convergent infinite sequence among the L's.

Now came a further step ; for Dedekind himself seems not to have been entirely satisfied with his definition, useful though it remains in many contexts. Cantor (1845–1918), a Jew convert, starting (1871–2) from Fourier's series, claimed to have proved that non-algebraic, " transcendental," numbers are infinitely more numerous, not merely than the rational numbers (themselves of course infinite in number), but than the " infinitely more numerous " algebraic numbers. Kronecker, denouncing Cantor for giving no method by which he could actually find even a single one of these transcendentals, took a further step in the development of one characteristic mathematical philosophy—that which insists on performable operations. Cantor (1874–84) went on to a series of papers which alienated even Weierstrass and gained scarcely any support until the end of the century. In these he tackled the questions raised by the old Greek philosophic objectors to mathematics, as well as certain new ones which needed answers. Such was : is $x^2>x$, as $x\to\infty$, since both can be made as large as one pleases ? Or again, are the rational numbers more numerous than the integers ?

He asserted that the only criterion which can yield definite answers

* The connection, in each case, with periodicity should be noticed.

to such questions is: can the two series in question be brought into 1 : 1 correspondence * with each other, an "infinite" quantity being one with parts capable of being brought into such correspondence with the whole. This idea of a "standing" infinity seemed unsound to the orthodox school, but Cantor was able to guide it through its difficulties by inventing a systematic method of counting two or more dimensions of countable assemblies like integers. Thus he showed that any countably ("denumerably") infinite number of such dimensions remained in correspondence with the original simple series of the integers; but that the continuum of real quantity did not—was of a higher "power." By this essentially sceptical proceeding of denying apparently obvious distinctions he solved some of the ancient and modern problems which beset mathematics.

With Hankel and others he went on to prove that a (discontinuous) set of points can have, like a continuous line of them, metrical properties. Their definitions give a zero measure for *denumerably* infinite sets. But that there are sets with a non-zero measure, if the term is suitably defined, appears from the fact that if we divide a finite continuous line into two intermingled sets, each interrupted infinitely often by the other, they have between them a finite measure, namely, the length of the line. We cannot describe the highly technical work by which Borel (1897) and Lebesgue (1903) assigned definite measures to such sets.

Cantor's work roused bitter controversy; and certain paradoxes were deduced from it which caused it to merge on one side into mathematical philosophy. But in spite of this, Cantor would have performed a great service even if he had merely brought questions of rigorism in many fields—numbers, functions, geometry—under one discipline, thus freeing other branches of mathematics from a potential plague. Just when extreme rigour was becoming a conscious process, a related one was coming into play, the deliberate use of hypothesis, as in physical science. Riemann made an hypothesis about the zeros of a certain function, and Hardy, Littlewood, and others have (much more recently) gone on to explore its consequences in that part of the theory of numbers which can only be reached by the *continuous* complex variable.

The development of the theory of sets of points was practically the assertion that continuity is not the best ultimate concept for the finest sort of analysis. Thus, part of Kronecker's contention was admitted: that we must build up everything from unextended units—points though not integers.

Just as progress in the superstructure of science lies in adding unit after unit of complexity, so progress in the foundations lies, in this view, in paring away, one after the other, needless, or mutually inconsistent, basic units or postulates. For long ages mathematicians had in practice used both continua and points, as it suited them. Now they decided to forgo the former, as raising more problems than they

* It will be recalled that Dedekind's re-definition of divisibility involved the correspondence or otherwise of two infinite sets of numbers. The idea was plainly ripening for expression.

solved. It seems, however, a little doubtful whether a science founded *strictly* on one concept could have any problems left. It is undoubtedly sources presenting at least the appearance of continuity which have furnished mathematics with most of its problems.

Such questions merged, towards the end of last century, into a systematic attack on the philosophy of mathematics. We have spoken of the mathematical logic of Boole and others. But is mathematics a branch of logic—are its truths such that thought is impossible without them—or do we found it only by laying down axioms which may be false? That some branches of mathematics require axioms which may be denied was the whole burden of non-Euclidean geometry. But is the whole structure of the totality of alternative geometries a necessity of thought? Frege (from 1884), Peano (from 1894), put forward the view that logic and mathematics have exactly the same status; but they altered that status. There is, they suggested, an independent realm, different both from the "matter" of the physicist and the "mind" of the psychologist.

Geometry, unlike other branches of mathematics, had always tried to set out its assumptions clearly at the start; but no one had been able to *prove* that these assumptions were consistent—would not ultimately lead to self-contradictions. Hilbert (1899) tried to do this; but he only succeeded in reducing the question to that of the consistency of arithmetic—Kronecker once more.

Peano was anxious to secure for this study of elusive concepts the mind-fixing advantages of symbols for which exact rules of handling could be strictly observed. He invented a complete "language" of symbols ("Peanian"!). Russell's and Whitehead's elaborate treatment of the bases of mathematics as a generalisation of ordinary logic (1910–14) was written almost entirely without words. Thus did Boole fructify. And yet it is far from clear, historically speaking, that it is rigour which is the advantage of symbolism. Even in this case of "Peanian" and its later enrichments it appears as if the advantage gained lay not there but, as with previous algorithms, in the ease of handling complex subsidiary or superstructural questions which arose. The later works of (for instance) Wittgenstein (1921) and Van Orman Quine (1934) are written mainly in words.

Russell and Whitehead avoided some of the paradoxes of the subject by classifying propositions into "types," confusion of which had to be carefully avoided. But other difficulties remained. These men needed, after all, certain existential propositions, one of which (Axiom of Reducibility) seems to raise the old difficulties of the continuum in a more complex form; but their work will probably long remain the principal monument of the subject. Quine has lately generalised it, and simplified its algorithm, by introducing the asymmetric 2-term relation of ordination.*

The whole subject is still controversial. Typical ideas include the one revived by Hilbert, "formalism"—the "game with symbols."

* A similar asymmetric 2-term relation, "before and after," has been used to replace the more usual symmetric 3-term one, "between-ness," in Robb's recent (1936) axiomatics of relativistic space-time. See Chap. XXII.

Unfortunately this needs to be related to questions of number and space by a separate science of axioms. Then there is " intuitionism " —or Brouwer's view that we can define a mathematical quantity validly only in terms of operations by which we can actually evaluate it. It is difficult for Brouwer to avoid denying validity to large branches of mathematics. Wittgenstein brilliantly envisaged mathematics as " the syntax of all possible languages." For any language, he contended, there are truths which can be *stated* in that language and others which can only be *shown*; though he did not prove that a further language could not be invented in which these latter could be stated. Gödel's theorem (1931) that it is self-contradictory to suppose that mathematics can be *proved* free from self-contradiction—that, in fact, there must always be true but unprovable theorems—seems to be a development of this.* Such work clearly puts the old question of the relation of logic and mathematics in a new light: mathematics is not logic but the algorithm or language (see Chap. IV) of logic.

The " logical positivism " which has developed from Wittgenstein is of interest beyond mathematics. Its view is that the function of philosophy is purely negative—lies in showing the illusoriness of difficulties raised for themselves by philosophers who try to impose their private abstractions upon reality. This is evidently a refinement of Comte's positivism, and of one whole age-long attitude of scientists.

There emerges the basis for a new classification of the sciences. Kempe long ago (1886–90) proposed to reclassify concepts, not only in mathematics, on the basis of their " logical structures." Thus, a hexagon, as usually understood, has only one kind of vertex, but has three kinds of pairs of vertices—that is one way of stating its " structure." Euclidean geometry recognises only one kind of straight line, but two kinds of pairs of straight lines. Projective geometry recognises only one kind of pair of straight lines.† Robb's relativistic space-time recognises three different kinds of single lines. These are only rough instances, and Kempe did not deal with the infinite classes which have since become the whole difficulty; but a similar view occurs in Whitehead's " Principle of Extensive Abstraction," that science deals with things not " in themselves " but only as revealed in the structure of their relations with other things. Experimental and observational approaches in science have evidently a very different logical structure.

Since, in several of these views, and especially in Brouwer's, there has appeared an anxiety to keep a " practical " outlook, a contact with the workable, it is perhaps not unfitting to end with a few notes on actual computers' mathematics as developed in modern times to meet the vast floods of figures put forth by science and business. This was one of our earliest mathematical topics, and it shall be our last.

* It appears that the convenience of using imaginary quantities or points, ∞, umbral symbols, and the like (which extend the current meanings of terms while ultimately making the extended field easier to understand) may be a further exhibition of the same " formal property " of the subject.

† All, of course, admit of many " kinds " *defined by reference to further lines, points, etc.*

There is a certain rough correspondence of practical and theoretical *fields*, coupled with a violent contrast of *methods*. For the theory of functions of several variables, practice substitutes nomography (D'Ocagne, etc.); for integration it substitutes planimeters and photo-electric and other integrators; for crude Fourier theory, harmonic analysers; for formal solution of differential equations, electro-magnetic apparatus; for the automorphic functions which formally solve algebraic equations, mechanical calculators. In each case simple parts of the theory are needed, such as Sturm's and Horner's procedures for algebraic equations, while in some cases the resources of higher mathematics are needed to justify the practical devices used. This is especially so in electrical engineering, where use is made of complex variable methods and of "operators." A simple operator is D^n, used as if it were an ordinary power, for $\frac{d^n}{dx^n}$ in differential equations. Heaviside's extensive use of operational methods was at first distrusted; but for most functions of use in electro-magnetism it has since been justified from function-theory by Carson, Bromwich, and others. It substitutes for the original functions a "symbolic image" which is usually much simpler.

As to calculating machines, mechanical or electro-magnetic, their surprising feature is the fewness of the fundamental units or movements, out of all the wealth disclosed by physics, suitable for dealing with numbers. Few machines are simpler in principle (or more intricate in repetition!) than a modern calculating machine; and by a few such movements as repeated adding, automatic "moving up one," and mechanical trial-and-error, much of the colossal variety of known functional relations can be dealt with, including some, in number-theory, which can still be dealt with in no other way.

CHAPTER XIV

MAINLY ORGANIC CHEMISTRY

In the next two chapters, we take up again that line of experimental physics, chemistry, and physiology which we followed in Chaps. VIII and IX. We show how the travail of the fifties over atomic weights had issue in the doctrines of valency and of periodicity. We show how the old interest in crystals led to stereochemistry and so to partial constitutions for the chief classes of organic compounds. We end the first chapter with notes on apparatus.

Turning to physical chemistry, we develop the thermal and related sides of reaction from where Berthollet left them, up to their fruition in Mass Action and the Phase Rule. Then we take up electrochemistry where Davy and Berzelius left it, and show it growing into ionic theory. This leads up to colloids and to catalysis.

Chronology would have led us to intrude thermochemistry between inorganic and organic work, but it has seemed best to keep all the physical sections together, since long overlaps are in any case inevitable.* The last two physical sections are decidedly later than any of the chemical ones.

Throughout, there was close give-and-take with physiology: the ideas of cells and of ferments, themselves newly brought into relation, gave men a start both in stereochemistry and in catalysis. Cells provided vital links in both ionic and colloid history. On the other hand, physiology was enriched by the physical advances. This was so especially as to Bernard's great concept (1855) of the "internal environment." Before his time the idea of a bundle of half-separate agencies had been to the fore in envisaging the animal organism. Bernard showed how largely the body is built round, and unified by, the effort to maintain the constancy of the internal environment.† We shall see elsewhere how this led to endocrine doctrine, and how, simultaneously (1859), Darwin was relating the organism to its external environment. The developments in the present chapters gave content to Bernard's concept, showing the organism as made up of buffered ionic solutions and of thousands of semi-permeable membranes enclosing colloids.

In much of this development the idea of a membrane, more generally, of a surface of separation, was vital: osmotic experiments led up to both ionic and colloid theory. Both in colloids and in heterogeneous equilibrium such ideas were central.

Improvements in apparatus were, as usual, vital: improved

* And often reflect the isolation of the subjects.
† That is, of the blood or tissue fluid.

calorimeters, for instance, not only in thermochemistry, and so chemical thermodynamics, but also in that nutritional research which led up to vitamins.

Chiefly Inorganic

The confusion in chemistry was due partly to theories which should long before have been forgotten. For the flank of dualism and of the oxygen theory of the acids had in reality been finally turned as early as 1833 by Graham's work on the phosphoric acids, followed by that of Liebig in 1837. Here we must note an important point: a matter apparently of the merest technicality and detail proves to be the pivot on which entire massive theories are overturned. Such facts as this gravely limit the usefulness of books like the present, for the tangled nature of the phosphate question makes it impossible to treat it here.

The phosphoric acids, with their acid salts and hydrates, had long been puzzles, largely because up to the thirties chemists were in a fundamental confusion as to the parts which water could play. Dualism required them to view every compound as essentially made up of *two* parts. Graham's and Liebig's doctrine of polybasic acids drew a line between combined water, leading to different acidities and really water no longer, and mere "water of crystallisation," which brings a *third* element into the question.

But such facts were not assimilated, and chemists continued to make play with their theories of radicals and of types. It was as just another such theory that "valency" first appeared on the scene. Like them, it had to go through a period of chastening before reaching a permanent place in chemical theory.

It may seem surprising that the idea of valency had not come to the front before, but that is precisely the point: until our present atomic weights had been fixed, the pervasiveness of the idea was far from obvious. We treat these two subjects together, and follow them by a discussion of the periodic law; for it was one of the germs of the latter which enabled Frankland to gain the notion of valency (1852).* For one germ of the periodic table was that a number of elements fall very obviously into families. This fact had become prominent when in 1826 Balard discovered bromine to insert between chlorine and iodine, and when ten years later Dumas classified other non-metals into groups: sulphur, selenium, oxygen; boron, silicon, carbon; nitrogen, phosphorus, arsenic. And it was the compounds of this last trio which brought Frankland to the point. One atom of any one of these, he saw, always combined with three or five other atoms unless these latter were themselves of the kind which combine with two (or more) others, in which case *two* (or more) atoms of nitrogen (etc.) are needed.

There were, from the first, many exceptions to this rule. Frankland himself supposed valency to be variable up to a maximum fixed for each element. But about 1857 several chemists began to see that there is one element with a particularly definite valency, and that is the element carbon itself. Kekulé, particularly, spread the doctrine of the

* More vaguely put forward by Williamson in 1851.

four-valency of carbon, and with it the general valency-idea. Organic substitutions and (especially at the hands of Gerhardt) homologous series narrowed the possible valencies and atomic weights of elements which could combine with carbon or with elements combining with it. And yet the conflict of criteria remained.

A conference was called at Karlsruhe (1860) to thrash the matter out. Here Cannizzaro circulated a pamphlet, advocating Avogadro's and Ampère's hypothesis. The conference was inconclusive, but through it Lothar Meyer was inspired to write a treatise on chemistry (published 1864) which adopted the old hypothesis. This was important, not because it swept away all the difficulties, but because it decided what the difficulties were. One of them had been that vapour densities * were not always simple multiples of values predicted from empirical formulae. Deville's revelation (1864–5) of the dissociation of gaseous compounds at high temperatures showed the reason for this. As regards valency, Frankland's view of it as variable and Kekulé's view of it as fixed became bones of contention for the rest of the century; but we leave its later developments to Chap. XXIII.

The periodic law became prominent rather later, and then amid much scornful scepticism. It was, at the time, much less important than the foregoing developments, because it did not relieve any such urgent strains. But we treat it here both because it follows logically and because the way will then be clear for organic chemistry.

Natural families among the elements were a sober beginning for the periodic law, but numerical relationships among atomic weights were what brought it into being. As early as 1839 Döbereiner called attention to several "triads" of similar elements with atomic weights almost in arithmetic progression. Careful workers might deny that the weights were reliable, or that they were integers, but the game went on. In 1850 Pettenkofer pointed out further arithmetic series. In 1852 Dumas urged the importance of such relationships even when not exact; and in the next five years several workers put forward more extensive schemes. Then came the settlement of the atomic weight table; and almost simultaneously, in 1863, Chancourtois began a series of studies which revealed the essential fact, the all-pervasiveness of *periodicity* among the elements. But he used Proutian whole-numbers and ascribed gaps not to missing elements but to varieties of existing ones. His (helical) table only aroused attention after the subject had been settled by others.

In 1864–5 Newlands (1838–98) stated that the elements run in octaves. He allowed no gaps in his scheme for further elements to be discovered. These were a definite feature of the largely final scheme published by the Russian Mendeléeff (1834–1907) in 1869 and 1871. These gaps relieved strains in the table and above all made prophecy possible. Most of Mendeléeff's prophecies proved surprisingly accurate, when, later on, the gaps were gradually filled. We may suitably recall here that each advance in technique (such as spectrum analysis) brought about a run of discoveries of new elements. We lack

* Much relied on in connection with atomic weights.

space to recount these effectively, but, as always, they were one of the main ambitions of chemists.

Recognition of the *fundamental* place of the periodic law was slow in coming. At the end of the century in England, for instance, it was still a novelty, and a doubtful novelty, to organise teaching around it. The law could not fail to suggest that all the atoms were built up of one of two fundamental units according to some simple rule. But Mendeléeff himself frowned on such suggestions, and important schemes based upon them had to await a new century's advances in atomic physics.

Organic Chemistry

Crystals give us one clue to the coming of stereochemistry and of the modern type of constitutional formula. As we have seen in former chapters, crystals had long had a fascination for scientists, and had already provoked vital advances in optics. We mention here only one special line of work connected with them. Haüy had noticed that crystals of quartz fall into two classes differing only by two tiny faces. The two classes are mirror images of each other. Biot knew that some classes of quartz crystals rotate the plane of polarised light one way, some the other. Herschel (1821) showed that the two facts go hand in hand. It will be recalled that in 1830 Berzelius had discovered the isomerism of tartrates and racemates. In 1844 Mitscherlich stated that, though these have exactly similar crystals, only the former rotates the plane of polarised light. Then, in 1848, Pasteur, under the microscope, was able to separate the racemate crystals into two kinds, mirror images of each other. These were equal in quantity, opposite in rotation. One was tartrate, the other was new. One of the few marked differences between them was (see next chapter) that a certain ferment acted only on one.

Now Pasteur stressed that this phenomenon occurred *in solution* as well as in crystals. The difference in crystalline form—in the relations between neighbouring molecules—was not the essence of the matter. The essence must lie within each molecule. He concluded that the molecules must form " dissymmetric " figures (such as helices and certain tetrahedra) of which the mirror images would not be superposable. But he did not (1860) identify such irregular tetrahedra with the valencies of a carbon atom with four different atoms attached. To explain this, the next idea, we must examine the development of the notion of a chemical formula.

Dualism and radicals both implied a long-rooted habit of writing formulae split up in various ways into groups connected together. Valency gave this connection a numerical definiteness. But it did not give it a spatial or physical definiteness, even though ordinary isomers were explained by splitting up a formula into different sets of radicals.

Then in 1858 Kekulé did suggest that the valency bonds were physical and geometrical realities. The physical idea was not immediately operative, and we keep to the geometrical one. In the same year Couper, also, began writing what we should call ordinary graphic formulae. These were not applied to the Pasteur phenomenon, but in 1863 Wislicenus found this latter in lactic acid also. He concluded, from the extreme similarity of all but the optical properties, that these cases were due to different arrangements of the *same* radicals in the formulae. But the problem was still not solved, for so long as the formulae were written in one plane it was not possible to see what such different arrangements could be.

It was 1874 before Le Bel and Van't Hoff independently pointed out that formulae, different arrangements of which were mirror-images of each other, could be got, when an atom had four different atoms attached, by going into a third dimension and boldly taking the formula as an actual picture of the molecule. This view was bitterly attacked, but as all the active and none of the inactive cases known had in fact got just such an " asymmetric " carbon atom, it forced itself into acceptance.

The idea of constitutional formulae in general was severely tested about this time by another case in which Wislicenus was concerned, that of aceto-acetic ester (1877). In this case the crux was " tautomerism "—the apparent capacity of the ester to behave to different reagents as if it had two different constitutions (in this case the " keto " and the " enol " forms, of which the essential linkages were :

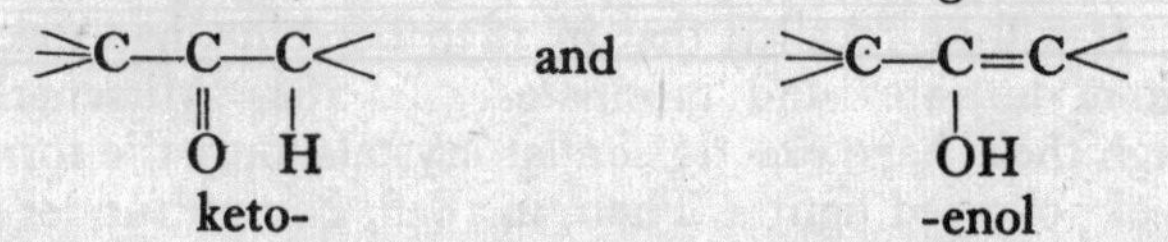

a hydrogen and a valency wandering). It was finally admitted that both forms, in reversible equilibrium, are normally present in the ester, keto usually predominating, but either being formed from the other as soon as any of the latter is removed by reagents. This type of explanation has become of enormous importance in 20th-century chemistry.

A difficulty was also met with in explaining the case of maleic and fumaric acids even by spatial formulae. These isomeric acids ($C_4H_4O_4$) differ in properties only in ways which suggest a different disposition of similar groups. They are not optically active and, using the tetrahedral space-formula for their carbon atoms, it is hard to find two forms which are different. The clue, worked out by Van't Hoff, Le Bel, and Wislicenus (1887), was that the carbon atoms are unsaturated. If we assume that the double bond which connects them defines a plane or cannot rotate freely like a single bond, we can conveniently write the two forms thus :

H—C—COOH ‖ H—C—COOH	H—C—COOH ‖ HOOC—C—H
Maleic (" Cis " form)	Fumaric (" Trans " form)

In 1846 Dubrunfaut observed " mutarotation " in certain sugar solutions, the fact that the optical rotatory power may alter with time to a new steady value. This was ascribed as early as 1855 to the formation of an isomer, an ascription which, as we shall see, affected the view taken of the constitutions of the sugars. With Mass Action, the change came to be viewed as a case of reversible reaction.

Another, more puzzling, phenomenon in stereoisomerism was the "inversion" discovered by Walden in 1893. Optically inactive reagents usually cause no change in the sign of optical activity when they react with active compounds. Walden's discovery that they sometimes do so has been important in confronting chemists with the fact that they have no theory of the relation of rotation to the nature and relative positions of the atoms concerned. In the last few years a little progress in this direction has been made, but the subject has proved exceptionally inaccessible to theory.

In 1899 Le Bel and others found that carbon is not the only element which gives rise to stereoisomerism: quinquevalent nitrogen can give it. The following decade saw enough new instances to enable chemists to conclude that *any* asymmetric atom will give it. This emphasised the atom. Then, from 1909–11, they found reason to return to the earlier emphasis and say generally that any molecule having a non-superposable mirror image will give the effect, whether this condition is caused by an asymmetric *atom* or not. This "dissymmetric molecule" is the present unit of thought in stereochemistry.

The constitutional formulae of the main classes of organic compounds must be our next concern. We have seen that by the fifties a very large number of organic compounds were known, often falling into "homologous series," such as those of the fatty acids, the paraffins, etc. There was still a tendency, however, for the activities of the chemist to be analytic rather than synthetic. Synthesis from actually inorganic materials remained exceedingly rare, Pasteur even saying that it was still unknown except as to excretory products like urea, which were not fully organic. To remedy this became the set aim of Berthelot and others from about 1850. We cannot describe their work, but we must remember that synthesis has been a vital factor in the finding of constitutions. It began, too, not very much later, to be a great factor in chemical industry. And chemical industry was important in another way: it gave the chemist finance and a motive, one of the two great motives of organic chemistry henceforward. The other was, of course, physiology.

Industry provided, especially from the sixties, many examples of a type of compound which was also important in physiology, "aromatic" compounds. As compared with "aliphatics," like fats or paraffins, these were stable, and so rather inaccessible. They included benzene, the essential oils immemorially used in cosmetics and medicine, and above all the aniline dyes which were destined to give Germany her great position in chemistry. In 1843 Hofmann (1818–92) had shown that certain substances prepared by Unverdorben in 1826, by Runge in 1837, and by others later, were identical: and aniline had come before the world. A last kick of the dynastic internationalism mentioned in Chap. V, namely the influence of Prince Albert, called him to England. In 1856 his pupil Perkin discovered the aniline dye "mauve." *

Of all these "aromatic" compounds, the very simplest, benzene, was a challenge for Kekulé's view of the fixed four-valency of carbon: since its empirical formula was CH. The idea which Kekulé advanced was that aromatic stability was due to the atoms being in closed rings as against the open chains of aliphatics, unstable because they have vulnerable ends. He asserted in 1865 that benzene consists of a hexagon of carbon atoms, each with its hydrogen attached.

While the general truth of this has never been in doubt, and while it has in fact proved possible to regard hundreds of thousands of aromatic compounds as combinations or modifications of these rings, it still left Kekulé with a valency over in each of the six carbon atoms. To dispose of this, he suggested that alternate bonds were double; and the argument against this is typical of many used in organic chemistry. For, if the bonds are of two kinds, we might expect that a pair of adjacent hydrogens could be replaced (by say chlorine) in two ways—according as they were joined by the single or the double bond. Of course, such substitutions might be only possible for one kind of pair, but this and other arguments led Claus in 1867 to suggest that the extra valencies formed diagonal braces for the hexagon, and Armstrong (1887) and Baeyer (1888) to suggest more generally that they are simply directed towards the centre of the ring ("centric formula"). Such a formula, while allowing only one kind of di-substitution product with adjacent hydrogens, obviously allows a small but definite number (three) when the hydrogens are not necessarily adjacent. How distinguish these? The answer given by Kekulé's pupil Körner is another instance of the simple argument used above. All three kinds exist; but they are found to admit of different numbers of ways in which a *third* substitution can be made. This has become a standard type of argument. It is, however, too elaborate to form, like the simple reactions of qualitative inorganic analysis, a routine method of identification. In many cases the examination of physical properties has become the normal substitute for this analysis in organic work, and tables of such properties are as indispensable a part of an organic laboratory as any piece of apparatus.

Towards the end of the 19th century, the synthesis of hundreds of thousands of compounds became the distinguishing feature of organic chemistry, an often pointless process for which the German chemical industry was partly responsible. One fact this prodigious output did confirm. This was the striking fewness of the types of organic compound, their falling into a few apparently endless series, each member obtainable by identical processes from the last. Before we examine how the constitutions of a few of these series were found, we must glance briefly at the physiological source from which, as we have said, many of them came.

* We cannot follow the subsequent migration of the dye industry to Germany, nor the researches of Griess (1858–66) on the diazo-(·N : N·) and related linkages, which made possible the innumerable dyes, explosives, etc., now known.

We will consider one point only, animal metabolism. Liebig in 1832 had found lactic acid in meat extracts, and in 1850 Helmholtz showed that an acid is formed in muscle contraction. Liebig had thought that it is protein metabolism which is the direct source of muscular energy. But in 1867 Fick showed that the direct source is the oxidation of carbohydrate, later work (1881) showing protein to be structural rather than active. Fick (1829–1901) was a pupil of Ludwig's and indeed this great man and those he inspired were largely responsible for all this work. For in 1871, another pupil of his, H. Kronecker (1839–1914), showed that the carbohydrate oxidation produces the lactic acid—which must have been the acid noticed by Helmholtz. As we shall see in Chap. XVII, this is by no means the last word on muscle contraction.

We turn to another point, namely the relation which this carbohydrate oxidation bears to the changes (in colour and other respects) which the blood undergoes during circulation. It was known in Lavoisier's day that the lungs are the site of the re-oxygenation of the blood and of its giving up carbon dioxide; but where did it *lose* its oxygen?

Poiseuille (1799–1869) and the brothers Weber (1795–1878 and 1806–71) of Wittenberg had gained much new knowledge of the circulation of the blood (and of the lymph), and the former's viscometer and law of fluid flow in tubes are as much physiological as physical in origin.* But the crucial improvement in apparatus was again due to Ludwig. His mercurial blood pump was in the direct succession from Boyle and Hales. Indeed the former had got as far (1670) as using a similar apparatus to separate the gases dissolved in the blood, though he went no further. By 1872 Pflüger (1829–1910), using Ludwig's device for the same purpose, had shown that the oxidation takes place in the muscles and other tissues, the blood serving only as a carrier of fuel and wastes.

In leaving physiology and reverting to chemistry, we note that *plant* physiology also laid stress on the carbohydrates, since in plants a main structural element is cellulose, and a main storage compound, starch. Thus it was not remarkable that towards the end of the seventies a strong effort was made to elucidate the structure of these bodies, which had in common empirical formulae containing hydrogen and oxygen in the proportions of water. But the processes of analysis took endless time because many sugars and derivatives crystallise—can be separated—with such extreme difficulty or slowness. It was only E. Fischer's (1852–1919) discovery (1875) of phenylhydrazine which made easy crystallisation possible, one of many cases where the invention of a single reagent has opened up whole new fields of work. We mention another case almost at once.†

* See Chap. XXIV for its physical connection.

† A further one was the use of organo-metallic compounds for synthesis. In 1890 Lohr isolated magnesium diethyl, and in 1899 Barbier showed that magnesium powder and an alkyl halide could be used instead. In 1900 Grignard completed the method known by his name by showing that ether catalyses the reaction.

CHART III

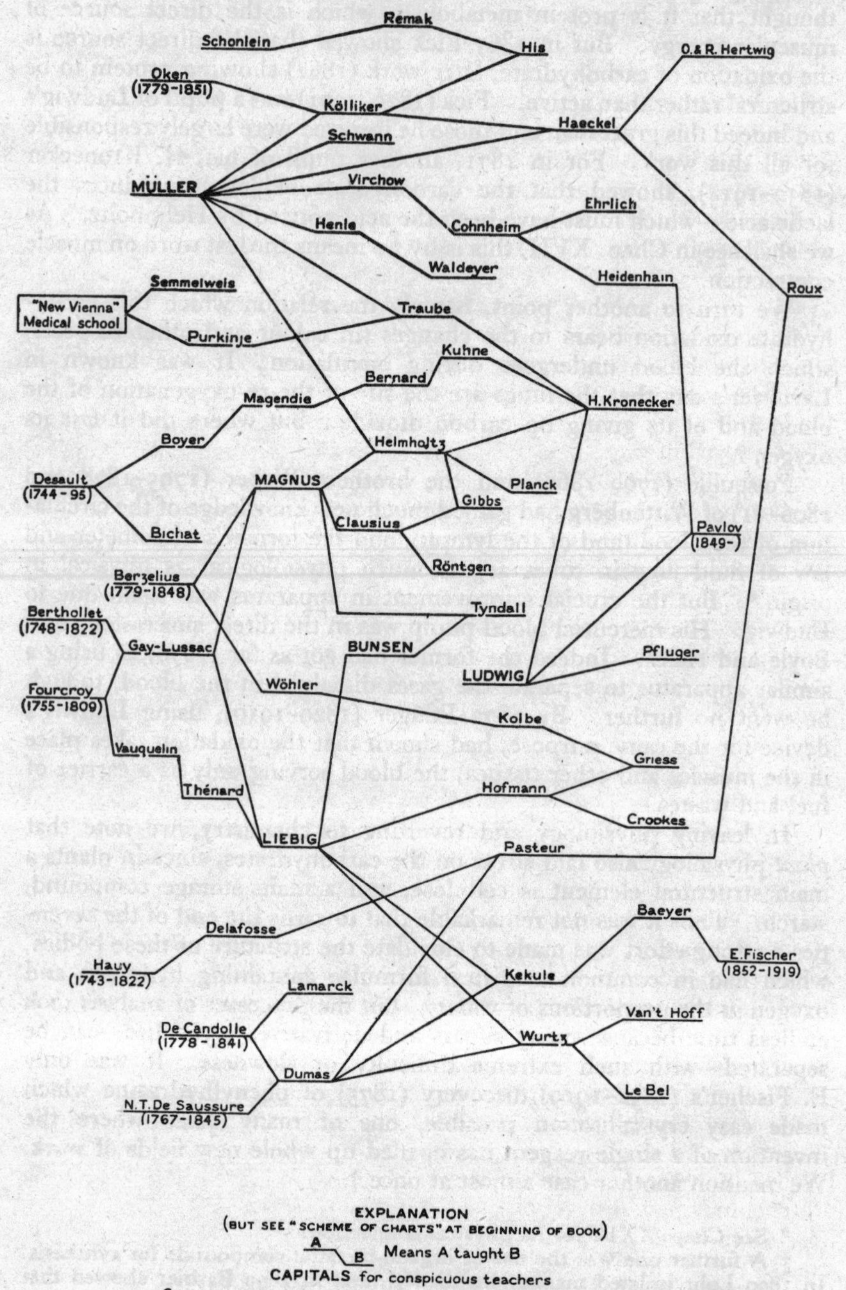

EXPLANATION
(BUT SEE "SCHEME OF CHARTS" AT BEGINNING OF BOOK)

A
B Means A taught B

CAPITALS for conspicuous teachers

Contemporaries are roughly, but only roughly one above the other

THE CONNECTION OF MASTER AND PUPIL. *Mainly Nineteenth Century.*

In 1880 and 1885 Kiliani found constitutions for some of the simpler sugars. Fischer began his main work in 1883. He found that sugars have the general properties of either aldehydes or ketones $\left({}^{H}_{P}\!>\!C{=}O \text{ or } {}^{R}_{Q}\!>\!C{=}O\right)$ and also of alcohols (—OH groups). He made bold use, among other facts, of the differing number of stereoisomerides offered by different formulae proposed, and concluded that the main feature of the molecule was an open chain of carbon atoms.

As early as 1883, however, Tollens had suggested a ring formula in which four of the carbons and one oxygen were involved. Ring formulae provided a further asymmetric carbon atom and so further optical isomers with which to account for the mutarotation of the sugars. They also suggested a reason for the production of the ring compound furfural when these sugars were distilled with strong acids. They caused, however, long controversy ; and it was 1926 before Haworth and his group produced satisfactory proof of the ring constitution. The basis of their method was a further advance in technique, namely, methylation. The methylated compounds can be broken down into others of known structure affording evidence of the structure of the parent sugar. The original methods of methylation, later improved by Haworth and others, were due to Purdie and Irvine, 1903 and 4.

The constitution of the higher carbohydrates, starch and cellulose, has proved a much more difficult matter, though methylation procedures were worked out for them by Denham and Woodhouse in 1910. Like the proteins, they are colloids of enormous molecular weight, offering therefore almost limitless possibilities in the arrangements of the atoms and groups concerned. We defer them to Chap. XXIII. Like Fischer himself from 1899 to about 1906, we turn to the proteins.

By 1899 it had long been known that hydrolysis of proteins always produces amino-acids, a class of substances of which a number of examples were by then familiar.* It was known, too, that these were fatty or substituted fatty acids, in which a non-acid hydrogen is replaced by the basic " amino " group $—NH_2$ (or a group containing it). These compounds act both as weak acids and as weak bases. Drechsel in 1889 introduced phosphotungstic acid to separate the more basic types, but only Fischer's device (1901) of fractionally distilling the esters under reduced pressure made his subsequent advances possible.

Fischer showed that these amino-acids could join in long chains which easily split up again. These " polypeptide " chains grew to have more and more of the properties of proteins as they grew longer. Fischer built up an eighteen-member chain, but the typical natural proteins may have from five hundred to a thousand, all of them selections from some twenty-five or thirty known amino-acids. We consider certain less chemical attacks on these huge molecules in Chap. XXIII.

Fischer, especially from 1894, had also been investigating the

* Glycine discovered 1820, tyrosine 1846, alanine 1849, etc.

" purines," uric acid and all the others, formed in the body when nucleins are broken up. In 1892 he succeeded in proving the formula suggested by Medicus (1875), based on the complex double ring :

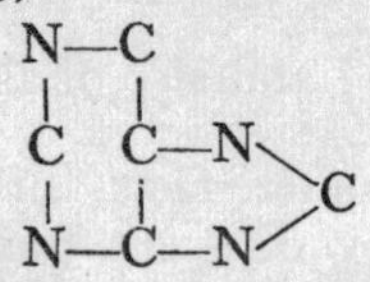

with hydrogens and oxygens attached in various ways and some linkages double. The number of such complex rings now known is great. For instance, the very important dye-stuff indigo, on which Baeyer did such brilliant work, consists of two

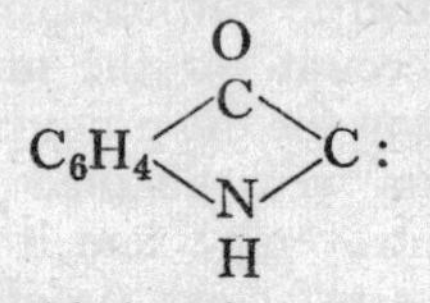

double rings joined by a double bond.

This use of constitutional or spatial formulae did not mean the full " physicalisation " of the concept of valency. For that we must await a later chapter (XXIII). But the idea was played with. By 1892 Kekulé, developing his earlier idea, had realised that a valency bond must really be a path of vibration, not the stem of a dumb-bell. In fact he defined valency as proportional to the number of collisions per unit time of an atom with its fellows in a molecule. On this basis he gave reasons why his formula for benzene need not allow the undesired second di-substitution product. Laar, similarly, supposed that the tautomerism of aceto-acetic ester was due to the actual vibration of the hydrogen atom between its two positions. But such ideas were not yet the leading ones, and the first effect of the new physics of the 20th century was to make them harder, not easier, to develop.

The *geometrical* concept of the atom, however, did take a further step, with which this section may close. If a carbon atom really does lie at the centre of a tetrahedron with its four valencies directed to the vertices, the angle at any one of these latter is about 110°. In 1885 Baeyer pointed out that this is in good agreement with the stability of the benzene ring in which successive valencies make an angle only 10° different from this, 120°. The benzene ring thus gradually came to be envisaged as an actual flat ring of atoms. In 1897 Collie showed that on this view not all the proposals for benzene's formula really conflicted. Similar considerations applied to sugars show that an " open " chain of six members would actually curl round very nearly into a ring, thus making Haworth's formulae, and also the natural prevalence of six-membered sugars, entirely natural. Towards the end of the 19th century Wegscheider suggested an explanation on the same lines of the slowness of certain reactions involving alkyl groups. These occupy a larger solid angle than hydrogen atoms and offer " steric hindrance " to access to the interior of the molecule.

Chemical Apparatus

We have at all points tried to stress the importance of instruments and of technology, and at certain stages we have paused to consider the state of standard equipment, particularly in chemistry. We now do so again. But the subject deserves a book in itself. For by the 19th century large-scale production had made many things easy for the instrument-maker which would be impossible for the home constructor; and there are a dozen such considerations which make instruments follow, to a degree, laws of their own. To avoid even the appearance of systematic treatment, we shall hang our notes on certain pegs, and in particular on the names of Bunsen and of Stas, and on the key-words, platinum, gases, and analysis.

Bunsen (1811–99) was the universal provider of the middle of the century. He worked in close contact with industry and economic need. In connection with blast furnaces he developed (1838–45) the definitive methods of gas analysis. When town-gas spread to his university in the middle fifties * he gave the world his burner. Until then Berzelius' spirit-lamps or Argand burners had been used for small purposes, charcoal furnaces for large ones. By the seventies, makers' catalogues in England were quoting Bunsen burners as standard products. It was economic considerations, too, which prompted Bunsen (1841) to invent his cell. This involved finding methods of making carbons not disintegrated by nitric acid, and so capable of replacing the costly platinum plates of Groves' cells. Large-scale primary batteries were thus made possible, and became standard practice until town current ousted them in the present century.

We have already discussed Bunsen's work in spectroscopy. His grease-spot photometer came in 1844, his ice-calorimeter in 1870. From his filter pump we may trace the vacuum pumps later so important. His iodine and sulphurous acid method was a great step in volumetric analysis.

Vacua call to mind corpuscular physics (Chap. XX), the liquefaction of gases and (1894 onwards) the separation of the rare constituents of the atmosphere. We mentioned pumps in Chap. XX. By 1898, the work of Dewar, Hampson, Linde, and others brought in the *large-scale* production and handling of liquid gases, Dewar's vacuum container being one vital advance here. A different side of gas technique is illustrated by Victor Meyer's method for vapour density (1876–7).

Let us glance aside for a moment at platinum, the standard resource for resistant vessels in Berzelius' and Faraday's time. By the second half of the century its alloys with its fellow "noble" metals, and especially with iridium, were being studied, for their even higher melting-points (Victor Meyer), for their suitability (Stas and Deville, 1877–9) for international standards. Standards, and units, ought to be further key-words of that time, preparatory to the present reign of hyper-accuracy.

* Much later than in England. Faraday could say in 1827 that "most large towns" have gas.

Hyper-accuracy brings us to Stas and also to the handling of gases. As to the latter, the name of Regnault was mentioned in Chap. X. His counterpoised globes for gas densities date from 1845, but Stas was the first to apply all the corrections in full. The date of Stas' paper on accurate atomic weights was 1865. Occlusion of gases by reagents (e.g. silver) and vessels was one great concern of his. He often substituted platinum for glass, but he also investigated the latter with a view to improving or controlling its adsorptive tendency and its vulnerability to reagents. Any and every precaution became his study. He prepared his reagents in several different ways (thereby anticipating the methods of rare-gas and of isotope research), he used extra large quantities. And this last may bring us to our concluding note. For, more recently, an extraordinary importance has come to attach to accuracy while using very *small* quantities, to micro-analysis.

But with analysis we can hardly deal. Fresenius (1841, 1846) systematised it. A later tendency has been the introduction of physical methods on the quantitative side, electrolytic, colorimetric, and the like. These are used especially for swift routine work, as in industry, medicine, or agriculture. One aim of modern analysis is to find a coordination compound of each element suitable for its detection and estimation.

As to micro-technique, we have noted Marggraf's use of the microscope on sugar beet in the 18th century. Raspail did chemical tests on starch-grains under the microscope in 1825. With the serious biochemistry of the last thirty years of the century came a great demand for, and some supply of, micro-methods. Later, from 1900–11, Emich perfected the technique of handling a drop or so of liquid and precipitate and of making quantitative measures upon them. The classical methods of quantitative organic analysis were definitely extended to minute quantities by Pregl (1910–16 and later), with the help of Kuhlmann's micro-balance. They have been greatly refined in our day by Linderstrøm-Lang and the Copenhagen School. But, as usual, the all-important details cannot be given here.

CHAPTER XV

SURFACES AND IONS

We now turn to physical chemistry.

Chiefly Thermal

It will be recalled that phlogiston tried to treat the thermal and chemical aspects of reaction together, and that when Lavoisier clinched the latter aspects, he overshadowed the attempts of men like Berthollet on the former. Other non-gravimetric sides of the subject shared the same eclipse. These included " affinity." Wenzel, in one of the old affinity tables (1777), had tried to correlate the " strengths " of acids with their *rate* of attack on metals, and he, and also Berthollet, observed roughly that the rate of reaction varies as the mass of reagent present. They did not see that concentration, not actual mass, is the vital factor.

Berthollet's interest in what we now call balanced reactions led him to that actual denial of fixed proportions which discredited his line of interest for over half a century. During that half century chemists were fully occupied in substantiating their claim that fixed proportions could in fact bring order into the subject ; but when (*c.* 1860, see last chapter) they at last succeeded in doing this, they soon turned to the other subject. By that time, physicists had given them kinetic and thermodynamic ideas. The concepts of dynamic equilibrium and of energy relations between compounds at last found their atmosphere. We treat the former first.

It was known early in the 19th century that in the presence of acids, cane sugar " inverts " into the simpler sugars dextrose and laevulose. Since the acid is only a catalyst (see later section) there is really only one molecule, the cane sugar, taking part if we ignore the water, the concentration of which does not change appreciably in dilute solution. This is one peculiar feature about the reaction. Another is that it takes a very considerable time, getting slower and slower as it goes on. In 1850 Wilhelmy observed that the rate decreases, in fact, logarithmically with the time. This suggested that it depends, essentially, on the number of molecules of cane sugar left unchanged.

Now entered the idea of dynamic equilibrium. In 1850, also, Williamson saw that in balanced reactions, of which a number were known by then, what happens is that the products of reaction are formed at a certain rate dependent on the current quantities of the initial substances, and then proceed to recombine into these latter at a rate similarly dependent on *their* current quantities. These opposite reactions both go on, the first at a decreasing, the second at an increasing, rate, until equality of rate, and so balance, is reached. This is a bimolecular, as Wilhelmy's was a monomolecular, case.

This was a new type of reasoning in chemistry, and as such was by no means accepted at once. Nor did the more systematic work by Berthelot and St. Gilles in 1862, on the balanced reaction of acids with alcohols to form esters, gain more attention. It was 1863, and especially 1867, before the Norwegians Guldberg and Waage stated, and extensively tested, the general " Law of Mass Action " that if [A] represents the concentration of A (and so on) in the general balanced reaction at any given moment

$$nA+mB\ .\ .\ \rightleftharpoons pP+qQ+\ .\ .\ .$$

balance ensues when

$$[A]^n\ [B]^m\ .\ .\ .\ =k[P]^p\ [Q]^q\ .\ .\ .\ .\ .$$

Of this, Wilhelmy's result is a particular case, for the expressions on either side are proportional to the rates of the forward and back reactions at any given time.

We now turn back to the thermal side of the question, which is bound up with that of physical changes, such as melting and vaporisation. In these there had been more continuity of development since the beginning of the century, and it is necessary to glance at thermometric and calorimetric technique since 1800. This, in turn, is bound up with the gas laws.

It had long been known that the equation $pv \propto T$ is in no respect exact; while Dulong and Petit's comparison (1815) of mercury and gas thermometers, and investigation (1822) of creep of zero and other errors in the former, brought the whole question of exactitude in this vital matter to the front. The man who symbolises the whole urge towards precision in physics is the Frenchman Regnault (1810–71). Regnault was an inspirer of William Thomson. Power industry was a main stimulus of both. While the master adopted the gas thermometer as his standard, the pupil endowed it (as we have seen in Chap. X) with an exact relation to the new thermodynamic theory.

Regnault showed that pv varies with p, and Natterer (1844–55) showed that it has a minimum value at each temperature. In 1870 Cailletet, and later Amagat (1878–84), made more and more exact investigations with greater and greater pressures. Meanwhile Andrews had correlated the pv, p curves with the liquefaction of gases (by Faraday and others) and with the prevalent concept of the continuity of nature. He showed (1863, 1869) that as temperature falls or pressure increases this variation assimilates the behaviour of a gas more and more to that of a vapour and so ultimately to that of a liquid.* One point to note is that the behaviour of a substance in all three states, liquid, vapour, gas, began to be represented on one p, v, T graph.

Regnault had made great improvements in calorimeters as well as in thermometers; and with these better instruments Berthelot (1827–1907) revived the thermochemical interest of his similarly-named fellow-countryman Berthollet. He determined, in the sixties, many heats of reaction, and examined their bearings. One useful fact in the chemical thermodynamics thus begun had already been pointed out by

* For the work of Van der Waals, see Chap. XXIV.

Hess in 1840—that the heat evolved in obtaining substance B from substance A is the same by whatever stages or "path" the reaction is carried out. This was seen by the sixties as merely a case of the First Law of Thermodynamics, not enunciated in 1840, but it already possessed one feature of laws in this subject which called for great caution in use: it was strictly true only if the resulting A's were the same in *every respect* (for instance, in crystalline form) in the cases compared, the same applying to the B's. Equally characteristically, it retained, for all that, much practical use.

The central hope and aim of chemical thermodynamics was that of deducing the course of a chemical reaction from its thermodynamic constants. At first Berthelot, and Thomsen of Copenhagen (who, 1853 onwards, was working on very similar lines) assumed simply that heat of reaction measures affinity, and that the course of an action is the one which evolves the maximum heat. It was some time before this error lost influence; though its fallacy could have been seen from the much earlier work of Berthollet on balanced reactions. When it did so, with Horstmann (1869 and 1873) and especially Willard Gibbs (1839–1903), the subject entered on maturity, and the realisation grew that no simple relation could be found completely governing chemical by thermodynamic factors. The subject has indeed become too complex to be followed here, and the work of (for instance) Nernst will not be treated.* There follow a few words on some of the work of Willard Gibbs.

This referred not to the homogeneous systems concerned in the Law of Mass Action, but to heterogeneous ones, like solutions in the presence of excess of solute or of their own vapour. That is, it referred to systems separated either physically or chemically into "phases." The physical cases of Gibbs' Phase Rule may be simply illustrated from the p, v, T graphs mentioned above; though it was the essence of Gibbs' own methods that he did not present them in this simple light. Instead he deduced them from an elaborately abstract presentation of the general principles of thermodynamics, which he placed on a new basis, free from that association with heat engines which had been their birthmark. It was already well known by Gibbs' time that the vapour-pressure of a liquid when no other gaseous "component" (C) is present depends only on temperature. Thus, these two homogeneous "phases" (P) can be in equilibrium at any point on a line in the p, T plane—such a system has one "degree of freedom" (F=1). The same is true for one C if the P's are (for instance) solid-vapour or liquid-solid, but it is no longer true if there are two C's; while three P's with one C can be in equilibrium only at a point (F=0). This simple "topology" of the p, v, T diagrams was crystallised by Gibbs into the very simple "Phase Rule" (1877) $F=C-P+2$.

Towards the end of the century, p, v, T curves were being increasingly used, for instance in chemical industry, to predict qualitative

* One point may, however, be mentioned: the Le Chatelier-Braun principle (1888) that if one factor in an equilibrium changes, the equilibrium shifts in such a way as to tend to annul the effect of the change. This is a principle following a well-defined thermodynamic pattern.

features of reactions involving several components, in ignorance of Gibbs' very abstract work,* which had thus made detailed plotting needless and brought an extreme simplicity into the question. The abstractness and the simplicity were apt in marginal cases to vanish. It is for this reason that we have not attempted watertight definitions of F, C, and P ; for so many provisos have to be made in these that no small empirical knowledge of a given case is often necessary, after all, before the rule can be used. One important point is that it applies—like all thermodynamics in the strict sense—only to " true equilibrium," and that there are " false " equilibria which look very like true ones.

Chiefly Electro-Chemical

The electrochemical interest of Davy and of Berzelius will be remembered. In 1832 Davy's assistant, Faraday, made a series of investigations on electrolysis by which he showed that unless side reactions set in, the weight of a substance deposited or dissolved at an electrode depends only on the *quantity* of electricity which has passed, not on either time or current separately. He also found it generally proportional to the equivalent weights currently assigned to the substances concerned.† He therefore concluded that a certain fixed amount of electricity is associated with each " ion."

Meanwhile Grotthuss' old mechanism for electrolysis was wearing thin. It required the two " ions " to move at equal speeds, and in 1859 Hittorf found that they do not do this. The concentrations round the electrodes change at different rates. Grotthuss himself had already pointed out (1806) another difficulty in the implied notion that the E.M.F. tears the molecules asunder. For in cases where the complexities of back-E.M.F. are avoided, as with copper electrodes in a copper sulphate solution, the least E.M.F. starts the process, though of course slowly. This would certainly not be expected.

In 1857 Clausius, with the full impulsion of the new energy-theory behind him, stated the trouble more clearly. In the case mentioned, Ohm's law was obeyed, which from Clausius' new point of view meant that no new " sink " of energy was involved. Hence he suggested that some of the molecules must be already dissociated before the E.M.F. is applied. Such dissociation, in the face of the powerful attractions at work, did not make much appeal to chemists, who were not conversant with the then new kinetic viewpoint. Nearly a generation passed before they had to admit its relevance.

Already, however, the position had ceased to be simple. For instance, Hittorf pointed out that the ordinary chemical tests of stability of a compound bear no relation to the ease or difficulty of its electrolysis. Strongly bound salts like copper sulphate were indeed the most natural choice for electrolytic demonstrations. This recalled the difficulties felt in the old ideas of affinity and of the strength of

* Maxwell had called attention to some of Gibbs' results in 1876, Roozeboom to others a decade or so later.

† The contemporary difficulties over this subject should be borne in mind. See preceding chapter.

acids. In the sixties, as just mentioned, interest in these ideas was reviving.

Hittorf had the idea that conduction consists in processions of charged particles moving towards their respective electrodes at different rates, instead of in series of consecutive interchanges of partners, as in the pre-kinetic Grotthuss theory. Unfortunately conduction in the body of electrolytes is hard to examine: we must insert electrodes, and they bring side-reactions and unknown back-E.M.F.'s. It was 1879 before Kohlrausch had perfected the technique of using alternating currents to avoid these latter, using the then very new telephone * as an A.C. "galvanometer." He developed the Hittorf theory and found that, if it gave the mechanism correctly, the velocity of each procession for a given E.M.F. could be regarded as characteristic of each ion, and as unaffected by the other: conduction was additive.

We must now turn to two other lines of research which, hitherto independent, were destined to converge with this one. It had long been known in practice that sodium chloride lowers the freezing point of water; and as early as 1788 Blagden had shown that for any one salt the lowering varies as the concentration. In 1881 Raoult showed that it also varies, in a large number of cases, inversely as the molecular weight of the salt dissolved. This latter point was widely taken up, for it afforded an addition to the all-too-few means of finding that vital number.

Unfortunately the law, though true for organic compounds, soon began to fail completely for inorganic ones—that is (as we *now* see) for the very ones with which the ionic theorists were concerned. Many of these seemed to behave as if they consisted, in solution, of twice the right number of molecules of nearly half the right molecular weight. Sooner or later the likeness of this to the Deville (1864–5) phenomenon of gaseous dissociation could not fail to be seen.

The second line of research was one of those which we have mentioned as concerned with membranes: osmotic pressure. In 1748 the Abbé Nollet, immersing in water a membrane enclosing a sugar solution, noticed that the water diffuses in. It will, in fact, exert pressure in order to do so. By 1826 Dutrochet had recognised the importance of this phenomenon in connection with the (then very new) cell theory, for an imperfect form of which he was himself responsible (see Chap. XI). He began to make quantitative measures, but to do this was difficult. The phenomenon implied that the membrane stops the solute molecules but not those of the solvent ("semi-permeability"). And this has proved a difficult ideal to combine with the strength necessary to withstand the pressures developed.

Both Nollet and Dutrochet had used animal membranes, and in 1867 this gave Traube his clue for the crucial advance in technique. In 1854, as we shall see later, Graham had stated the resemblance of animal to "colloid" substances. Traube then developed the method of depositing a colloid in the pores of a porous pot, which method enabled the German botanist Pfeffer to make (1877) the first fairly

* Reis, 1861; especially Bell, 1876.

reliable measures of osmotic pressure. Real exactitude has proved very difficult to achieve, but certain points vital in our present connection were established without it.

Early in the eighties the Dutch botanist De Vries, true again to the physiology-physics-chemistry connection, began in effect to use plant cell-walls as semi-permeable membranes and the cells themselves as indicators of osmotic pressure. The solutions in which they were immersed had the same osmotic pressure as that within the cells if these latter neither shrank nor swelled on immersion. He connected this up with Raoult's work. He found that such " isotonic " solutions, whatever the salts dissolved in them, had the same (lowering of) freezing-points. In fact both osmotic pressure and lowering clearly depended rather on the number, than on the nature, of the molecules.

In 1884 he mentioned the matter to another Dutchman, Van't Hoff (1852–1911). Now on the kinetic theory, the ordinary pressure of gases is also a matter of number, not nature, of molecules. In 1885 Van't Hoff succeeded in uniting these several lines of thought. He made the capital advance of showing that the relation of osmotic pressure p to concentration, if expressed in terms of the volume v " occupied " by each solute molecule, takes precisely the form of the gas laws :

$$pv = \mathrm{RT}$$

In many cases R actually has the same value as for gases, so that the solution with the membrane behaves as if the solvent were abolished and the solute molecules formed a gas.* But in many, R had to be multiplied by a factor ; and this again happened primarily in the case of the simplest inorganic salts. Van't Hoff did not solve this problem, but he tabulated the factors. The situation was thus that the exceptional cases irresistibly called to mind the electrolytic line of research, to which we now turn once more.

In 1881 Helmholtz, using the greater precision which ideas had acquired in the interval, gave Faraday's ions the interpretation that each molecule carries a certain small definite number of " atoms of electricity." About five years later the young Swede Arrhenius (1859–1927) entered the field. He pointed out that the conductivity of salt solutions is only sometimes as great as would be expected from Kohlrausch's transport numbers. If a table of conductivity factors was prepared, it bore an unmistakable resemblance to Van't Hoff's table. The question thus became : how this was to be explained if ions—if dissociation—only occurred when an actual E.M.F. was applied, and had no relevance to osmosis or freezing ? To answer this question Van't Hoff was led, in 1887, to put forward the idea that electrolytes in solution are permanently dissociated into pairs of oppositely charged ions.

The number of fields in which this theory was successful forced its

* pv, indeed, was not quite constant ; but then neither was it for gases (see Chap. XXIII). The gas-picture does not at once show why the solvent comes in from outside. The point is a little complex (see, e.g., Guggenheim, " Modern Thermodynamics ").

acceptance in spite of its obvious handles for attack. The attractions between the ions was the greatest of these; and such attractions have, in fact, proved to be of importance in its more recent developments, though not in the way urged by objectors. The theory explained why strong acids and bases, neutralising each other in dilute solution according to the formula (say)

$$H_{2n}A + B(OH)_n = BA + nH_2O$$

behave thermally as if the only reaction occurring were the formation of the water from hydrogen and oxygen. For if C is as much dissociated as were A and B, this *is* the only reaction. It also enabled advances to be made in the old question of the strength of acids. Degree of ionisation was seen to be closely related to "strength"; but by this time it was, of course, realised that strength in a particular reaction is a matter of many factors, such as solubility or volatility.

Concentrations of ions now took the place of those of molecules in some Mass Action equations, even a weak acid dissociating further if (for instance) formation of an insoluble salt takes up the ions already present. Water in the pure state, being a very poor conductor, was assumed to be very little dissociated (10^{-7} grs./litre of hydrogen ion). This application of the Law of Mass Action to ions was a great advance, and few quantitative procedures can have had so many applications in chemistry. As an instance of the type of new result in which this and similar arguments issued, we will mention the "Donnan equilibrium," though it belongs to a rather later time.

In 1890 Ostwald pointed out that unforeseen effects may be expected when a system of electrolytes is divided by a membrane permeable only to *some* of the ions present; but the point did not rivet attention until in 1911 Donnan and Harris used it to explain an anomalous value obtained by Bayliss in 1909 for the osmotic pressure of Congo Red. Bayliss' method had involved the use of sodium chloride, the ions of which, unlike those of the dye,* diffused through the membrane which was used. The essence of Donnan's effect was that the sodium chloride ions came, in equilibrium, to a lower concentration on the side containing the dye than on the other.

Donnan was able to show that this followed from thermodynamic principles, and to erect a quantitative theory. It was later found that, in essence, the case had already been dealt with by Willard Gibbs. Though the effects are often referred to as membrane equilibrium, Donnan's argument is not especially concerned with membranes, but with any case of ions differentially restrained from assuming a uniform concentration.

Since the War, Donnan has made many applications of his idea, especially in physiology. For the ion, and especially the hydrogen ion, concentrations of the salt solutions of the internal environment have become of manifold importance. In particular, as is easily seen from the law of mass action, weak acids and bases can act as "buffers," playing in fact a large part in securing that constancy of the internal environment which Bernard had proclaimed. Carbonates, phosphates,

* For Congo Red is an electrolyte.

and proteins themselves, are prominent in this capacity. The same principles have been applied to the aqueous solutions which pervade soils, and which are in immediate osmotic relation to the tissue solutions of plants.

Colloids

These advances drew attention to " colloids," already mentioned. The properties of these had been unconsciously used in several immemorial industries, such as that of leather. Moreover, the alchemists, by working with body-fluids, accumulated data which could have been systematised by anyone with the necessary line of interest. It was just this latter which was lacking until Graham's work in the fifties. The idea of a " colloid solution " could not, in fact, develop until that of an ordinary (" crystalloid ") solution had been so clearly defined that certain cases appeared as exceptions.

Berzelius, however, following up work by Richter (1802), on Purple of Cassius, did begin to define a line of research. He noted that some substances, such as some forms of sulphur and of silicic acid, which are insoluble in the ordinary sense, could be got into a condition mid-way between fine suspension and actual solution. Selmi who, 1844 onwards, investigated sulphur and Prussian blue, called this state pseudo-solution (1847). Baudrimont (1844–6) was another early worker.

Faraday (1857) prepared and investigated gold in this condition. Gold is not normally a colloid, so that by implication Faraday had reached the later point of view which speaks of colloidal states rather than substances. Neither he, however, nor Graham, who was equally aware of the point, always recognised it. In fact, they had not reached the outlook which makes it a vitally important one.

Graham's distinction, which " made " the subject, related, like osmotic pressure, to membranes. He found that crystalloids do, and colloids do not, readily diffuse through parchment membranes. All degrees occur, but this point, again, was not adequately recognised by Graham.

It was found that many agents, thrown into colloidal solutions, precipitate them, presumably by causing them to coagulate into much larger particles. Some solutions needed much less of such agents than others to effect this ; and in some cases the process could be reversed, in some not. The subject, then and since, has proved to be one in which it is much easier to find facts than to coordinate them under simple rules. Hence the treatment of the subject on our present scale must be especially unsatisfactory. Certain simple rules have indeed been found, such as that resulting from Schulze's work in 1882 on the relation of the coagulative powers of different substances to the valency of the atoms concerned. But the main advance in the subject only began at the end of the century when, as we have seen, a general theory of solutions became possible.

In this advance, two particular realisations helped, both of them cited in the title of this chapter. One was that solutions were shot through and through with electrical effects, colloidal solutions no less than others. It began to be seen that the particles of colloidal (as indeed of coarser) suspensions, are electrically charged. In 1892 Linder and Picton showed that under an applied potential difference colloid particles move to one electrode (" cataphoresis "). This charge was in fact the reason why they did not come together and coagulate. It began to be seen that it is as *ions* that the coagulants act, neutralising the charges on the particles. That, when neutralised, the particles should coagulate into larger ones followed from the second realisation, that with colloids we are essentially dealing with the phenomena attendant on enormous specific surface.

A surface always has energy associated with it, and thus coagulation reduces potential energy. The train of thought which had enabled Galileo to explain the difference in shape of fleas and elephants, and Le Sage to explain the relation of size to (surface of) evaporation in plants and animals, thus found a new application. Colloids exerted almost no osmotic pressure, a fact which pointed, as we have seen, to enormous molecular weight. The question, in fact, arose, whether their particles might still not be (as in ordinary solutions) actual molecules, though of enormous size. This point was tested in 1903 in a way which brought colloids directly into relation with kinetic ideas.

As early as 1827 the botanist Brown had noticed a " perpetual motion " prevailing among the tiny particles of a solution which would later have been called colloidal ; and in 1879 Ramsay pointed out that this was but a visible proof of the kinetic theory for the case of relatively large and slow particles. He was met with doubt ; but in 1888 Gouy showed that other explanations, such as unnoticed vibrations of the support, could not explain the motion. In 1903 Siedentopf and Zsigmondy, followed the motions, not indeed of the colloidal particles, but of the diffraction patterns which they give when intensely illumined (" ultramicroscopy "). They thus assured themselves that the particles could not, at least in the cases examined, be actual molecules, but must be aggregates of them.*

By 1903 the idea of state rather than substance had gained ground among such workers as Zsigmondy. It was especially stressed, with the idea that all gradations occur between the colloid and crystalloid states, by von Weimarn (1905–8).

Another property of colloids was found to be their ability to take up large quantities of other substances. This, again, associated them with other states characterised by fine subdivision and enormous specific surface. Adsorption (and, see later, catalysis) by charcoal or finely divided metals like platinum had long been a well-known phenomenon. Such considerations led Ostwald (1907) to bring in the idea of treating all such surface phenomena together, classifying them by whether gas, liquid, or solid occupied the " disperse phase " or the " dispersion medium " (the former discontinuous, the latter continuous). Thus " mists," " dusts," and the like became particular cases of a large class, colloids but another.

* This " Brownian movement " was used by Perrin (1910) to estimate the numbers and sizes of atoms and molecules.

It was, however, more important and more difficult to classify colloids than to generalise them. Many classifications were proposed, but we shall confine ourselves to one, that of lyophile and lyophobe proposed by Perrin 1903 to distinguish the fairly uniform class in which a tiny ionic concentration irreversibly coagulates the solution (lyophobe) and the more varied class in which a high concentration is needed and the effect can at least sometimes be reversed (lyophile). These latter are usually organic, and it is they which give the subject its enormous importance in physiology. Proteins themselves are colloids, as are also certain vital constituents of the soil (see later chapters). Hardy in 1900 showed that by gradual addition of acids to such colloids an "iso-electric" point is reached when the charge on the particles changes sign. The point is one of instability, and coagulation tends to occur. This iso-electric point is of great importance in distinguishing proteins from one another. It is marked in each protein by a fairly constant and characteristic value of the hydrogen ion concentration.

In concluding this section, we may mention that the state of matter within the colloidal particle was early the subject of speculation. We shall see in Chap. XXIII that there has lately appeared definite evidence for von Weimarn's view (1907) that this state is crystalline.

CATALYSTS

More consciously than colloids, *small quantities* of certain substances have been used by man since very early times. The classical civilisations, for instance, were well aware of both acetic and alcoholic fermentation. The general connection between science and fermentation has, as we saw in Chap. XI, been close. Alchemy, too, abounded in ideas of the great effects of small quantities of "quintessences." But the *science* of substances which can produce their effects in minute amounts can hardly be said to begin much before 1800. Spallanzani's work of 1783 dealt, as we should now say, with enzymes; but a better initial date is 1794. In that year Mrs. Fulhame showed the necessity of traces of water for several reactions, among them the oxidation of carbon monoxide. Another point later taken as essential emerged in 1812 when Kirchhof showed that the acid which promotes the change of starch into sugar is unaltered at the end of the reaction. The same point was implicit in the use of the idea of an intermediate compound in explanation of the action of the nitrous oxides in the chamber process for sulphuric acid (Clément and Désormes, 1806).

Thénard's demonstration in 1818 that hydrogen peroxide is stable only in the presence of acids afforded a new type of case. Yet a further type emerged when Davy (1817) and Erman (1818–19) showed, at increasingly low temperatures, the catalytic action of finely divided platinum on oxidation, the climax being Dobereiner's proof (1822) that it will enable hydrogen and oxygen to combine at room temperature. Dulong and Thénard showed (1823) that many other finely divided metals, especially noble ones, are only less effective. By 1825 another modern point was being hinted at: Henry was investigating the "poisoning" of the platinum by many agents, ammonia, sulphur compounds, etc. By 1831, Phillips had patented the platinum catalytic process for commercial sulphuric acid manufacture. But this case of Phillips gives us the key to the history of catalysis at this time: it was still in the stage when the workers are not aware of each other. The actual use of the Phillips process was held up *until about 1900* by neglect of Henry's point of view.

In 1830 the original fermentative connection was taken up when Dubrunfaut showed that extract of malt had the same action with starch as have acids. And in 1833 Payen and Persoz isolated the active substance "diastase" from the malt. In 1830, too, Robiquet and Boutron-Chalard show that extract of bitter almonds hydrolyses amygdalin. In 1833 Faraday suggested one of the mechanisms still held to underlie such catalytic effects as those of platinum, surface adhesion of the reactants to the catalysts. But still a conscious and embracing viewpoint was wanting. This was supplied by Berzelius in 1835; and this date is therefore the real beginning of the subject.

Berzelius' idea of a special "catalytic force" was opposed by Liebig, who supposed that vibratory mechanisms already not unfamiliar in science could account for the effects, thus may small troops of soldiers break down large bridges by walking over them in step. Later workers have been forced to recognise different explanations in different instances, as when Williamson revived the intermediate-compound idea in 1854. The cases which, about then, were causing most argument, were those connected with living creatures. Such were Pasteur's micro-organisms. Each new class of chemical agent discovered in living matter seems destined to raise up the old vitalistic hope of a sacrosanct region *within* the material world. It happened here. Only from about 1897 did Buchner and Hahn show that cell-free extracts can produce the same effect as the organisms originating them; but fortunately Kühne's definition of "enzyme" had, much earlier, cleared up a good deal of the difficulty.

By the eighties studies both of these and of inorganic catalysts were becoming very numerous.* We cannot cite them in detail, for, as with colloids, they are too varied; but in 1888 Ostwald raised a point of principle by introducing the then coming concept of reaction velocity into the subject and stating (without universal assent) that catalysts and enzymes do not initiate reactions nor alter their final equilibrium point, but only change their speed.

Fischer emphasised (1894–5) how exceedingly specific most enzymes are in the "substrates" upon which they will act, contrasting them in this respect with most inorganic catalysts. Faraday had already suggested that catalysis by platinum and the like was a phenomenon of high specific surface, adsorption bringing the reactants into close proximity. In 1906 Bayliss urged this as a primary feature of enzyme action also, since enzymes are colloids. Later developments, while not diminishing the significance of the surface idea, have tended against a purely physical interpretation of it.

* For instance, in 1877 Friedel and Crafts showed that aluminium chloride catalyses the alkylation of aromatic compounds. In 1901 Perrier isolated an intermediate compound in this case.

One consequence of the colloid nature of enzymes is the high importance of ions in relation to them, as revealed by Cole (1904), but especially by Michaelis, and also Sörensen, 1907 onwards. That ions themselves act as catalysts had been contended by Arrhenius as early as 1887. For, in cases like the inversion of cane sugar, the effectiveness of acids is in the order of their strengths, that is, of their hydrogen ion concentrations. The rise of these latter to importance may indeed be conveniently dated from about 1907.

It should be noted, as a point already realised by Faraday's time, that effects like those of catalysis are not confined to chemical change. The effects of "seeding" on crystallisation, and of scratching the relevant surfaces on boiling or efflorescence, are not recent discoveries.

Before turning to enzymes, we mention certain sensational catalytic effects of moisture. In 1869 Wanklyn showed that violent reactions like that of chlorine and sodium will not occur at all in complete dryness. In 1880 Dixon reaffirmed Mrs. Fulhame's case of carbon monoxide, and from 1886 Baker began to accumulate cases. His later work on changes in *physical* constants under extreme dehydration was positively alarming, since it seemed to invalidate much-relied-on analytical constants like boiling points. The explanation that two or more components with different physical constants were present, the proportions of which were altered by the water, was little comfort. More reassuring were the strenuous efforts needed to exhibit the effects. Baker's work had other sides. For instance, about 1912, he secured evidence that certain long-standing puzzles about the sulphur oxides were due to the fact that they had always been examined when imperfectly dry. Such facts suggest the importance of catalysis in inorganic nature. Other catalysts in this region which have received attention lately include alkali and alkaline earth chlorides and phosphates in their effect on silicate formation.

We have seen in the last chapter that Pasteur knew of an extreme case of Fischer's enzyme specificity, only one of a pair of optical isomers being acted on by a particular enzyme. Further cases had accumulated by 1900. On this and other subjects, a keynote of much of the work since about 1895 was the relation of enzymes to the idea of balanced reactions then taking this, like other branches of chemistry, under its dominion. For, given Ostwald's definitions, an enzyme must catalyse *both* directions of a reversible reaction, and so, in proper conditions, be capable of synthesis, as well as of hydrolysis. In fact, in 1898, Croft-Hill showed that maltose can synthesise the sugar it hydrolyses; and in the next fifteen years the general doctrine came to be accepted. The same dynamic point of view led Fajans to suggest in 1910 that an enzyme really acts on both optical isomers, though at different rates, an assertion supported by work (1909, 1910) by Neuberg and Pringsheim on bacteria and fungi. The specificity of enzymes is still a puzzle. It is much greater in some cases than in others, and may relate rather to linkages than to actual compounds or radicles (see Chap. XXIII).

We refer in conclusion to a feature which, long known as to inorganic catalysts, was found by Magnus in 1904 and by Harden and Young, 1906, among enzymes also. This is the fact that their action may not only be poisoned by traces of certain substances, but may be promoted by traces of others. Thus the catalysis of the " water-gas " reaction

$$CO+H_2O \rightleftharpoons H_2+CO_2$$

by iron, nickel, or cobalt, is promoted by (for instance) chromium oxide; while Baker in 1902 found that *perfectly pure* water will not enable hydrogen and oxygen to explode together. The phenomenon is so widespread that the search for promoters is now a regular activity in industrial catalysis.

The terms coenzyme and antienzyme were criticised by Bayliss in 1914 on the ground that the substances in question were not enzymes; but here we need only note that it was not in dispute that certain substances or groups, such as the sulphhydryl group (see Chap. XVIII), and certain heavy metals like iron, are necessary to some enzyme actions.

CHAPTER XVI

CYTOLOGY AND GENETICS

THE next four short chapters are biological. It is hoped that the chapter divisions will not obscure the close interrelations of the subjects. Another division might equally well have been chosen.

We take up cell theory where we left it, at the point where cell-division ("mitosis," 1882 *) had become the crux. Its clearing up (*c.* 1880–5) depended on the new microtechnique already described under bacteriology. Among the chief names were Flemming, Van Beneden, and Strasburger.

We have seen that attention had already begun to centre on the nucleus. The rest of the cell ("cytoplasm," 1882) was the seat of the more variable factors, and so was less suitable as a starting point. Kölliker had early investigated what happens to the nucleus when the cell divides. At first, it had seemed to disappear, reappearing in the daughter cells. But presently, even with the older micro-technique, it had been seen dividing, in some cases at least, like the cell itself (Leydig, 1848; Remak, 1852).

With the newer technique, it was found that there first occurs a change in the "chromatin" network (1879) in the nucleus, so-called from the deep stain which it takes with suitable dyes. This network first becomes, or becomes capable of being seen as, a number of fine coiled "chromosomes" (1888), which, on closer inspection, are each seen to be double, consisting, along their entire length, of two similar parts. These two may next coil upon themselves to look like one, and undergo shortening and thickening. Chromosomes were found by Van Beneden (1887) to be normally of a definite number in each cell of a given species. The number runs from a few units up to a hundred or so, being forty-eight for man. This constancy has proved to be of enormous genetic importance.

Chromosomes, with certain important exceptions to be examined later, always go in pairs. The members of each pair look alike. In 1901, the American, Montgomery, found that the chromosomes differ somewhat in shape and size in any one nucleus. The corresponding chromosomes can be recognised from nucleus to nucleus of one species—another very important point.

To return to mitosis, in the cytoplasm of many higher animal (but not plant) cells, there is found to be a tiny point, the "centrosome" (1888) which, by the above preliminary stage of division, has become two points, one on each side of the nucleus. Even with no centrosomes,

* Dates following inverted commas show when the names were given; usually, when the ideas crystallised.

there appears round the nucleus a spindle-shaped region looking like the lines of force between opposite poles and in some ways acting like them. For the chromosomes are found to collect at the centre of the spindle, in a plane at right angles to its lines, as if repelled thither.

Their two similar parts then reappear and (Flemming, 1879) separate, like mediaeval tally-sticks ; so that precisely corresponding parts drift off, now *along* the " lines of force," as if now attracted by the centrosomes. Each group then becomes a fresh nucleus, the cytoplasm following them into fission. Presently, it is hard to see the individual chromosomes in the nuclei, and the net-like appearance recurs. Rabl asserted (1895) that the chrómosomes do not actually lose their identity. This is a vital point in the modern view of them as carriers of heredity. It is supported by their reappearing at the start of a new division in their position at the end of the last one.

All this was observed whenever, in growth, cells were seen dividing. But many of the observations were of such fineness that doubts early appeared whether they might not be mere optical effects. Some alleged ones were, but these, for clearness and brevity, have been omitted.

We now proceed to the opposite process, the sexual *union* of two cells, cleared up in part by the same group of men. This work, too, was complex ; for the cellular histories of the ovum and the sperm are at first sight very different. So are the processes in question in the lower forms and in higher plants and animals. Let us examine how research converged on the unified scheme put forward by Boveri in 1892.

Amici, with his new high-powered microscope, had summed up in 1846 certain important aspects of the sexual process in higher plants. There followed a period of controversy. By 1870, however, when Hanstein initiated plant embryology (that is, the subject of what happens *after* fertilisation), it had been established that the anthers and the pollen grain and tube are the male organs, the pistil, the ovule and the embryo-sac, the female ones. And their functions had been disentangled.

In 1830 Amici had traced the course of the pollen tubes into the ovary, and seen one find its way into the ovule. About this time, it will be recalled (Chap. XI), von Baer was tracing the mammalian ovum further and further back. In 1855, Pringsheim saw the spermatozoids of a lower plant—an alga—enter the female cell.

The next line of research refers to higher animals. In 1861 Gegenbaur showed that the vertebrate ovum is a single cell. In 1879 the Swiss Fol saw the entry of the sperm into the ovum. As Hertwig had foreseen in 1875, this was seen to be followed by the union of the single nucleus of the sperm with that of the egg—a phenomenon seen in plants by Strasburger in 1877.

The subject now received vital help from the study of heredity. We treat this as a whole later ; but it is convenient here to consider one point. Kölliker was, as we have seen, interested in heredity, and held (as most now do) that the nucleus carried hereditary factors from generation to generation. At any rate, it seemed clear that either the cytoplasm or the nucleus of the male and female sexual cells must do so. Now the French revival of Lamarckian evolution, with its inheritance of modifications acquired during life, had provoked the German Weismann (1834–1919) to the definite formulation of the opposite view. He taught that such modifications are *not* inherited, a part of the protoplasm of each creature being transmitted immortally from generation to generation in the germ cells, unaffected by environmental accidents. He derived his idea from protozoa and other primitive forms of life, where reproduction may be by cell-division and separation of the daughter cells, or by rejuvenescence of the cell as a whole.

This " continuity of germ-plasm " was an abstract, almost mystical theory, independent of any particular ideas as to what bodies carried the germ-plasm ; and it made evolution much harder to explain. Yet no one has ever succeeded in producing the kind of heritable acquired characters which Weismann's generation expected. Weismann was led to a conclusion very germane to our present subject.

In fact, if in each sexual act germ-plasms unite, the germ-plasm will *not* remain the same, but will double its complexity with each generation. Hence (1887) Weismann foresaw that, prior to the union, there must be a halving of each germ-plasm. This proved to be the case (" meiosis "). During the development of ovum and sperm to sexual maturity preparatory to their union, each was found to undergo a halving process. Unfortunately this process is embedded in others irrelevant to the essence of the transaction, and so was hard to elucidate.

By 1824, in animal cells, certain " polar bodies " in the ovum had been seen extruded by the latter during maturation ; and in 1877 Hertwig found that this extrusion involves a division of the nucleus. The division is repeated, but the essential point is that at the first of these divisions the chromosomes were found not to divide. It will be remembered that the chromosomes normally go in pairs. What now happens is that, instead of *each* chromosome splitting, the members of each pair separate and go different ways. A second division, of the usual type, follows, issuing in the egg nucleus and three small polar bodies (which usually degenerate). Hence, afterwards, the ovum, nucleus, and polar bodies each have only *half* the usual number of chromosomes, though one of each kind.* Thus a " haploid " cell generation sets in, in contrast with the ordinary " diploid " ones (Van Beneden, 1887).

In the case of the male cells, the similarity of this process of meiosis to that in the ovum was obscured for some time by the absence of the contrast in size and fate of the nuclei resulting from the divisions, four equal sperm-nuclei being formed (instead of one large egg nucleus and three small polar bodies). Hertwig demonstrated the similarity in 1890. The production in both cases of bodies not all of which will ultimately function should be noted as a case of trial and error or natural selection in an internal process. Van Beneden (especially around 1885) was responsible for much of this work, including the proof that the

* An important exception is noted later.

chromosomes of the offspring come in equal numbers from the egg and sperm, that is, from the two parents.

The haploid and diploid cell generations were now seen to be the alternate generations of Hofmeister, the spore-bearing generation being diploid, the other haploid. In 1894 Strasburger showed this in non-flowering plants, and soon after 1900 he brought the sexual process in the higher animals under the same scheme by viewing the polar bodies and ova and sperms as the remnant of the alternate generation. Thus in both higher animals and higher plants, the haploid generation is tiny and parasitic ; though the males thereof may be viewed as " free-living " if not self-supporting.

Heredity

The work of Zirkle, Roberts, and others has lately shown how far back we must go to reach the first deliberate use of heredity by practical men. The Stone Age crossings by which our early cattle and dogs were produced may have been accidental. But the artificial pollination of the date palm * must have suggested deliberate selection of desirable strains, and this is apt to imply some unconscious theory of particulate inheritance. One does not " select " between infinitesimally different specimens. Aristotle and many others among the ancients were greatly interested in heredity, and Hippocrates held a form of the theory elaborated by Darwin under the name of " pangenesis." But even the greatest of the theorists, as well as the practical men, also held absurd ideas as to what crosses were possible. They even assigned sex, like gender, to inanimate objects.

The Chinese must have been selectively breeding garden flowers long before 1000 A.D. ; while Columbus found that the Central American Indians had enough practical knowledge of breeding to engender and preserve different colour-varieties of maize. The mediaeval European religious houses, which were everywhere forward in clearing swamps and forests for cultivation, also had some such practical knowledge. The Cistercians, by greatly improving the wool-bearing quality of English sheep, laid one foundation of this country's economic supremacy, in which the first factor was wool. Mendel's was thus no new activity within monastic gardens. We have been reminded by J. B. S. Haldane of the forbidding lengths of time needed on present methods for acquiring genetic knowledge. No way of life is more fitted than the monastic to afford the necessary continuity.

We have noted the breeding activity of the 18th-century agriculturists like Colling and Bakewell, improvers of the cattle and sheep of their time. In the same century Koelreuter (1733–1806) was establishing the reality of plant hybridisation and doing artificial pollinations. The practical breeders must long have known that startling " sports " occur, and also that they are usually sterile. Usually, but not always. In 1788, a fine individual oat plant was found to transmit its superiority to its offspring. In 1824–5, a cow-herd of Aberdeenshire found another good strain in a similar way. In 1819 began the experiments of Shirreff, of Scotland, on hybridising wheat.

* First recorded 2400 B.C.

Shirreff was the first considerable worker with cereals to realise the importance of "pure lines." By the date of the "Origin of Species" he was well-known, and was having heads of wheat sent to him from many places.

As we have seen, it was not the "discontinuous," almost chemical, aspect of this work which most interested the evolutionary thinkers from Erasmus Darwin onwards. And in the first flood-tide after 1859 the impulse was rather to stress that the new theory had no *need* to assume variation by large heritable jumps than to investigate whether such jumps actually occurred. So far as theories of heredity proper were put forward, they were in one sense particulate theories. Thus, in 1863, Spencer came out with the idea of atom-like carriers of heredity present in each individual; and we have seen that some such ideas must have been present in the mind of Kölliker. In 1868 appeared Darwin's pangenesis, designed to furnish a mechanism for the inheritance of acquired characters.* Each organ was supposed to pass particles to the sex glands, so that any modifications of an organ came to be represented in what was passed to the next generation. No direct evidence of the existence of these particles was adduced, but particles they were. This theory, however, was not a "particulate" one in the modern sense of a quasi-chemical view of heredity. Darwin's epoch believed in blending inheritance; if there were particles they were so numerous that, like the atoms, they gave virtual continuity.

Galton, in this tradition, laid down, on the basis of his statistical work, a "law of ancestral inheritance," that an individual is made up of contributions from each generation of its ancestors, one half from its parents collectively, one half of the remainder from their four parents, and so on. Here again, the particulate element was blanketed by the impression of continuity given by statistical apparatus.

But in all this there was no impulse actually to search for heritable "mutations" (that is, jumps). It was the middle eighties before the Dutchman de Vries began to search for these. Presently he began to find them; though much criticism was levelled at each instance he adduced. In the nineties Bateson (1861–1926) of Cambridge began to do similar work.

Weismann's germ-plasm, with its fixity, and so definiteness, now did a second service to science. It gave de Vries the basis for his theory (from 1889) of "pangens" carrying particular characters through the generations. Primarily fixed, these occasionally underwent mutations.

With all this work, the question of mutations was now in the air; and in 1900 de Vries and others took one of the steps which changes an eccentricity into a "subject." They began to search past literature for light on it. It was thus that they came upon that work of Mendel (1822–84) which might have saved them so much trouble.† For

* This notion, though not stressed, was not fixedly doubted. Pangenetical ideas had been held by a long line of thinkers from the ancients down through Ray and Buffon.

† For further precursors of Mendel himself, see Hogben, L., "Science for the Citizen," 1938.

particulate theories, and the discovery of actual mutations, were not enough. What was needed in addition was a touch of logic to work out the slightly subtle *combination* of several particulate factors. And it is not wholly fanciful to suggest that, outside mathematics, no atmosphere was so likely to furnish this logic as that of Thomistic theology.

In 1866 and 1869, Mendel, a monk of Brünn (Brno), with a long peasant family tradition of gardening behind him, had published the results of certain researches which he had made between about 1857 and 1868 on the crossing of plants. The sweet peas, with which he worked, normally fertilise themselves. He examined the effect of crossing a tall and a short variety. He found that the results were not intermediate in height, as many then current theories supposed would necessarily be the case. In fact, all were tall. But when these fertilised themselves one quarter of the offspring were short—shortness, though eclipsed for one generation, had not suffered *dilution*. This "segregation of characters" was Mendel's first law.

These shorts, self-fertilised, always bred short. Of the three tall quarters, only one always bred tall. The other two—half the crossed generation—gave, again, three T's and one S. And so on. It is thus obvious that tallness and shortness do not behave alike. A similar contrast was found by Mendel for other characters. Mendel said that tallness was, relatively to shortness, "dominant." Shortness was "recessive." He showed that every case could be explained if a pea always has two "places" for factors related to height, one for the factor from each parent. To take the case given, in the generation after crossing, each pea must have S+T, and in the self-fertilised generation following there would be either 2T or 2S or (in two ways) S+T again. If now dominance means that tallness is exhibited as long as *either* place is filled with the dominant, this gives T in all the S+T's of the first generation. It also gives it in the S+T's and in the 2 T's of the self-fertilised one; that is, in three cases out of the four. Half these latter—the S+T's—would behave like the first when self-fertilised in turn. It was a neat piece of reasoning, neatly confirmed by experiment.

In 1902 the American, Sutton, pointed out that the mode of splitting and union of the chromosomes in mitosis and meiosis is formally the same as that of Mendel's characters, and that these latter must, in fact, like the "factors," etc., of other writers, be carried by the chromosomes. Since 1900, in fact, the two bodies of theory and experiment have converged, and the resulting "chemical"—and so experimental—attack has more and more filled the horizon of studies in heredity. This is not to say that the idea of "blending" inheritance has been dropped. Galton and other older workers strongly maintained it. In 1930 it was discussed by Russell. It must be remembered that, from the nature of the case, the variations which conspicuously obey Mendelian laws are abnormalities if not lethal qualities. It has been suggested, therefore, that Mendel's laws do not apply to normal inheritance. The Mendelians, however, seem to have shown that no case yet unearthed positively demands blending inheritance.

Inheritance carried by non-nuclear parts of the germ cells (and, therefore, from the nature of spermatozoa, by the *mother*) has also been investigated; but it cannot touch the main problem. Here we confine ourselves to Mendelism.

One important preliminary to the new phase which now set in was the Dane Johannsen's (1857–) work with pure lines (1903, 1908, etc.). He had at first worked with Galton's statistical methods, but, having been forced to think in terms of unit characters, he saw that Galton's methods might easily confuse the effects of several causes. He therefore chose a particularly simple type of case, namely self-fertilising plants, of which garden beans are an example. Having secured pure lines of these, he showed that in any one generation of any one plant there is variation which follows the normal error law, but that the averages obtained by breeding from the largest and the smallest bean of such a generation are the same. That is, these variations (which are purely environmental) are not inherited. On the other hand, the differences of size between *different* pure lines *are* inherited. He was thus led to envisage the existence of a "genotype"—of a complex of features of any living thing which are not affected by environment—and of a "phenotype"—of features which are—the latter including most crude visible characters. In fact, the important distinction grew up between the genetic constitution of a creature and the resultant of this and of environment and time of growth, which resultant alone is observable. Heredity and environment are always hard to separate, and neither the exponents of this modern theory nor those opposing them always avoided claiming too much. Hogben points out that only when one of the few and definite possible patterns of Mendelian inheritance is present in family trees can environment be certainly treated as secondary. It can never be treated as absent. But such a realisation belongs to the last decade or so. We must retrace our steps.

Another important preliminary had been the discovery by Castle (1867–), in 1906, of a peculiarly suitable experimental subject, the fruit fly *Drosophila melanogaster*. This has short generations and few (four) chromosomes; and by now millions of genetically-controlled individuals of it have been examined. Certain other highly suitable species have also been much used, such as the amphipod *Gammarus chevreuxi*. Certain characters, too, such as pigmentation (e.g. of the eye), are prominent in genetical literature, but are of course pointers to deeper ones.

Under the concentrated attack of the American school of T. H. Morgan (1866–), and of other workers, certain difficulties began to emerge in the first simple form of Mendel's theories. Many of these proved to be explicable on lines due to Bateson and Punnett (1905, 1906), who experimented on the combs of fowls. Briefly, characters were found not to behave, in all cases, independently. They might behave as if linked. Thus, a character might be "sex-linked"—might run with one sex. Or, again, an apparently single character might turn out, genetically speaking, to need analysis and an allowance of more than one causative factor. Linkage was easy to envisage if the causative agents of the character (soon to be called "genes," from "gens," 1909) were located on the same chromosomes, and if characters (in the visual, phenotypic sense) were normally the resultant of more than one "gene." But even this was found to be only a beginning. Characters normally linked might become unlinked, new linkages succeeding the old.

This also, however, could be explained; it was indeed a triumph of the chromosome theory. It was explained as due to the crossing of chromosomes at the thread-like stage, both chromosomes breaking in two and reassorting partners. Plough (1917) was able actually to demonstrate this.

Small entities, which stain more deeply, were in certain cases distinguished on the chromosomes. These "chromomeres" were arranged in a single line along the length of the larger bodies, and their linear arrangement naturally intensified the ambition to "map" the genes. Surprisingly good results have been obtained by supposing that the probability of "crossing over" of any particular pair is directly as their distance apart on the chromosome, so long as this distance is small enough for the possibility of the intermediate crossings to be negligible. Crossing over normally occurs at least once in every pair of chromosomes at meiosis. Mutations themselves are now viewed as minute rearrangements.

Other types of heritable change of genic constitution may result, especially in plants, from failure to divide at meiosis, so that diploid ova or sperms may be produced. We cannot pursue this point here; but polyploid individuals show differences which adapt them to a different, often a more northerly or alpine or otherwise harder, environment. Their degree of variability is also different. Tetraploidy, partial or complete, appears to be the commonest, and optimum, form of this variation. Further studies in ecology may show that very often in the spread of a species out of its optimal region hardy tetraploids form the van.* But the point for our present purpose is that the types of inheritance here discussed, being all *particulate*, will not be gradually drowned by crossing-back to unchanged specimens. Thus one great difficulty in Darwinism was removed.

But other difficulties remain, for the deeper sort of mutations are unpromising raw material for a theory of the formation of new species. It is true that, after many failures to produce mutations artificially, it has at last been found possible enormously to increase the mutation rate by means of X and γ radiation. It has even been possible (Oliver, 1930) to show that this rate varies as the energy of the applied radiation and is independent of its wave-length if the latter is short enough. But this is no great help, even if cosmic rays modify the conclusion of Muller and Mott-Smith (1930) that naturally occurring radiation is not strong enough to produce the desired amount of mutation. The difficulty is of another kind: mutations are nearly always lethal, and nearly always recessive. To counter this difficulty geneticists have

* J. S. Huxley's conception of "clines" brings out the existence of gradations of characters among animal and plant populations. See *Nature*, 30 July, 1938.

developed the idea of linkage of genes to the highest possible point, in the theory of the "gene-complex."

According to this theory which is highly susceptible of mathematical development, the phenotypic effect of a new gene depends on the existing genes, and vice versa. We thus cannot tell, merely from observing farmyard sports, when a new gene appeared, what its first effects were, or what its "final" ones will be. Dominance and recessiveness themselves evolve, no less than everything else, and are in fact phenotypic, not genotypic.

The application of this theory to natural selection depends on the observed fact that mutations tend to recur. It thus becomes possible to envisage (and envisage mathematically) selection operating on the gene-complex (R. A. Fisher, 1928). Those individuals on whose existing gene-complex the new gene reacts badly will be eliminated, and a line gradually established in which the gene is recessive, and ultimately, as an individual, of no phenotypic effect. The new complex will, however, for that very reason be different from the old. The change will not be sudden; so that one objection urged (for instance by MacBride, 1930, etc.) against cruder mutation theories of evolution ceases to hold. At the same time, it must be granted that the less abnormal a variation the more it will *appear* to be a case of blending.

Any rare favourable mutations will, it can be shown, rapidly become dominants in the same circumstances; but the above argument as to initially unfavourable ones still needs support. We refer to it again under ecology. Here the main point to note is that the gene-complex is an "environment" for mutations in the organism. The new theory in fact furnishes, and must consider, *two* adaptations to environment instead of the single one of the old ecology and evolutionary doctrine. The subject becomes two-complex instead of one-complex, like so many previous theories in the history of science.

Once a mechanism is established by which a species, or some lines of it, may change and become two, it is easy to imagine that hybridisation would give us further species. On the other hand, "true" species used to be regarded as mutually sterile; and even when it has proved possible to effect crosses, one sex of the offspring is usually sterile or even fails to occur. It has, however, been shown that in plants, which are genetically much more flexible than animals, hybridisation, followed by doubling of the chromosomes, can give rise to new species. In this connection Lancefield (1929) has made a discovery highly disturbing to the systematist. He has found that races of *Drosophila* which, as tested by fertility of crossing, have nearly attained specific rank, may yet be morphologically indistinguishable—unless in future the systematist concerns himself with the micromorphology of the chromosomes.

We may here mention a not unrelated matter which is modified by particulate genetics, namely, the important concept of "homology." Souèges * has indeed shown that there were always varying uses of this term and of "analogy," but all are altered if we can, within wide limits,

* *Act. Sci. et Ind.*, 381, 1936, pp. 11–12

speak of a modification being "the same" in different species though its arising is due, not to common ancestry in the old sense, but to the same mutation's having occurred more than once.

In conclusion, we mention briefly the history of a complex subject, that of the determination of sex. As early as 1891 Henking noticed that in one case of sperm formation there was a chromosome ("X") which did not divide, but went whole in one direction. In 1902 Sutton noticed in insects (and in 1905–8 Strasburger confirmed in plants) the fact that the chromosomes do *not* always go in pairs, one being commonly unpaired. In 1902, also, McClung suggested that the X chromosome might be the determiner of sex, since, from the above, half the sperms would be without it. Later studies showed that in some species the X chromosome may have a mate, "Y," but different from itself; and from this there grew up the simple doctrine that one sex is produced when sexual union brings together X and Y, and the other when it brings together X and X. The XX sex is in some species the male, in others the female. Cases occur, though they are not the rule, when closely allied species are different in this respect. This has been found to be correlated with a comparative ease in the upsetting of the sex determination in these cases.

In fact, as was shown by Bridges in 1921, and as the doctrine of the gene-complex suggests, the mechanism is only in extreme cases as simple as this, the other chromosomes having normally some influence. This more balanced, less decisive, condition among the genes lets in environmental factors, especially in plants (Schaffner, McPhee, etc., *c.* 1920–5). In higher animals hormone production usually confines alterability to very early stages.*

Haldane (1922) successfully explained the above-noted case of interspecific crossing. Sex-determination depends, evidently, on a right balance of X and Y factors; and though this may be right in each of two species, it might not be right between them—especially, it is clear, for the XY sex. Now Haldane was able to show that in the cases mentioned it was in fact the XY sex which failed.

* The genic factors appear to act through metabolic rate, maleness having the higher rate.

CHAPTER XVII

THE GROWTH AND UNITY OF THE INDIVIDUAL. I

As growth proceeds, organs become differentiated and might become actually isolated were not further organs to develop, the express function of which is to keep the others connected. Hence in the next two chapters we couple together the growth and the unity of the individual, and follow up a section on embryology with sections on the nervous and endocrine systems.

EMBRYOLOGY

Embryology has run on two main lines since 1859. In the first the development of the individual has been followed in a great number of species in increasing detail, with evolution in mind. In the second, *experiment* with the living embryo, as opposed to observation and post-mortem dissection, has been called in, to assign causal factors to successive stages in growth and, ultimately, to reduce these causes to physics and chemistry ("experimental embryology," "Entwicklungsmechanik"). We consider the two lines in that order.*

The keynote of the earlier stage of embryology (Chap. XI) had been the germ layers. Remak, working (1850–5) on vertebrates, especially frogs and chickens, established the organs which develop out of the three different layers. But it was generally held that no similar layers can be traced in invertebrates, until in 1871 Kowalewsky showed that the doctrine could be extended to all metazoa. The middle layer might be absent, but this served as a basis for classification.

We have already seen, in the case of the notochord and the gill-arches, how discoveries in embryology threw valuable light on animal classification, revealing unsuspected relationships between groups. This remained a large part of their interest for the generation following 1859, when the impulse (for instance of Haeckel) was strongly evolutionary. Moved by recapitulation theory, F. Müller did detailed work on the development of crustaceans. This was taken up by Haeckel, under whose aegis Kowalewsky went on to revolutionise the systematic position of certain lower animals hitherto placed generally and vaguely with molluscs. Two of these, in particular, the tunicates (1866–71) and *Amphioxus* (1866–77) were shown by him to possess notochords and other vertebrate features during at least a larval stage in their life

* Only animal embryology is dealt with. In Chap. XVI we dated the start of plant embryology from Hanstein, 1870. In the next fifteen years, certain stages followed by all flowering plants were laid down, and certain laws relating the successive divisions. The first few of these latter escaped the available instruments until about 1895. Since then the subject has become a large one.

histories. In other respects these animals would not be suspected of vertebrate relationships. This resulted in the recognition of a new major division, chordata, of which vertebrates were only one, though numerically enormously the preponderant, part.

The sixties and seventies were a period when "brilliant" theories of stages or phases which every evolving entity must go through prevailed in many subjects from anthropology to embryology. Generally the theories were based on the feeling that origins must be essentially *simple*. Since then, as we have seen in Chap. XI, a more historico-geographical viewpoint has prevailed.

In embryology Haeckel (e.g. 1877) rushed in with a scheme of stages for the early development of all metazoans, namely, anucleate cell, nucleate cell, many-celled sphere, hollow sphere, hollow sphere "pressed in" so as to be double-walled over most of its surface, elongate double-walled figure with small opening at one end. The two walls of this "gastrula" stage were of course the endo- and ectoderm. The next phase was the development of the mesoderm * between the other two, at one side of the tube, and then, according to Haeckel (1872), the budding of the hollow body-cavity, or "coelom," from the endoderm.

The anucleate beginning was a piece of imagination, based, as we have seen, on the idea of simplicity. The next stages, however, have worn fairly well; although, for instance, the opening in the gastrula stage may be blocked with yolk overflowing from inside. The coelom proved to be controversial and complex (Huxley, 1875, Balfour, the Hertwigs, Ray Lankester). But its presence or otherwise has ever since been a valued criterion in classification.

While subsequent attention has centred rather on comparative than on common features of development, no one in the early period contributed more than Haeckel to vitalise comparative studies also, and to convert the old formal Cuvierian comparative anatomy into the rich, dynamic comparative embryology of F. M. Balfour (1851–82) and O. Hertwig (1849–1922).†

At the time, however, there was a reaction from Haeckel's "recapitulatory" tone of thought. Hertwig, Keibel, Sedgwick, and others stressed rather that embryos are *not* alike at the start; and since then the idea of recapitulation has dropped into a very secondary place, except as one of a more general group of notions which will be dealt with at the end of this section.

We now turn to experimental embryology, with its ultimate ambition of formulation in terms of the laws of inanimate matter. This ambition, when expressed by W. His (1888) and others, was but coldly received by the workers in the other line, Hertwig and the rest. For Hertwig, if two living shapes were similar this was due to similarity of heredity and growth, not to common submission to (say) the laws of physics, in spite of the resounding triumphs which these laws were constantly achieving.

W. Roux (1850–1924) in 1881 distinguished a "pre-functional"

* Where this exists.

† Whose great works appeared in 1880–1 and 1886 respectively.

MAP V

BIRTHPLACES OF SCIENTISTS

Please consult Explanation on p. 10 before studying the Map

***South and Central Germany and Eastern France since about* 1800**

Number on Map	Name	Birthplace *
1	Abbé	Eisenach
2	Bessel	Minden
3	Dedekind	Brunswick
4	Dirichlet	Düren
5	Driesch	Kreuznach
6	Einstein	Ulm-a-d-Donau
7	Fischer, E.	Euskirschen
8	Gauss	Brunswick
9	Gegenbaur	Würzburg
10	Gerhardt	Strassburg
11	Gerlach	Zelle
12	Henle	Nurnberg
13	Henry	Marche
14	Hermite	Dieuze
15	Hertz	Hamburg
16	His	Basle
17	Hittorf	Bonn
18	Hofmann	Giessen
19	Kaufmann	Elberfeld
20	Kekulé	Darmstadt
21	Klein	Dusseldorf
22	Koch	Klausthal
23	Kohlrausch	Rinteln-a-d-Weser
24	Kühne	Hamburg †
25	Laue	Coblenz
26	Laurent	Langres
27	Le Bel	Soultz-sous-Forêts
28	Liebig	Darmstadt
29	Lorentz	Arnheim
30	Ludwig	Witzenhausen
31	Mayer	Heilbronn
32	Meckel	Halle †
33	Mohl, von	Stuttgart
34	Müller	Coblenz
35	—	Göttingen
36	Ohm	Erlangen
37	Oken	Bohlsbach
38	Pettenkofer	Neuberg-a-d-Donau
39	Pfeffer	Grebenstein
40	Pflüger	Hanover †
41	Plücker	Elberfeld
42	Poncelet	Metz
43	Riemann	Dannenberg
44	Röntgen	Lennep
45	Schleiden	Hamburg
46	Schültze	Freiburg
47	Schuster	Frankfurt-a-M.
48	Schwann	Dusseldorf
49	Schweigger	Erlangen
50	Siebold	Würzburg
51	Spemann	Stuttgart
52	Staudt, von	Rothenburg-a-d-Tauber
53	Waldeyer	Brunswick
54	Weierstrass	Münster †
55	Weismann	Frankfurt-a-M.
56	Wöhler	Frankfurt-a-M.

* In a few cases, work-places, or the nearest larger place, have been substituted. Göttingen has been inserted for reference, owing to its importance, though it was not he birthplace of any of the scientists mentioned. It has been given the number 35.

† Hanse town. Many prominent 19th-century German workers, Helmholtz among them, were of course born to the East. See (e.g.) Map IV.

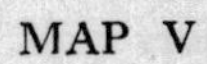
MAP V

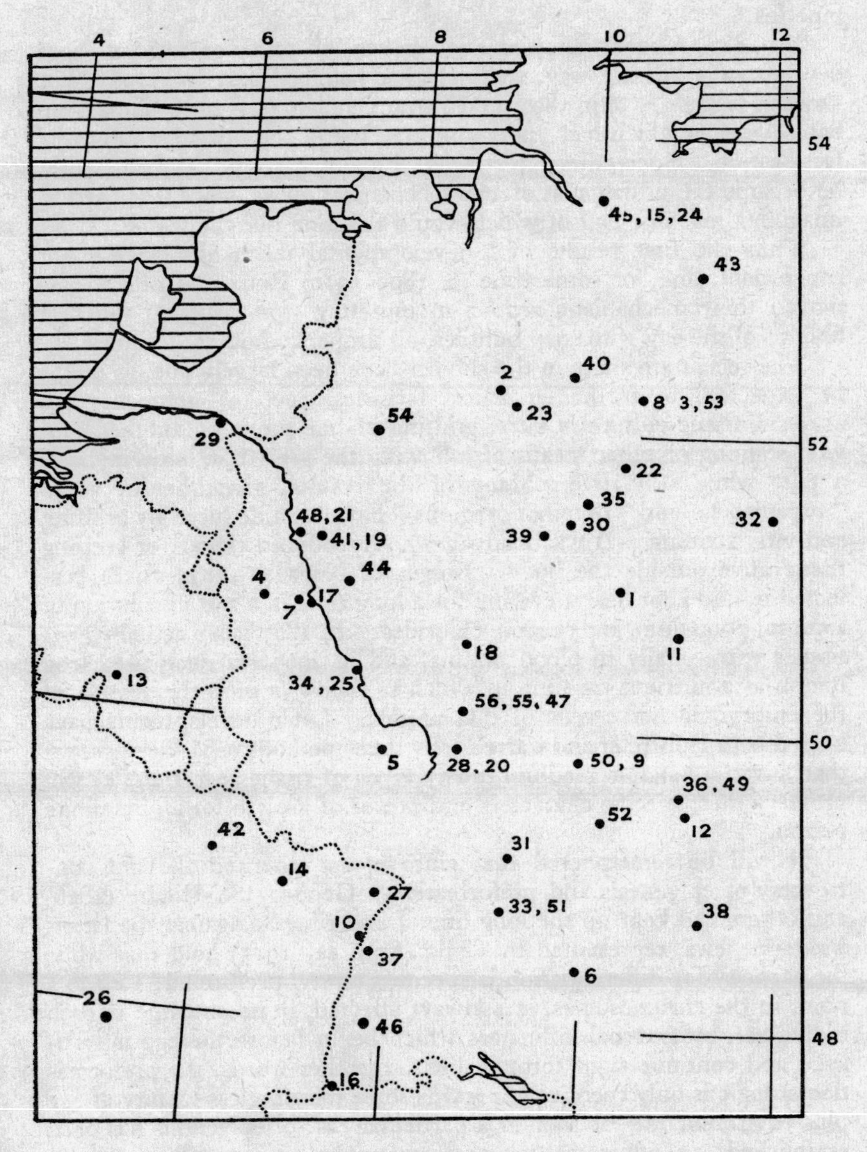
4
6
8
10
12
54
52
50
48
45, 15, 24
43
2
40
23
8 3, 53
54
29
22
35
32
30
39
48, 21
41, 19
44
4
17
7
1
18
11
13
34, 25
56, 55, 47
5
28, 20
50, 9
36, 49
52
12
42
31
14
27
33, 51
38
10
37
6
26
46
16

stage in every creature's growth and a " functional " one in which use influences form. In early years (and largely still) the latter defeated mechanistic analysis. It was, therefore, the earlier which most of the work concerned. The experimental appliances were as usual extremely simple : needles, knives, and simple salt solutions. Even now, much is done with fine surgical scissors and fine glass needles, loops, and pipettes.*

In 1887, Chabry showed that halves or parts of ascidian eggs develop in a mosaic way, that is, give half- or part-embryos. But Driesch (1867–) in 1891 reared complete larvae of sea urchins from halves of eggs which had undergone the first of the divisions following fertilisation. Succeeding afterwards with a quarter and even an eighth, he became the protagonist of anti-mechanical views, since it is hard to imagine a machine half of which would function like the whole.

Thus the first results of " developmental mechanics " were not concordant, and for some time (*c.* 1895–1910) Roux and others employed their mechanistic ardour in imitating (for instance) the first fissions of the egg with soap bubbles, oil drops, and other " models."

The recent advances in the subject have been largely due, as might be expected, to further technical facilities, such as microchemical analysis, tissue-culture *in vitro*, grafting of one tissue on another, and vital staining or other means of following the growth or movement of a part while still alive. Many of the results, given below, as to " organisers " and " dominant regions " have been deduced by grafting and vital staining. It was Ludwig who introduced the art of keeping tissue alive outside the body. Ringer of Norwich (1835–1910) perfected it, and kept hearts beating for a long time in a mixed solution of sodium, potassium, and calcium chlorides. In 1897 Loeb actually *grew* tissues extra-vitally in blood plasma, and in 1907 Harrison used frog lymph as a nutrient medium in which to follow *in vitro* the growth of the embryonic nerve-cells of that animal. Later developments have been due to Holtfreter and Carrel. By their methods it has been shown that, given a suitable medium, many types of tissue (e.g. muscle) will preserve their specific characters and power of growth for an indefinite period.

It will be remembered that embryology inherited the old controversy of epigenesis and preformation ; Geoffroy St.-Hilaire (1826) and others had kept up the long line of experiments against the latter. Modern views, represented by Child (from say 1915) hold that while the capacity for differentiation is specified (and " preformed ") up to a point in the chromosomes, it is always affected, in its outcome in each individual, by external influences which begin before the egg is fertilised and continue right through life. In other words, the preformationist logic is only coercive as regards some meaningless totality of egg plus environment. As soon as a particular case of epigenesis has been established, an effort, so far preformationist, is naturally made to

* His, it will be recalled, was a chief progenitor of the microtome. But this is used in every branch of biology ; and the effect of the experimental school of embryology was, rather, to take the embryo out of the hands of the microtomist and keep it alive.

explain it in terms of prior states and events. Thus, ideas of isotropic eggs, and proofs that in early stages parts can be interchanged (Roux, Pflüger, Hertwig) are "epigenetic," those of anisotropic, mosaic, regionally determined eggs (Roux, His), are in a sense preformationist. But we shall drop the use of these words.

It is more important to note that a biological explanation must usually precede a mechanistic one. For instance, workers were concerned to discover "why" the spherical egg "chooses" to divide or invaginate in one direction rather than another.* As early as 1850–4, Newport related the plane of segmentation of the amphibian egg to the point of entry of the sperm, a purely biological explanation. This entry normally determines the future planes of upper and lower and of bilateral symmetry; but in some insects and cephalopods this determination precedes fertilisation, while it can often be disturbed experimentally.

But from an early date in this work it was realised that the *lack* of bilateral symmetry, and in particular a helical formation, may be the matter requiring explanation. Even in mammals symmetry is partly superficial; while in (for instance) snail shells helical asymmetry is plainly shown. It was, however, early known that the successive cleavages of an egg may themselves take place helically, and in 1894 Crampton showed that opposite helical cleavages resulted in oppositely twisted shells. Harrison (1921, 1925) has suggested that all this must be correlated with a "space lattice" of protein (see Chap. XXIII). It is considerably easier to evolve bilateral symmetry from helical forms (for instance, by placing two opposite ones side by side) than the opposite; and in view of the generally spiral arrangement of plants, it may turn out best to regard the helix as fundamental and the symmetrical shapes as the ones to be "explained" in terms of it. It is a natural speculation to connect this with symmetrical and antisymmetrical spins in quantum mechanics.†

Even more important is the determination of the main axis of front and back. Such work as that of Lillie (1906), Jenkinson (1911), and many others since, has suggested that the determinant here is the egg's point of attachment to the parent. This axis seems to be determined by factors external to the egg, and it is possible by experimental application in early stages to affect it.

It early seemed clear that visible differentiation must be preceded by invisible (for instance chemical) differentiation; and as early as 1901 Boveri brought in the idea of gradients of activity from place to place in the embryo. The serious development of the idea began with Child about 1915.

* It may be mentioned that early workers in this subject were concerned to lay down laws about cells in general, not confined to egg-cells; hoping that these latter could be viewed as particular cases. Thus Hertwig laid it down that the nucleus tends to be central, and that in division the "spindle" lies on the longest axis of the cell. Boveri (1895–1905) asserted that the size of the cell varies as that of the nucleus. Such rules, however, were found to be only rough.

† See Chaps. XX and XXIII.

These gradients seem to exist before the egg divides and to determine the later visible differentiation into organs and planes and axes, while the actual cell-cleavages are often unrelated to them. Thus the fertilised egg retains its undifferentiated spherical form while cellular division has proceeded up to perhaps a thousand cells, but gradients already exist within it. So long as they are only matters of degree, these gradients do not disturb the capacity of a part of the embryo to develop on removal of the remainder into a normal adult, but it may happen that before there is visible differentiation, these gradients and other invisible differences have reached a new, " mosaic," stage when chemical differences of kind are involved. In this stage, passed through by all higher animals, the embryo is strikingly lacking in unity, and can no longer regenerate from a part of itself. Grafts from part A on an injured remnant of part B result in A, not B, growing there. Many experiments of this sort were carried out from the early days of this century, but it was long before the above-mentioned doctrine emerged, because it was well known that in many forms, like newts, regeneration will take place even in the adult. Here, however, it has been shown that the primitive power is lost, and another one is acquired later. This new one seems to be associated (Morgan 1902, Schotté 1926) with the coming of the nervous system.* This, with the endocrine and other circulatory systems, restores to all higher animals something of the " organic unity " originally possessed by the egg, and often taken as the universal mark of life.

Different eggs reach their determination points at different times, some before fertilisation (hence mosaic eggs), others at gastrulation.

We have seen that the course of events after the shape ceases to be spherical was known in Haeckel's time; but quite lately, by vital staining and other methods it has become possible to see what movements take place when these changes of shape come about. The invagination is caused by the disproportionately rapid extension and thinning of the outer layer of cells, particularly on one side, relieved by an inward turn along a certain line.

From the lip of the opening, as the invagination begins, spreads an " organiser " (Spemann, 1869–   ) which determines the fates of the parts. This was an important discovery, and close research has fastened on it. Organisers are roughly specific as to the class of structure they induce, but quite non-specific as to animal groupings. They seem to be substances, not living entities, for cell-free extracts work just as well, and so do many pure compounds. Organisers are, in fact, morphogenetic hormones. They are an important point in that chemical embryology which became a distinct subject after the War. The first experiment on them was by Spemann and H. Mangold, 1924.

As the nature of a gradient suggests, the later stages of growth have been found to be dominated by one part, in most animals the front part. Differentiation and growth are often dependent on influences

* A complex stage with which we cannot deal in strict sequence. See, however, next section.

coming from more apical levels. If an earth-worm is cut in pieces, the front parts of each segment tend to develop these apical properties.

In the later pre-functional, and in functional, stages, the variety of courses taken by growth in different groups of animals becomes prodigious ; so does the number of parts of which the growth and relationships have to be followed. Such variety has confined most work to early stages or special aspects of later ones. We consider only two such aspects.

One of these is metamorphosis. The ancients had of course been interested in the metamorphosis of insects, which had also been a favourite subject with the 17th- and 18th-century microscopists. The army surgeon Thompson had shown (1823–30) that extreme forms of it also occur among marine forms such as crustacea, where it often goes with fantastic forms of parasitism (see " Ecology " later). One result of his work was a re-classification of certain groups, such as the barnacles : a further instance of that dependence of advances in systematics on embryological and physiological discovery which we stressed in Chap. XI. In this connection it will be recalled that completeness of metamorphosis is one of the main bases of insect classification. From the evolutionary point of view one of the most significant features of marine life, as revealed by the work of a century since Thompson, is the vast proportion of even sessile forms which pass through a larval stage as free-swimmers in the plankton.

As our second aspect, we will take the laws of growth in the functional stage, investigated by J. S. Huxley and others. Such growth is often highly geometric (e.g. in spiral shells and in radiolaria), and as such had attracted many workers from the earliest days. This early work was summed up by D'Arcy Thompson (" Growth and Form," 1914).

Huxley's line of attack again involved the concept of gradient, this time of rate of growth. The non-spherical form of a creature or of an organ is regarded as due to different rates of growth in different directions and regions. One case examined was the asymmetric growth of crabs as regards the size of their claws. It is found that the ratio of these sizes increases as growth goes on. The simple law of " heterogonic * growth " : $y = ax^k + b$, is found to have wide application, a limb usually presenting " growth-centres "—regions where growth is fastest—surrounded by regions of slower growth.

Such changes of proportion, though they spring merely from increase of size, reach a point when they would be taken as constituting new species. J. S. Huxley has (1924, 1927) pointed out that in fact they may have an evolutionary significance, especially where characters appear to be non-adaptive.

In general, among physiological genes, few are more significant than the "rate genes " investigated by J. S. Huxley and Ford (e.g. 1927, 1929). Whenever the rate of a process (for instance, of growth, or of pig[illegible] deposition) is correlated with time of onset and of final equilibrium, a [illegible]tation which increases the rate will cause recapitulation,

* Huxley (*Nature*, 9 May, 1936) has since proposed " allometry " instead of " heterogony."

while one which decreases the rate will cause the opposite—an embryonic character spread on into later stages instead of being forced back into earlier and earlier ones. New light is thus thrown on this old idea by recent concepts in growth and in genetics. Recapitulation is shown, in fact, to be only one of a general class of alternatives which, more probably than not, are relevant to every living creature.

NERVES (see last, Chap. IX)

We preface our account of this subject by the usual remark that advances in it were largely conditioned by advances in instruments and technique, sometimes made especially with nervous physiology in view, sometimes not. We may roughly distinguish two lines of advance, which analysed the nervous system respectively by the technique and theory of cytology, and by examining the electrical and other responses of individual units such as fibres to electrical and other stimuli. Before treating the first, it is convenient briefly to mention the earlier phase of the second, made possible by the first great group of electrical and other devices.

We have mentioned earlier some of the instruments due to Helmholtz. At about the same period (1840–50) Du Bois Reymond and the Webers were also active in this way. Du Bois Reymond showed that the electrical state of the nerves changes when an impulse passes. He investigated nerve-muscle preparations, and the " tetanus " produced in them by a rapid succession of stimuli. In 1845, the Webers made a discovery which started a new era in nerve classification. They found an efferent nerve (the vagus) which is inhibitory, not excitatory. It had hitherto been held that excitation is the one function of efferent nerves.

In 1871, the American Bowditch discovered a principle which was to be of great importance for the future : the " all-or-none principle." If an impulse is big enough to make a muscle-nerve unit respond at all, the response is independent of the size of the stimulus. Each response is followed by a " refractory period " during which no stimulus is effective. It was in the heart that Bowditch discovered the law, and for long it seemed inapplicable to other units. In fact, with the available instruments he would not have discovered it at all but for the fact that the refractory period is unusually long in the case of the heart. Generalisation of the principle awaited the modern ultra-sensitive electrical equipment.

About 1835, when cells and the new microscopes entered biology generally, they also entered the question of the nerves. They made it clear that while both white and grey matter contain an enormous number of fibres, the grey enjoys the distinction of containing also a large number of cells. In 1838 Remak claimed that in the vertebrate sympathetic system the cells are connected with the fibres ; but it was 1849 before Kölliker proved that all fibres are connected to cells.

Embryology, the question of how things *grow to be*, was al[illegible] the air at this time, and in 1839 Schwann suggested that the fibres themselves arise by the junction of chains of cells. At this time the cell-idea itself, it will be recalled, was not clearly understood.

In 1851, Waller made a discovery of vital importance in that it gave means of tracing the course of an individual fibre through a dense mass. He found that on cutting a fibre, the part remote from the cell will degenerate, while the part still connected will not. Later Marchi discovered the stain, osmium tetroxide, most suitable to show up the degeneration.

In 1864 and 1868 Hensen suggested that nerve fibres are formed by functional stimulus from bridges of protoplasm between cells. But it was a third theory which was destined to hold the field. We may find the germ of this in the old comparative anatomists like Carus and Cuvier. They had shown how complete is the dominance of nerves even in invertebrates. In these, it is possible to see nerve cells as separate entities before their branches grow into an inextricable net. That all fibres have in fact grown out of cells was suggested by Bidder and Von Kupffer in 1857, and especially by His (1886, 1887). But the full "neuron" theory, that the entire nervous system consists of nothing but cells and their outgrowths, is the work of many hands, Kölliker's among them. It was first definitely stated by Waldeyer in 1891. The point that the branches are outgrowths was experimentally confirmed by Harrison in 1907 and 1910, using the technique already mentioned to grow embryonic nerve-cells *in vitro*.

The next question was, why do the fibres grow out of the cell in one direction rather than another, what is responsible for the existing distribution and branching of the nerves? Cajal * and Forsmann stressed (1893–4) the effects of chemical stimuli on nerve growth, Kappers (1917, 1921) and Child (1921), those of electric forces. What is remarkable is the extent to which nerves grow simply along lines of *mechanical* force. This was urged in early days by His (1887), later by Harrison (1910), and was recently (e.g. 1934) confirmed by Weiss from study *in vitro*.

A further question raised by the neuron theory concerned the interconnectedness of the cells, via "synapses" between extremities of branches. Unless branches from different cells positively grow into one another, discontinuity of nerve-substance should occur at these points. Golgi was able to demonstrate that in fact such discontinuity does occur. As to the interconnectedness of *responses* resulting from the interconnectedness of cells, every research has confirmed its importance. Localisation of function has been found to be never more than rough. Nor is there complete autonomy in the involuntary nervous system. Gaskell (1886–9) and Goltz (1891) investigated this. The work of Goltz (1869 onwards) had moreover shown that the effects of (say) decerebration are relatively the more serious the higher the animal in the evolutionary scale. It is Sir Charles Sherrington (1861–) who is associated with the integrative action of the nervous system (especially 1893 onwards). Our power of tracing nerve paths has far outrun our power to assign them functions, but Sherrington has given simple examples of coordinated action. One is the relaxation of one muscle to make possible the contraction of another. Such integrative action increases as we go up the scale.

* One of the few modern Spaniards of scientific note.

This book does not touch psychology, but the work of Pavlov (1849–1936) was relevant here. From about 1898 he investigated "conditioned reflexes," and showed how many of our actions they can account for without the intervention of "mind" or of the wider integrations just mentioned. Yet even in the case of animals like dogs, he needed years of experiment before he perfected his technique; that is (in part) before he succeeded in getting rid of this wide integration and in isolating suitable reflexes. So far does coordination go.

In effect, Pavlov's conditioned reflexes are *induced* integrations, linkages, by repeated coupling of stimuli, of one part of the nervous system with another hitherto separate. Give a dog a book and then food often enough, and eventually his mouth will water at the book alone.

Messages are more easily stopped or distorted at synapses than in the body of fibres. Moreover, when stimuli come by indirect paths, that is, across synapses, delay may occur in the motor response. This may be explained on lines due to Forbes (1922). Individual impulses across synapses may be too weak to give the (all-or-none) response; but reinforcements may come by more and more indirect routes until they build up the needed level of stimulus.

Such considerations bring us back to the more physico-chemical approach with which we started, and shall end, this section.

In 1904, Elliott hinted that the effects of the sympathetic nerves were similar to those of the substance adrenaline found (Oliver and Schaefer, 1894) in the adrenal glands (see next chapter). In 1921, Loewi showed that the vagus * and sympathetic nerves of a frog's heart liberate two different substances. In 1926 he showed that the "vagusstoff" closely resembles a compound, acetylcholine, first prepared in 1867.

The exceedingly small quantities involved made investigation difficult. A large enough scale to show the widespread formation of acetylcholine was only secured, in 1929, by Dale and Dudley. It has been even harder to show conclusively that the accelerator due to the sympathetic nerve really is adrenaline; and the whole subject is still *sub judice*. It is, however, widely held that the autonomic nerves, at least, may be classified according to which substance they liberate.

The key fact about recent work on nerves and muscles is that, owing to the extreme delicacy of modern physical instruments, it has become possible to deal with single fibres, so that we need no longer be content with unanalysed averages. This has cleared up many of the difficulties as to the all-or-none rule. In 1909 Keith Lucas showed that this rule applies to the action of muscles attached to bones, and also to the nerve impulse itself. Different intensities imply different frequencies of impulse, not differing strengths. For investigating the electric accompaniments of nerve impulse, he used the capillary electrometer. In recent work this has been superseded by a technique involving the valve amplifier, the cathode ray oscillograph (for instance, Adrian and Matthews, 1928) and ultra-sensitive galvanometers. Adrian and Matthews in 1929 were able to follow a single impulse in a

* The vagus, it will be recalled, is inhibitive.

single nerve fibre. It had long been thought that there is no heat emission accompanying nerve impulse, but this has now been disproved, and the emission followed, by A. V. Hill in England and by Gasser and Erlanger in U.S.A. Hill, Gerard, and Downing's thermopile for this type of work (1920, improved 1932, 1933) involves hundreds of silver-constantan junctions. One important point elicited has been that, as in muscle, heat is evolved not only during the impulse but during recovery.

The inaccessibility and overwhelming complexity of the central nervous system, and especially of the brain, have made the analysis of its action on these electrochemical lines very difficult, but suggestive facts have come to light. Towards 1928 Berger discovered a roughly sine-like oscillation of potential between two points on the scalp, apparently due to the resting activity of the ocular region of the brain.

To find an electrochemical model for nerve impulse is one of the current ambitions. According to Hill, one of the less unsuitable is that of a wave of charge and discharge along the immensely long thin condenser constituted by the inner and outer coating of nerve fibres.

There is the same ambition as to muscle, where the foundations of recent work were laid by Hopkins and Fletcher. And there is the same difficulty, as yet, in forming a connected picture. We have mentioned the muscle protein myosin. Myosin chains which make up the muscle fibrils undergo reorientation, with contraction, under certain conditions involving energy transformation. But the old idea of carbohydrate oxidation as the source of this energy has been found to need much modification. The year 1927 was an important date here. In that year the Eggletons discovered an unstable compound, phosphagen,* in vertebrate muscle. Energy is obtained by a cyclic series of changes in which phosphorus is continuously transferred from one molecule to another. The subject is too complex to be pursued in this book, but it will be seen that here, as in other protein researches, contractility has come to occupy a central place.

Research on the chemistry of the nervous impulse has not gone very far. But it has suggested that a parallelism may be traceable with muscle: nerves can function for a time without oxygen, lactic acid is produced, phosphagen occurs and is broken down and rebuilt.

Viewing muscle and nerve together as to the all-or-none rule, an obvious suggestion arises that the quantum may play its part in the mechanism concerned, as in that of photosynthesis. Hill has considered this suggestion and has found no reason to negative it.

* Later shown to be a compound of phosphoric acid and creatine.

CHAPTER XVIII

THE GROWTH AND UNITY OF THE INDIVIDUAL. II

THE ENDOCRINES

PRIMITIVE peoples consume bodily organs and secretions to remedy their own deficiencies in them, or in the corresponding human qualities, digestive, warlike, amorous. In early days mystical potencies were assigned where function was unknown. As gross physiology became familiar, it was to the smaller organs that mysticism clung. In particular, curiosity was early excited about certain ductless but otherwise gland-like structures found in the body. Galen assigned to the pituitary the production of nasal phlegm, and thought that the thyroid lubricates the larynx. Descartes, as we have seen, credited the pineal body with the " secretion " of thought.

Such views depended on the readiness with which men believed in minute ducts or pores where they could see none, as with the pre-Harveian heart. Haller, in a more modern spirit, emphasised in 1745 that in fact no ducts are visible, and in 1766 classed the thyroid, thymus, and spleen together as pouring special substances into the circulation. In 1775 Bordeu (1722–76) suggested that each part of the body gives off its emanations. He stressed the tonic effects of the secretions of the sex-glands, as evidenced by the effects of castration. His more general assertion was before its time and was unsupported by detailed examples ; but in the early 19th century some such doctrine began to gain ground, and was even (1840) held proved, by some workers, where the adrenals were concerned. In 1849 came Berthold's proof that transplantation of the testes of a castrated cock annuls the usual effects on the comb. It has, however, been characteristic of the early stages of this subject that, as the first evidences were scattered and the facts complex, bursts of interest have been followed by periods of discredit ; * and most of the precursors of genuine discoveries have, like Berthold, only been unearthed afterwards.

The idea of internal secretions as links in the mechanism for keeping the internal environment constant was clear in Bernard's enunciation of 1855. This year marks a beginning in endocrinology. It saw Addison's (1795–1860) description of the constitutional effects of disease of the suprarenal capsules ; while in the following one Brown-Séquard asserted the fatal results of excision of the adrenals, and Schiff, of the thyroids. Now appeared another characteristic of the history of

* Discredit due to repulsive or exaggerated claims, such as those of Brown-Séquard as to the rejuvenating effects of testicular grafts or extracts, which, however, ended (1889) by concentrating interest on the subject.

endocrinology.: luckier or more skilful men asserted that they had carried out these excisions without causing death, provoking the retort that their excisions had been incomplete. This reminds us that the subject, no less than others, waited on technical advances ; and, in particular, in the fifties and sixties, on Lister's antiseptics. In fact thirty years passed from Bernard's pronouncement before the subject was effectively resumed.

Then, for some time, the thyroid was the chief subject. In 1882 Reverdin produced the (already known) morbid state called myxoedema by thyroidectomy. In 1888 Horsley proved that both this and cretinism are due to thyroid deficiency (hypothyroidism) ; and in 1890 successful treatment by grafting or (1891) by thyroid extract was begun. Here appeared yet a third characteristic of endocrines : the improvement lasted only as long as the application continued. The methods available treat the symptoms ; they do not remove the cause.

Then in 1891 Brown-Séquard and D'Arsonval brought out the wide interest of the subject by describing a non-nervous integration of the cells of the body. This made it natural to suppose that glandular excess as well as defect might produce morbid results. The previous view had rather been that these glands merely removed or made harmless specific toxic substances apt to get into the blood ; so that their excessive action would not be a serious matter. The present century had dawned before exophthalmic goitre was accepted as hyperthyroidism. The later, intensive work which set in after about 1910 gave the thyroid the rôle of maintaining normal oxidation, and so growth and body-temperature. We reserve the chemical side of the picture, the connection with iodine metabolism, until later.

We have described (Chap. IX) Bernard's work on the pancreas. Secretion by this was known to be occasioned by the arrival of the digestive acid in the duodenum ; and in 1897 Pavlov suggested that this was a case of nervous reflexes, on which he was working. But in 1902 Bayliss and Starling showed that the pancreas is stimulated not by a nerve but by a " hormone " or " chemical messenger " (" secretin ") poured into the blood stream by the stimulated duodenum. This was a crux. To leave the pancreas for a moment, the earlier theories of cell emanations and interrelations now assumed definite form, and the blood was seen to be constantly receiving and delivering highly specific chemical substances. The identific[illegible]on of these became the ambition of a new generation. Th[illegible]rliest chemical success had just been obtained—adrenalin[illegible]om the adrenal glands (1894–5, see Chap. XVII). This [illegible]ws the heart and raises the blood pressure. It was prep[illegible]ystalline, and its formula ($C_9H_{13}NO_3$) was given by Ald[illegible] The adrenals seem to control the ionic concentration of the [illegible]ssue fluid, especially as to sodium and chlorine.

In 1880 the parathyroids had been described by Sandström ; but for some time they were generally viewed merely as potential thyroids. Then in 1903, and again in 1907, the state called tetany was found accompanied by changes in the parathyroids. In 1909 it was further correlated by McCallum and Voegtlin with error of the metabolism of calcium. The effects of this would naturally be wide, for calcium is among the most important elements both in animals and in plants. It shares with silica the responsibility for the hard parts of both classes; but lately (e.g. Heilbrunn, 1936) it has been suggested that its part really lies much deeper, since the plasma membrane of the cell is at least partly a calcium gel. Its metabolism, too, affects that of phosphorus, so fundamental in muscle and nerve.

To revert to the pancreas, Langerhans had, in 1869, revealed in this organ certain "islets" which have since borne his name. In 1889 the excision of the pancreas was shown by Von Mering and O. Minkowski to result in diabetes. In 1893 the islets were recognised as having an "internal secretion" of their own, which was apparently the anti-diabetic factor, since deprivation of the pancreatic juice alone did not produce diabetes. In 1913 Schafer called this secretion "insulin"; but it was 1921 before Banting and Best succeeded in isolating it.* The long interval was due partly to the difficulty of estimating small quantities of sugar and partly to the fact that insulin is destroyed by trypsin, the secretion of the pancreas proper. Since this work, the relation of insulin to carbohydrate metabolism has become well known.

There was no doubt, by the time that the hormone doctrine was established, that sex hormones or some equivalent of them, must exist in men and animals, but the subject is complex and we shall only note that "androsterone" (isolated 1931) is important in the male, while "oestrone" (1926), in the female, is concerned with producing the oestrous cycle. We refer in Chap. XXIII to the chemical resemblance of these substances to the carcinogenic compounds (Kennaway, 1930, etc.). There are other male and female hormones, the ovarian hormones of the female being necessary for the development of the secondary sexual characters and for the early stages of pregnancy. In both sexes they produce, like other hormones, general secondary effects, and reciprocal effects on other organs which affect them, such as the pituitary.

But we must press on to consider the main collective character of the endocrines, namely, their forming a system each part of which affects, and normally balances, each other part. The organ to which Cushing (1869–) has assigned the dominant and unifying rôle in this interaction is the pituitary, and especially its anterior lobe. This, of course, implies that, originally, the specific and proximate rôle of the organ was especially hard to disentangle. As early as 1887 it was found that acromegaly and gigantism were correlated with abnormal conditions of it, but so they were with other things; while in 1898 the pituitary was found to be connected with the oestrous cycle.

A further difficulty was that until the arrival of the special surgical technique of the Roumanian Paulescu in 1908, complete excision was a fatal operation. It was, in fact, only in that year that Cushing definitely assigned gigantism to hyperpituitarism and dwarfism to its opposite.

Evans and Long in 1921 found a growth hormone in the pituitary,

* A contribution from Canada.

while from 1924 the effect of the gland on the oestrous cycle took definite shape, and in 1928 it was found to secrete a thyroid stimulator. To its effect on the thyroid was also traced (1930) the effect on metabolism which had been attributed to it in 1927. In 1927, too, its effect on the adrenals was shown, and in the nineteen thirties its specific action on most of the others has been proved. In particular, it affects carbohydrate metabolism, antagonising insulin. It also affects fat metabolism.* To complete the picture of interaction it should be added that many of these glands have also been shown to act directly on each other.

The view is held that the endocrine system was, in evolution, the earliest method of coordinating the parts of the body. Nerves came later. Hormones are chemically the same in widely different groups, though their effects may be different. Thyroxine from mammals is as effective as the home-produced article in causing amphibian metamorphosis. Adrenal extract from fish causes its usual effect in man. There are also " plant hormones " (1910). These are *not* the same as animal hormones, but they seem to be the same in all the plants considered. We can sketch them only briefly.

The responses of plant stems, leaves, and roots to light and gravity had been correlated by De Candolle (1832) with differential growth, as on the two sides of a bending stem. In 1910–11 Boysen-Jensen suggested displacement of a growth-stimulating substance as the cause. It will be remembered that this was a time of activity in the subject of endocrines and also in that of vitamins. The neat modern method of diffusion into and out of blocks of agar was developed by Stark (1921) and Went (1927–8). Much is now known about these plant hormones, or " auxins," including their constitutions, which are relatively simple. They seem to be formed in regions of most rapid growth, and by their diffusion away from such regions they " control " the growth gradient, much as the hormones in the blood stream control various activities in the animal body. Like some of the animal hormones, they can also cause tumours.

The auxins so far studied seem to cause cell-enlargement, not cell-division. The latter may possibly be accelerated by some compound which (like glutathione) contains —SH, and which may act as an oxygen carrier (see later). The subject of plant growth is very complex. Several of the vitamins seem to promote it. Ethylene, too, rouses plant tissue from dormancy. In fact, the subject is only just being opened up.

Vitamins

Scurvy, to early doctors, was not more evidently due to dietary deficiency than was any other disease. The American Indians had a cure for it, prepared from spruce needles, when Cartier touched Newfoundland in 1535, but over two centuries passed before Europe had one. The disease was a veritable scourge, especially where (as in ships

* Protein metabolism, to complete the triad of bulk constituents, has not yet been directly correlated with the endocrines.

or barracks) wholesale provisioning led to an artificial diet. In 1720 an Austrian army doctor Kramer asserted that a little lime or orange juice soon cured it, and in 1747 Huxham prescribed vegetables for the British fleet. But 1757 came before Lind's work gathered up the threads and made further outbreaks needless. Lind (1716–94) was a naval surgeon.

Even though prevention, and that by food rather than drug, thus succeeded cure, two-thirds of the next century passed before the connection with food was realised. Until then, even the gross factors in nutrition had not been clearly grasped.* Moreover, a stimulating draught of physics was needed to make nutrition a live subject for research.

It will be remembered that, in Germany, energy doctrine originated in physiology (Mayer, Helmholtz), with special reference to metabolism. Like thermochemistry, the exact calorimetry of metabolism came in with better calorimeters in the sixties. Von Voit and Pettenkofer began (1866–73) to examine the optimum quantities of the three already-"classic" types of foodstuffs, proteins, carbohydrates and fats. Then in 1773, Forster, one of Voit's assistants, showed that dogs and pigeons fed on a diet of these substances artificially purified died more rapidly than of starvation. This he traced to elimination of certain minerals, which thus merit the name of the first known "accessory food factors." But thermal investigation still dominated. It had, and has, a relevance to many diseases besides those of "deficiency." In the eighties and nineties, Rubner, Atwater, Bryant, and others produced refined calorimeters which made it necessary to distinguish the resting metabolism of the body when in a bath of its own temperature and when (as normally) it is giving heat to its surroundings.† A beginning having thus been made with defining basal metabolism, various physical aspects of the subject could be settled. Bergmann (1845 onwards) had asserted that such metabolism is proportional to the bodily area, and in 1883 this was verified by Rubner. Its value per unit area, very low in the new-born, was shown to be 50 per cent. above the adult's in infants of one year and to go on increasing for some years.

In 1880, Lunin and Bunge began work which showed that young animals, though they can live, cannot grow on synthetic milk, but this observation did not at once lead anywhere. The more sensational study of the deficiency diseases was to make headway first. Between 1882–6 Takaki attacked the disease beri-beri. He was able to eradicate it from the Japanese navy by means of a mixed diet, although he was wrong as to its origin, attributing it to protein deficiency. Thus the military or naval connection continued. Imperial interests also produced the next step. In 1897 the Dutchman Eijkman, produced beri-beri in birds by feeding them on de-husked ("polished") rice. He showed that this "polyneuritic" state could be cured by alcoholic

* In 1825, Blane had seen that the cure of scurvy could not be due to bulk factors furnished by the lemon-juice. But this passed unnoticed.

† Cold-blooded animals usually have a temperature differing somewhat from that of their surroundings; as John Hunter knew. It has lately been found that the same applies to plants.

extracts of the rice polishings. But it was 1901 before his co-worker Grijns interpreted this as a dietary, not a pharmaceutical, matter. He thus ushered in the viewpoint of a new century in which vast towns, like barracks of old, were being fed on synthetic and preserved foods.

Another disease, rickets, nearer home, illustrates this point. Towards 1890 one widely-held theory attributed this to deficient diet, and to diet deficient in quality rather than quantity. Many other theories, however, were current, and it began to be clear that fundamental research on the whole nutritional field was needed. To link data on deficient diet with that on the more sensational deficiency diseases became the explicit object of Gowland Hopkins in 1906. A unified view was being reached. In 1912, Hopkins concentrated attention by quantitative data on synthetic diets. Like Pekelharing in 1905, he showed that minute quantities of fresh milk will make up for the deficiencies of the preserved product. In 1912, too, Osborne and Mendel showed, following Lunin, that there are diets on which a rat may be able to live but not grow. The addition of a specific factor brought on growth.

On the other wing of the subject, the Pole Funk, in the same year asserted that scurvy, beri-beri, pellagra, and probably rickets, were to be classed together as due to deficiency of accessory food factors, which he called " vitamines " (final " e " now dropped).

The matter was now fairly under way, but it was still widely held that only *one* such factor was involved. By 1915, the Americans McCollum and Davis had made it clear that there are at least two, vitamins " A " and " B "; and the process of isolating such factors began. For this purpose a primary question turned out to be solubility in fat solvents or water. Between 1913 and 1916 McCollum, Davis, and Kennedy located fat-soluble " A " (responsible for growth) in butter-fat and eggs, distinguishing it from water-soluble " B." This latter was surmised, and later proved (Goldberger, 1925), to contain the anti-beri-beri factor in the rice-husks. But Goldberger soon afterwards proved that B is not simple, containing anti-beri-beri B_1, and anti-pellagra B_2, itself not simple.

In 1913, Holst and Fröhlich found the anti-scurvy water-soluble " C " in citrus-fruit juices. The anti-rachitic " D " was more difficult, as it was easily confused with " A." But in 1921 the American school made it clear that preparations of " A " are not anti-rachitic; and soon afterwards they separated an anti-rachitic fraction from cod-liver oil. In 1926 " D " was found to be produced by subjection of a definite compound, ergosterol, to ultra-violet light. In its absence the absorption from the intestines of bone-forming calcium and phosphates is lowered. Anti-sterility " E " was located by Evans of California in 1922 in meat, wheat germ, lettuce, and elsewhere.

It is practically impossible in a book of this size to give a significant discussion of the biochemistry of the hormones and vitamins. To treat them together is itself to adopt a fairly recent point of view (Abel, 1928, vitamins as "extraneous hormones"). The obvious common point is of course that, like catalysts and enzymes, they act in very small quantities. Each class is chemically highly various, but there are several ill-understood relations between members of opposite classes: the sex hormones have some resemblance to "D." "C" has been found in the adrenal cortex; which latter, with the anterior pituitary and "A," "D," and "E," have effects on growth. The parathyroid hormone has an unclear relation to "D," both affecting calcium metabolism. It is, however, a very different substance, probably proteinous if the present impure preparations are a guide.

The following is an almost arbitrary choice of items from this vast and rapidly moving subject.

We have mentioned that the thyroid concerns iodine metabolism, and in 1914 Kendall isolated an iodine compound, thyroxine, from it. In 1927 Harington and Barger found the constitution of thyroxine. The subject, however, is by no means closed. Only 20–25 per cent. of the thyroid iodine is in the thyroxine.

Early studies of "C" led Zilva to conclude that it was only stable in the presence of a strong reducing agent. Others thought that it actually *was* such an agent. It was not then known that "C" occurs in the adrenal cortex; but in 1920 Szent-Györgi was studying oxidation and reduction in the latter, and in 1928 he isolated a reducing acid which he connected with Zilva's requirement for the stability of "C." But it was 1932 before he found that it is itself anti-scorbutic. This "ascorbic acid" is now identified with "C" itself.

It has been said that the sex hormones, male and female, resemble in formula certain of the carcinogenic compounds isolated by recent medical research. So do "D" and the compound ergosterol from which it is derived. We refer to this again in Chap. XXIII. The common basis is the ring formation

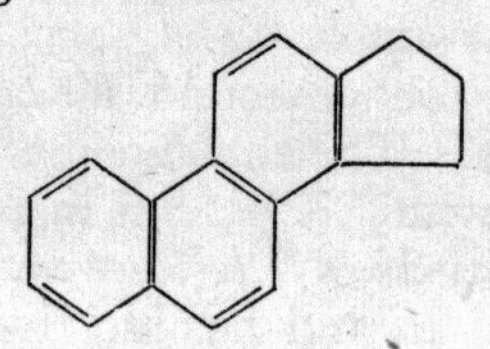

but, needless to say, the side-chains are not the same and such a common nucleus is compatible with very different chemical properties. "A" and "B_2" have been found to have in common a close relation to naturally occurring pigments, carotene (Steenbeck, 1919) in the former, lacto-flavin in the latter, case. The carotenes have two rings joined by chains of alternate single and double bonds. In spite of superficial contrasts they seem to be related in origin to the above ring-formation.

From the last paragraphs it will be clear that chemists are coming to be able to see in one picture many widely scattered biochemical groups. In the remainder of this chapter we show them making out one or two further patterns. We consider especially, first, respiration (" oxidation-reduction ") processes, then photosynthesis. Both these are closely related to pigments.

We have mentioned the early history of the photosynthetic pigment chlorophyll in Chap. IX. Haemoglobin, among respiratory pigments, has long been known, and the criminological importance of being able to detect minute traces of it was responsible for its being examined spectroscopically as soon (1855) as spectrum analysis was developed. In 1864 chlorophyll was found not to be a simple substance, and in 1882 spectrum analysis revealed that it is very largely red and violet light, especially the former, which activate chlorophyll in the plant. In fact, the absorption spectrum of the substance roughly follows the curve of energy of the solar spectrum.

Soon, however, with the coming of concentrated organic analysis, individual pigments began to be lost in the crowd of those occurring naturally or used in the dye industry. Functions other than oxygen-carrying and photosynthesis were found to be carried out by them. Their coloration seemed often to be quite incidental.

In 1913 Wieland suggested a general equation of respiration processes; but the equation was not supposed to give more than the barest summary of the complex reactions in the living cell. Keilin supposed that the hydrogen and oxygen which are the pivots in oxidations and reductions must both be " activated " in the cell; since reactions occur there between compounds which are stable to each other *in vitro*. Enzymes were known to be concerned here, for instance the " indophenol-oxidase " of Batelli and Stern (1912).

Then in 1921 Hopkins isolated a tri-peptide " glutathione," containing the sulphur group —SH, the respiratory activity of which had been observed by Meyerhof in 1918. Further work (e.g. Hopkins, 1929, Brand and Harris 1933) has revealed that this substance is present in all animal cells, and perhaps in all plant cells too, certainly in those of yeasts, fungi, bacteria. It acts as carrier both ways in oxidation-reduction reactions. E.g.:

$$2G{-}SH + X \rightleftharpoons \begin{array}{l} G{-}S \\ \;\;\;| \\ G{-}S \end{array} + XH_2$$

It was also known, however, that certain inorganic catalysts, and notably iron, are of importance in these reactions; and it was in this connection that the pigment theme recurred, iron being a known component of the blood pigments. For a long time (1912–28) Warburg had been investigating the effects, as respiratory catalysts, of certain iron pigments of the general type of haemoglobin. Then Keilin (e.g. 1925, 1929, 1933) urged the similar importance of the " cytochromes " discovered by MacMunn in 1884. Keilin found that these are widespread among cells, and that their concentration is roughly proportional to the respiratory activity of the cell. Before we discuss this, we note that meanwhile, but especially 1907–14, Willstätter and his pupils had been discovering the constitution of the chlorophylls. The remarkable fact thus emerged that nearly all the prominent pigments involved were made up of rings or chains of pyrrol

HC——CH
HC CH
NH

rings, and that in particular both chlorophyll and the chief " haem- " compounds consist of a ring of four pyrrol rings (P say) :—

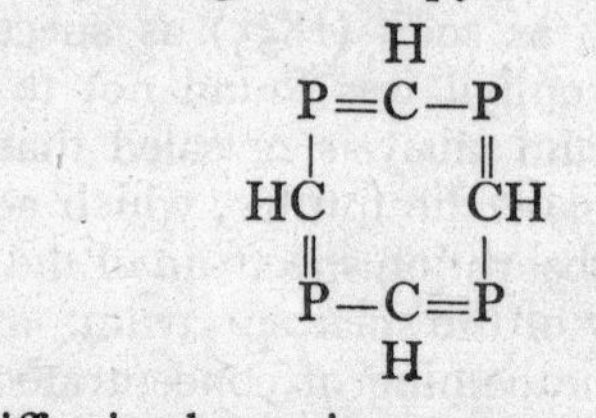

The side chains differ in the various cases, while the chlorophyll group have magnesium, and the other group, iron, as a kind of diagonal strut. Further, the latter are united with protein molecules,* and the iron seems to be, in the main actions concerned, freer than the magnesium. In fact, in the last few years, oxygen exchange has come to be viewed as a system of balanced catalysed reactions. Iron and manganese seem to act as antagonistic catalysts in systems where the ring of pyrrol rings (whether haematin or chlorophyll) intervenes in the oxygen transfer, while copper and zinc perhaps act likewise where a glutathione or ascorbic acid † mechanism is acting. The three bulk constituents of living things have been linked in a complex interchange where animals are concerned. Hormones are concerned in carbohydrate formation, and probably *vice versa*. Carbohydrates are formed in part from the amino-acids coming from protein digestion. They themselves are partly used as raw materials for fat formation.

Turning now to photosynthesis, we find that chemists are still out of their depth. The original view had been that formaldehyde is the first product :

$$CO_2+H_2O=CH_2O+O_2$$

and that this afterwards condenses to more complex carbohydrates. But, in spite of many attempts, no one has *proved* that this reaction occurs. By 1925, it was clear (Briggs, Willstätter) that enzymes and oxygen, as well as pigments, are necessary to the action. But the course of the latter itself is still largely unknown. One point to be noted is that lately much has depended on the mastery of *physical* instruments such as spectroscopes and colorimeters.

In 1925 Kluyver and Donker extended Wieland's general respiratory equation to cover photosynthesis too :

$$H_2A+B \rightarrow A+H_2B$$

and it has since (e.g. Van Niel, 1935) been pointed out that this covers

* Lubimenko, however, concludes that this is the case also with the photosynthetic pigments (1926–8).

† Found, it will be remembered, in both animals and plants.

both ordinary green plants and the green and purple sulphur bacteria, which contain photosynthetic pigments of closely similar type. In the first case

$$A=O, B=CO_2:$$
$$H_2O+CO_2 \rightarrow (CH_2O)_n+O_2.$$

In the second

$$A=S, B=CO_2:$$
$$H_2S+CO_2 \rightarrow (CH_2O)_n+H_2O+S$$

But these equations furnish only end products. One clue to the intermediate stages is based upon a fact of general photochemistry. The history of this subject cannot be touched here except to note that in 1912 Einstein applied the quantum theory (see Chap. XXI) to it, enunciating on thermodynamic, then (1913) on quantum grounds, the principle that one quantum of light transforms one molecule. This has turned out to be rather a guide than a law, but its relevance here is that in 1923 Warburg and Negelein found that four quanta are needed per CO_2 molecule; so that effort has been directed especially to finding four-stage models for the photosynthetic reaction. But no entirely satisfactory model has yet been proposed. Nor is this the only type of difficulty. Arnold and Kohn point out (1936) that the absorption unit of an alga contains perhaps 500 chlorophyll molecules, and even then quanta from several such units seem to be needed to secure the necessary four. A mechanism, based on surface effects, by which this might occur has been suggested by Weiss, but the whole subject is involved in obscurity. According to Thomas (1935) it still awaits the application of modern microdissection methods to disclose the minimum unit which can effect photosynthesis.

But photosynthesis and the phototropic effects mentioned earlier are by no means the only important effects of light on plants. Another is photoperiodism, with which we may conveniently couple "vernalisation." The latter discovery was the earlier. Indeed, before 1857 the American Klippart was working on the conversion of winter into spring wheat by chilling the germinated seeds, while in 1875 the Russian Gračev controlled the time of fruiting in maize and artichokes by similar means. Photoperiodism was noticed in hops by the Frenchman Tournois in 1912 and taken up in 1920 by workers in the U.S. Department of Agriculture. It was found, in brief, that some plants, to come into flower, need a day of at least a certain length, while others need one of not *more* than a certain length.

Vernalisation for wheat was seriously taken up again by Gasser in 1918. Many of the main facts of both phenomena are due to Klebs. We shall not, however, attempt to assign credit in detail. The Russian, Lysenko (e.g., 1932) envisaged the two phenomena as the earliest in a series of phases, each necessary for the full development and reproduction of the plant. In the "thermo" phase, a certain low temperature is needed for a certain time, in the "photo" phase, a certain minimum of light for at least, or at most, a certain time. This minimum, it may be remarked, is much lower than that needed for photosynthesis, so that this latter seems to be a distinct phenomenon. More recently, a more complex series of phases has emerged, in which the parts played by light, heat and other factors are not as clearly separable as in the first scheme. Lubimenko has emphasised (1933, 1934), as against Klebs and others earlier, that growth (a process of *accumulation*, by photosynthetic and other means) is quite distinct from development, which includes reproduction. Reproduction is not the result of growth's having reached a certain point. Its organs develop concurrently with growth from an exceedingly early stage. The fact, indeed, that they are to an extent rivals is one of the themes of contemporary biology; and it has been suggested that if we desire only the seed, we may, in future, be able to curtail purely vegetative growth almost indefinitely. Be that as it may, the economic possibilities of these discoveries are great. Many countries are re-examining the possibility of growing all sorts of alien crops. Countries within the Arctic circle, and in particular Russia, are especially interested. Hence the number of Russians concerned in the work. The facts involved must have profoundly affected also the natural distribution of plants; since in the tropics neither cold, nor very long nor very short days, may be available.

CHAPTER XIX

MAINLY ECOLOGY

Ecology concerns the relations of living things with their environment and with each other, whether described in biological or in physico-chemical terms. Before turning to it, we consider two related points. First, we return for a moment to pigments, the visual aspect of which leads us (for instance) to the ecological relation of insects to flowers. Second, we consider the embryonic subject of comparative physiology and biochemistry. It is in virtue of their chemical contrasts and similarities that living creatures dovetail into one another, so that ecology enters again.

How did plants originally come by the very complex chlorophyll molecules on which their main growth depends: how, in general, did life originate? At first sight the question might seem too speculative to receive much attention. But to suppose that science always observes due caution is to oversimplify history. Were an attractive and widely applicable interpretation of (say) certain micro-features of azoic rocks to present itself, it might be eagerly taken up.

Possibilities discussed have been, bacteria or other low forms capable of using light or other available energy with no catalysts at all, or with such simple ones as iron which might have occurred free in the early periods of the earth. Whole ages of rock formation may thus have passed, and it has been claimed that certain appearances in early rocks suggest algal action. We have mentioned that Tilden (1935) has correlated the colours, and therefore the absorption bands, of the blue-green, red, brown, yellow-green, and green algae still common, with the differing absorptive powers of the hotter, more opaque atmospheres of early periods. All these algae contain chlorophyll, their differing colours being due to their having other pigments as well. Now the universality of certain other pigments is one of the realisations of current biochemistry. It has been asserted (1935) that all biological photo-processes, including vision, involve carotenoid compounds. This is more than can be said of chlorophyll itself. It will be recalled that vitamin "A" belongs to the carotene class, while Wald (1933) found that the "visual purple" of the retina of the frog (and therefore probably of mammals) behaves as a carotene attached to a protein (compare the haematin compounds).

This brings us to the visual side of pigmentation. It will be recalled that Helmholtz had revived Young's theory that, physiologically, vision is due to the presence on the retina of three "elements" sensitive respectively to red, green, and violet. But it had always been realised that, psychologically, yellow is quite as fundamental; and this suggested a *four*-element theory, given definite form by Hering in 1874.

Viewing yellow and blue and red and green as complementary pairs each making white, Hering postulated three balanced reactions in the retina, corresponding to these two pairs and to black-white. These balances were upset, with stimulus to the optic nerve, by incident coloured light. This theory cannot be true as it stands ; for instance Fick in 1890 pointed out that red and green do not produce white, but yellow. In 1893, however, Mrs. Ladd-Franklin used the latter fact to give a new, evolutionary, theory.

Originally, she suggested, white alone was detected (by the retinal rods). Awareness of white next split, with the coming of the cones, into sensitiveness to the complementary pair blue and yellow (possessed by the bees). Yellow-sensitiveness next split, as above, into red-and-green sensitiveness. A modified form of Hering's chemical mechanism was postulated.

Such theories open up many avenues. Thus, early flowers are supposed to have been mostly white and yellow, the colours detected by the early insects. The comparative study of the visual and other sensitiveness of animal types has become quite a line of research. Something, too, is known of the physical basis of coloration in nature. Thus, the brown of young birds is the oldest colour in that phylum, then black, white, and grey. All these are pigment colours, due to melanin. Then come blue and green. These are interference colours, due to the structure of the feathers, and recall Rayleigh's tracing of the blue of the sky to dispersion. Finally come red and yellow, due to fat-soluble pigments in the feathers.

Comparative physiology and biochemistry, of which we have already noted some instances, are still in an early stage. The second depends on the first ; as may be seen, for instance, in the case of oxygen-transport in animals. Insects have a very different physiological device for this from vertebrates. Their extensively branched tracheal system enables them to transport oxygen to their tissues without—and here comes the biochemical contrast—using the chemical carrier haemoglobin.

One fundamental possibility which has been examined is whether chemical relatedness can be taken as evidence of common evolutionary origin ; and the conclusion has been that generally it cannot. A case in point would have been the development of the saline solution which now forms the internal environment of life out of that other environmental saline solution, the primeval ocean. Another would have been the evolution of haematin from chlorophyll or of both from the perhaps older and more ubiquitous cytochrome. But the evidence suggests rather that invertebrates have " tried " a number of respiratory pigments, and that several different phyla have hit independently on haemoglobin. Cases of such " convergent evolution " are frequently brought to light nowadays.

But the immediate difficulty of comparative studies is the surprising fewness of deep chemical contrasts even between plants and animals or vertebrates and invertebrates. Deep contrasts, of course, there are, which run with the great divisions ; but there are many more which run right across them. Moreover, individuals of a species may vary enormously, and an individual may vary widely from time to time.

Mez's serum taxonomy of plants is in one sense a triumph of comparative biochemistry; but to test sera one against the other is a very different matter from expressing their compositions in atomic terms. This latter process has gone very little way. Carbohydrates do present well-known broad contrasts running with the great groups: the starch and cellulose of plants, with their storage and structural functions, the storage glycogen of vertebrates. At the other extreme is the case discovered by Heidelberger (1925, 1927)—the existence of three different strains of pneumonia bacillus, enclosed in capsules of three different carbohydrates, distinguishable by their being hydrolysable by three different enzymes. Dissolution of the capsule allows the blood to deal with the protein within, so that attack on the disease becomes possible. The same has been found with other bacteria. In general, minute parasitic plants, such as moulds, furnish instances of extraordinary variability of physiology and biochemistry without morphological change to correspond. Hundreds, perhaps thousands, of strains have in the last year or so been distinguished in a morphologically homogeneous species of mould. Since some of these changes are inherited, this is an earnest of future biochemical contributions to systematics.

The best cases so far found of chemical contrasts running parallel to ordinary systematic characters occur among the simpler nitrogenous compounds. One is the case of creatine phosphate, in vertebrate, and of arginine phosphate in invertebrate, muscle.* Another concerns nitrogen excretion—purines. Most invertebrates and some lower vertebrates (e.g. bony fishes) excrete mainly ammonia. Most vertebrates excrete mainly urea, as do cartilaginous fishes. Other groups excrete mainly uric acid: such are insects, birds, and reptiles. Herbivora excrete hippuric acid. Men and the higher apes are deficient in the power of excreting uric acid. Needham (1929 and later) has correlated these contrasts with the environment of the *embryo*: a good instance of the importance of keeping growth-changes in mind even in chemical studies. Ammonia and urea can diffuse, uric acid cannot. The latter is thus less of a danger in closed-box eggs, while the former can readily be got rid of in an aquatic environment or where, as in mammals, the organism in its early stages remains attached to the adult and perfused by its body-fluids. The contrast of leguminous and other plants reminds us that there are also nitrogen contrasts among plants.

Another suggested contrast running with existing divisions was that animals make hormones, plants vitamins (or their immediate precursors). But there are many exceptions. Chicks can synthesise "C." Cod seem to make "A" out of carotene provided by planktonic diatoms; for the "A" is absent from the copepods which are a

* Bateson's identification of the echinoderms as the nearest relatives of vertebrates among the invertebrates received unexpected support from the finding of creatine phosphate in the muscles of the jaw (Aristotle's Lantern) of the sea-urchin.

main link in the diatom-cod food-chain. On the other hand, copepods seem to get ergosterol irradiated " for the benefit of " the cod, which lives below the depths to which ultra-violet light penetrates. Of the function of vitamins in plants little is known.

Ecology

We have said that there was something left, in the first generation after 1859, of the atmosphere of Cuvier and of the 18th century, some thing stiff with pigeon-holes or phases into which evolution was rigidly fitted. We have also noted how this broke down into a largely descriptive science. A few of the wider features of this, the distribution over the whole earth and over geological time, have already been sketched. This distribution was, of course, built up of countless local inter-relationships. It was these which Haeckel and others, towards the eighties, began to see as a single discipline, ecology.

Ecology considers the relations of the living creature both with its inorganic environment and with its fellow creatures. Plants look less dependent than animals on the latter (" biotic ") sort of relation, and less dependent on animals than animals are on them. Hence, plant ecology started first and has been able to keep close touch with physiology. Its first text-book, that of the Dane Warming, was published in 1895, and stressed mainly the non-living factors, to be followed in 1898 by that of Schimper with more stress on biotic influences. The first text-book of animal ecology was published nearly twenty years later. Animal censuses are a generation behind those of plants, chiefly dating from about 1910, and being still very scattered and local—a lake or river here, a spinney or heath or fishery there. We treat plant ecology first.

We have said that the ecological point of view as such is not strong before about 1875–80. But isolated ideas occurred much earlier. L'Obel in the 16th century noted that mountain plants of warm countries descend to lower levels in colder ones. According to Clements, King in 1685 recorded one ecological phenomenon, " development," a steady, progressive change in the vegetation of a region, in his case an Irish peat bog. Humboldt used ecological ideas, and in 1844 Lecoq hinted at the idea (not the word) " dominance," the frequency, among plant societies, of a vast predominance of one or a few species. On the non-biotic side, we have seen Liebig working, in the agricultural interest, on the effects of soil on plants, and have noted the general recognition of the carbon dioxide and nitrogen cycles in plant economy. Unger, for instance, in 1836, began to show the close dependence of the type of plants on the chemical composition of the soil where they grow, and in 1849, Thurmann began the correlation of soils with their geological substrata. Here is an instance of a recurrent feature in ecology, its enormous economic importance, which has greatly swayed its development.

Kerner (1863) is claimed as a precursor of the idea of " succession." In its full extension, this envisages the development from uncolonised earth through a series of types of vegetation, each preparing the way for its own downfall at the hands of the next. Soil itself, as a suitable medium for higher plants, is no naturally occurring product, but has to be built up. Wet coastal sand, say, is colonised by minute algae whose glutinous sheaths bind it in some measure against water-erosion. Dying, they are attacked by the bacteria of putrefaction. These, and certain fungi, leave an accumulating humus of rich organic remains, making the ground suitable for certain adaptable " higher " land plants, like mosses, which further bind the ground and retain its water until it may become firm enough to support the stronger root plants and perhaps ultimately trees. These forthwith are " dominant," the mosses and algae only living a dependent existence on them. The first man to realise the universality of this type of process, and to suspect that many of its instances fall into a small number of fairly rigid successional series, was Hult, in 1886 and 1888. He also asked the question which it naturally suggests, namely, does the process come to a " climax," thereafter stopping, or does it go on and on ? He favoured a climax,* but, not realising the above-mentioned soil-*forming* activity of life, he thought that soil fixes the point in the succession at which the process stops in any given region. Schimper (1898) realised that climate might also set the limit, and since then Clements (1916) and others have urged that it is the main final factor, since it alone really limits the soil-forming powers of the plants. The geological substratum has less *final* effect than might be supposed on the *features of soil important to plants*.

Still, the idea of climax has always had doubters, such as Raunkaier and Gleason ; and climate, also, is by no means universally given final place. The known successions have not been long in the scale of geological time : several may in places be traced since the ice age, the last major change in climate. This discovery we owe to an ingenious advance in technique which has also given us much other information. Lagerhaim of Sweden has shown that peat contains pollen from past ages which is recognisable when all other parts of a plant have gone. Statistical analysis of such pollen has become quite a regular activity.

But the point is not only that the evidence for climaxes is criticised. It is, rather, that, although Warming (1895) saw the universality of change, a working stress on the latter has been chiefly confined to America and England (Clements, 1899 etc., Cowles, 1901 etc. ; also Tansley). Continental work has been rather on the analysis of communities as found under various limiting factors. It is natural that, as in meteorology, Scandinavians should be foremost in this, for cold and other limiting factors are very familiar to them. Raunkaier (a Dane) classified plants (1905, 1918 etc.) into a definite number of groups according to their way of bearing the hard season ; and on this basis could give a " spectrum " (of the number of plants of the different kinds) as a quantitative description of any given community (1908).

* In undisturbed dry-land cases, trees tend to be the climax, but most of the earlier stages remain, only subordinate. This illustrates the ecological idea of a " plant community."

This is a convenient point to note a limiting factor in ecology itself. Field studies are very much hampered by the factor of instruments, always so important in science. Robust yet sensitive field instruments (for instance, for measuring sunlight or plant transpiration) are much needed.

We conclude plant ecology with a brief account of soil science, of which we have noted some beginnings, taking it as an instance of the universal interdependence of living things. It has long been realised that many plant and animal individuals are in fact colonies. In many species, from termites to men, intestinal bacteria and protozoa play a part in digestion, often a useful part provided that the dominant creature is able to keep them in subjection, but one in which the " balance of nature " is always liable to be upset, or rather, has never really existed.

Again, in 1842, Schleiden showed that the roots of certain plants always have fungi. Of more interest, both ecologically and in its bearing on an old agricultural problem already mentioned, was another case. In 1866 certain nodules seen by Malpighi (1686) on the roots of leguminous plants were shown to contain bacteria. Ten years later Berthelot took the subject up, and after a further decade showed that certain bacteria present in the soil can fix nitrogen from the air. In 1887 it was shown that leguminous plants fertilise the soil by increasing its nitrogen content, and in 1888 the Germans Helriegel and Wilfarth showed that the bacteria in the nodules transform atmospheric nitrogen into the nitrate form in which most plants take it up.

Since then the complexity of soil ecology has grown increasingly clear. Among plants, minute algae, as well as fungi and bacteria, play their part. Among animals, earthworms are influential.* Most insects spend some phase of their life in the soil. The water films in and on the crumbs of soil are teaming with minute protozoa which prey on bacteria. Micro-organisms of various kinds break up the cellulosic and ligneous compounds which would otherwise choke the soil when plants wither and die. Bacteria may, indeed, by secreting strong acids and weathering rocks, play an enormous part in soil-formation in the deeper sense ; while sulphur and iron bacteria may have had great rôles in earlier ages of the earth.

In 1845 Thompson observed the process of base-exchange in clay soils, and lately, after a long gap, the central importance of this for agriculture and ecology has been realised. In this connection, as in so many others, hydrogen ion concentration has been increasingly stressed (since about 1908–9 when Sorensen made plain its importance in connection with enzymes). Clay holds the ions against dissolvent water and stores them for plant nutrition and for keeping the ionic environment sufficiently constant. Viewing recent work on the crystal lattices of clay and protein (see Chap. XXIII), one would hardly exaggerate in suggesting that these two together, and not only the latter, should be viewed as " the colloidal complex " at the foundation of life.

The recent study of soil, and of the dependence of plants on it, has

* A perception we owe to Darwin, though Jenner had had it before him. More recently it has been realised that termites largely replace worms in the tropics. In either case the function is mainly mechanical.

been associated (1914 onwards) with Glinka and a Russian school. Here we can only note that, in line with general plant ecology, this school gave climate the final word in soil formation. It was very much the vast variety of their national climate which directed their interest. The unit for recent soil research is the "soil profile," that is, all the layers down to the mother rock to which they approximate, not only the superficial layers. No smaller unit enables us to follow soil dynamics.

All this shows again the close economic dependence of some sides of plant ecology. In animal ecology, to which we now turn, the dependence has been even closer; for, however many bacterial, fungus, and virus diseases crops may suffer from, animal, and especially insect, pests are their most conspicuous enemies. The later generations of the 19th century were busily occupied in trying to grow crops in altogether new parts of the earth. The native life of every sort soon took advantage of this; while (as with rabbits in Australia) undesirable imports from the original setting were also unwittingly made. The phylloxera of the vine, against which we have seen Pasteur at work, was a case in point. It came from America. New pests and diseases, from locusts to yellow fever, made their appearance. Some of these, like malaria, revealed parasitic life-cycles of extraordinary complexity. Others called for strange remedial measures, such as pests of pests. Yet others proved baffling from their apparent causelessness, such as the huge fluctuations in the numbers of fish or of fur-bearing animals, to which we return later.

Some of the categories of plant ecology can be taken over at once into the animal study. Dominance, in numbers or influence, is extraordinarily widespread among animals. Within the individual the nerves "dominate." Within the hive the queen, if she does not dominate, is the keystone of the arch. Zuckermann (1932) describes baboon families as dominated by the male; dominance is far commoner in human history than equality. The work of Howard (1920) has shown that male birds of many species establish "property rights" over a particular territory, from which other males and their dependents are excluded. There is no doubt that the apparent chaos of nature is in reality almost as closely compact of local "organisation" as our own Middle Ages.

Other themes of animal ecology had already been independently developed before the unified point of view arrived, parasitism, for instance of which the above pests furnish cases. More extraordinary ones were discovered by Darwin (e.g. 1851) and others—female fish carrying tiny degenerate males permanently attached to them, males which, in fact, in some cases actually cease to function, fertilisation being effected by sperm thrown into the ocean by hermaphrodite females themselves. This "fluid fertilisation" by the currents of the ocean ought to be coupled with another, wind fertilisation of many land plants. This, in turn, leads us to another topic, pollination (discussed, as we have seen, before ecology was recognised). Sprengel (1793) had viewed every feature of the flower as formed round fertilisation by wind or insects. Darwin saw the relevance of this to his theory, and extended it with special reference to the flower's adaptations to avoid *self*-fertilisation. The ecological side of the subject was summed up by the Müllers (1873 etc.), who concluded that self-pollination survives chiefly in sheltered situations. Insect fertilisation is one of the widest instances of the "interrelationships of living things," and reminds us that parasitism shades into symbiosis.

Such relations are stable because the parasite is not lethal to the host; but with powerful, especially animal, "parasites"—creatures of prey—the predator may kill the proverbial goose and then succumb himself; so that it is no wonder that instability, perpetual violent fluctuation of numbers, is the most significant novelty revealed by recent work. Before considering the predator-prey relation, we consider another cause of fluctuation so far as one place is concerned, namely, migration.

The migrations of fish and of birds were known to Aristotle. To the seafaring Greeks, indeed, with their infertile hinterland, the migrations of the tunny in and out of the Black Sea were of as vital importance as the migrations of the herring were to the fortunes of the Hanse towns and of England and Holland at the end of the Middle Ages. The regular to-and-fro movements of cattle between summer and winter feeding grounds were in fact familiar to men before history opened, and, as conditioning the migrations of the steppe and desert nomads, have played a vast rôle in human history. Archaeologists came to realise this at about the time when ecologists were revealing how widespread is migration in the animal world, from butterflies to lemmings.

The case of eels is well known. Schmidt found (1904) that both the American and European varieties breed in a small region of the Atlantic, from which the fry, as they grow, radiate in opposite directions until they reach the coastal waters and the rivers of their respective continents. Until the Italian, Grassi, cleared the point up (1896), the young eels were taken as a different species and the meaning of the observed facts could not be seen.

It is conjectured that the breeding ground may once have been a fresh-water lake. Vast but gradual land movements such as have led to its obliteration may also account for the course of (for instance) bird migration, though not for the ability to follow this course. The latter difficulty is less in the case of marine creatures. Sensitiveness to direction of current and to salinity and temperature may account for it. As mentioned in Chap. XI we are led by such studies to banish the vague notion of ocean life as a vast homogeneous thing. Rather we see the ocean, like the land, as an exceedingly definite structure. Its abiding currents, each with definite direction, rate, and chemical and physical properties, each carry (closely controlled by these factors) their own teeming planktonic and larger life. In this plankton, as elsewhere, dominance finds play. One dominant, the copepod (a minute crustacean) is a centre of study at the moment. There are also, of course, vast forests of sessile algae round the coasts, a vast surface spread of minute photosynthetic algae (whose seasonal periodicity governs the whole life of the ocean), a special life of the deeps, and, between these, up and down currents of matter in course of bacterial decay. We have stressed ocean life because it has a relative simplicity which has lately made it a more and more favoured starting point for ecological and physiological studies.

Turning briefly in conclusion to the other source of fluctuation, we come to another of the rather few branches of biology susceptible of mathematical analysis. If one species preys on another, its numbers increase and it preys at an increasing rate. It may overtake the "natural" increase of its prey, but before it actually wipes the latter out, it may have felt the pinch itself, and have lost numbers again, perhaps until the prey can start to multiply once more (since *its* food supply will have profited by the interval), and so on. It is needless to stress that a variety of assumptions have to be made to enable the mathematician to get to work, but it is precisely the value of the work to compare such assumptions with reality. It is true that in this case the assumptions are too numerous and closely linked for decision to be final; but already quite a small literature exists.

Some of it has arisen out of economic problems. Lotka (1920) examined the equilibrium of two species. In 1926, Volterra, from observations by Ancona on marine fisheries, worked out the simple case of predator and prey. He had to make, among other assumptions, some on the relation of reproduction-rates to food supply. Nicholson (1933), with insect pests in mind, had to make assumptions as to the effect, on the predator's chance finding the prey, of their densities of population. Volterra has compared the changes in such systems with those in mechanical systems, both with and without "hereditary" factors. These latter introduce "functionals," being functions not only of the "last" instant of the past, but of "every" previous instant—that is, of a continuous infinity of variables.

Elton (1924, 1930) pointed out that the widely observed fact that a species may swarm in one year, and a few years later be hard to find, is of significance for evolution. Timoféef-Ressovsky has remarked that genes have different chances of survival in different population densities. Elton's point was that after any heavy disaster to a species, the small group remaining may be far from representative of the genes at work in the former mass or of the usual limiting factors on survival. J. B. S. Haldane suggested (1932), after mathematical investigation, that the effect would be too slow to explain the non-adaptive characters especially in question; but Ford (1931) pointed out that the expansion phase after a bad period is one of less severe competition than usual, allowing normally non-viable genes time to interact with others and produce a new gene-complex in which their own original expression is favourably modified. He had been able to bring forward (1930) a case in which a species of butterfly had been under sufficiently continuous observation for nearly fifty years, during which it had undergone a great and prolonged decline in numbers followed by a strong and swift revival. It produced few variants, and remained stable in type, until its revival set in, when an extraordinary proliferation of variations was observed. As a new equilibrium approached, most of these ceased to appear and a new uniformity set in; but the new type was distinctly different from the old. Fisher and Ford had previously (1928) found that abundant species vary considerably more than scarce ones. To the geneticist, these cases are facts to be explained, not explanations; but this does not decrease their interest to the ecologist.

It has been remarked that, in a general way, the new genetics envisages *two* adaptations, one to the internal, the other to the external, environment, instead of merely the latter. This suggests a new "ecology," one which has not yet had time to develop.

CHAPTER XX

MODERN EXPERIMENTAL PHYSICS

THE next three chapters are devoted to modern physics. In the first, most of the experimental work centring round the vacuum tube has been brought together. This work forms so continuous a run that it has been thought best to carry it straight down to the present day. At several points the mathematical side of the electron and quantum theories is presupposed, but as the exposition of this also demands continuity, it has been necessary to sacrifice logic, and to defer quanta to the next chapter. It is, perhaps, remarkable how relatively slight the sacrifice of logic is: how self-contained the vacuum tube technique has been.

Faraday and other early workers had felt the fascination of the vacuum tube—that is, of the varied and colourful electrical discharges which replace sparks as the pressure of the gas is diminished. Faraday had noticed that at very low pressures these colourful effects recede towards the anode and leave the " Faraday dark space " before the cathode. Opposite the cathode, the glass begins to phosphoresce. About 1855–60 Geissler developed the technique of making glass tubes with inserted electrodes, and of reducing the pressure in them to very low levels. At such levels, glass is apt to reveal that it has adsorbed large quantities of gas.

In fact, the vacuum tube itself was only the centre of a large number of technical advances which conditioned progress. Geissler brought in mercury pumps, Sprengel's and Toepler's improvements came in the sixties. The late Lord Rutherford recounted that in the formative period of " projectile physics," about 1895, Daniel's cells were still used for small currents, Grove's cells for larger ones—this although dynamos had been built for decades and accumulators had, since 1881, been entering commerce. Since Geissler's time, however, an increasing number of experiments had been done with electric discharges both at ordinary and at much reduced pressures.

In 1869 Hittorf had shown that the phosphorescence of the glass, above noted, is due to something which, wherever the anode is, proceeds in straight lines from the cathode and is cut off by obstacles (1876, " cathode rays," Goldstein). Crookes, however (among others) deflected the cathode rays by a magnet (1879). He argued that they were not light, but rather streams of negative particles possessed of mass. They resembled the processions of ions by which the then new ionic theory was accounting for electrolysis. In 1890 Schuster used this concept to calculate the ratio of charge to mass $\left(\frac{e}{m}\right)$ from the magnetic deflection with a given field. The result—hundreds of times that of liquids—was not very encouraging, since in spite of the idea of "electrons" (then in the air, see next chapter) it was not taken as evidence of a very small value of *m*. Even larger ratios were obtained by workers on the continent.

In 1886, Goldstein had found "positive" (that is, oppositely deflected) rays in the space behind the cathode when the latter is bored with holes and placed opposite the anode. What was needed to disentangle all these effects was a means of finding *e* and *m* separately, also the velocity (v) of the particles. Early in 1897 Wiechert, by using the oscillatory discharge of a condenser to obtain bursts of known very short length, estimated the velocity of the cathode rays as $\frac{1}{10}$th that of light. A few months later Kaufmann pointed out that a similar velocity would be consistent with the view that all the energy of the cathode stream came from the potential difference between the plates. And yet the energy of the cathode particles must, if so, be gigantic.

The best way to find the velocity was to compare the magnetic deflection with one due to a strong *electrostatic* field. Several attempts had been made (for instance, by Hertz) to do this, but ionisation of the residual gas always interfered. Finally, J. J. Thomson of Cambridge succeeded, using very high vacua (and so very high voltages for the discharge) in order to reduce the ionisation. Late in 1897 he thus not only gave further values for $\frac{e}{m}$ and of v, but announced the crucial fact that the former is the same for all the pressures, electrodes, and gases which he examined, therein differing from the ions of electrolysis with which it had often been compared. Thomson deduced that the cathode rays were the same in charge as the electrolytic ions, but about a thousand times smaller in mass. This needed checking by independent determination of *e* and *m*. Meanwhile, further sensational discoveries had concentrated scientific attention on the subject.

In 1895 Röntgen (1845–1923) had discovered that the walls of the vacuum tube, under the impact of the cathode rays, give off rays which affect a photographic plate or a fluorescent screen *through other matter*. These "X rays" were found to be undeflected by fields and to travel in straight lines. They were found to make gases conducting at ordinary pressures,* and since this conductivity could be removed by filtration, it was suggested that it must be due to gaseous "ions" not unlike those of electrolytes. These ions recombined gradually of their own accord. In 1897, C. T. R. Wilson found that they can act like ions coagulating *liquid* disperse systems, and condense saturated water vapour into a visible mist of drops. Dust can do the same. Thomson (1898 and 1899) seized on this, for it gave him means of counting the number of such charged particles present in an enclosure, and so of estimating the charge in each. A formula of Stokes on the greatest velocity attainable by drops of given size in free fall enabled the size

* This temporary conductivity due to electric discharges had already attracted investigation.

to be determined for the mists produced by the ions; and thence, from the total weight condensed, the number of ions was estimated. Their total charge could also be found. The values of e thus obtained agreed well with those deduced by Townsend from the rate of diffusion of gaseous ions; but until 1910 * actual numerical values for these quantities were of an accuracy far below the standard of physics in general.

In 1898 Wien, and also Thomson, applied the two-field method to the harder case of the positive rays, and concluded that these involve charges equal (though opposite) to those of the cathode ray particles. Their masses, thousands of times greater, are those of the atoms of the gases left in the tube (less those of the "electrons" or cathode rays).

Meanwhile, curiosity had been fastening on the question of the nature of the X rays: which, according to various schools, might be new forms of light, or electric doublets, or other neutral particles. Then in 1896 Becquerel discovered that certain substances gave, in the natural state, effects similar to those of a vacuum tube. First certain uranium compounds, then all such compounds including the element itself, were found to have this "radioactivity," and thorium likewise. In 1899 the New Zealander, Rutherford, then at Montreal, announced that uranium radiation consists of two parts, a less and a more penetrating (α and β). The β rays, when examined by the two fields, proved to be like cathode rays but with a velocity not $\frac{1}{10}$th, but sometimes $\frac{9}{10}$ths, of that of light. The α rays were compared with positive rays, but for some time were hard to deflect. Soon afterwards Villard discovered that some substances give off still more penetrating "γ rays," which were compared with X rays.

In 1899, also, Rutherford noticed variations in the activity of thorium which he attributed to its evolution of a radioactive gas. In 1900, while M. and Mme. Curie were making the concentrated chemical analysis which unearthed radium and the other active elements, Crookes was discovering that uranium slowly produced another active substance, thereby losing its own activity. The second substance also lost its activity in due course. In 1902, Rutherford and Soddy, discovering a similar effect with thorium, propounded the general theory that radioactivity is the breaking up of one atom into a lighter one, with emission of charged particles. When the activity-time curves were analysed, they always proved consistent with the idea that the reaction is "unimolecular," that is, that the chance of a given atom's disintegrating is uninfluenced by its neighbours. Indeed, it was found to be uninfluenced by the most intense agencies which could be applied. The α rays, especially, of a given element were found to have a very definite and fixed range in any given gas, so that, once they are emitted at all, they are emitted with a definite velocity.

Intensive research with the tiny quantities available elucidated the lines of descent of the three series of elements which have been found to exist, lead being the suspected inactive end-point of all three.

* 1911 was the date of Millikan's great improvements in the falling-drop method of finding e and m separately.

Concordantly with theory, β and γ rays, like cathode and X rays, went together. Their emission did not change atomic weight. On the other hand, α rays, which were later identified (Rutherford, 1906) as charged atoms of helium, came off alone, and in doing so dropped the atomic weight by four—the value for helium. Methods were developed of counting individual α particles by their impacts on phosphorescent screens.

In 1903 had come C. T. R. Wilson's discovery that a single-leaf electroscope tilted to a suitable angle becomes extremely sensitive, the second great advance in experimental resources due to this worker. Later on, Wilson showed that the track of an α particle can be followed by the drops condensing on the ions which it forms in passing. In 1904 Dewar discovered the value of charcoal, properly prepared, as an absorbent in high-vacuum work. The subject was eminently one for brilliant experimental improvisation.

The nature of γ and X rays remained disputed. The effects of matter on them were found to be complex, secondary radiations being set up. The effect of ordinary ultra-violet light in ionising air had been noticed by Hertz in 1887, and in 1899 both Thomson and Lenard showed that this effect is due to the fact that the light causes matter (for instance, metals) to give off electrons. Some alkali metals will show the effect with visible light. X rays produce a similar "photo-electric" effect. But they also produce, if of high enough frequency, secondary X rays of wave-length characteristic of the substance on which they fall (Barkla, etc., 1905 onwards).

The idea gained ground that these X and γ rays were light of extremely short wave-length due to sudden stoppage of cathode and β particles. They could not be refracted or reflected, but this was theoretically not unlikely in view of their very short wave-length. And then Laue (1873–) conceived the idea that they might be diffracted by the regular arrays of molecules in a crystal, acting as a "grating." In 1912 Friedrich and Kipping, at his instance, showed that this was so; and thus the wave-like nature of the rays came to be accepted. By 1913 the Braggs, W. H. and his son W. L., were using these facts to analyse the structure of crystals and to measure their molecular distances and the actual wave-lengths of X rays.

These methods revealed that an X ray bulb gives a "continuous spectrum" of X rays; but that if a metal anti-cathode is put in the path of the cathode rays to produce X rays, the continuous spectrum has lines on it characteristic of the metal used. This line spectrum is simple: generally each metal shows only two lines. And Moseley (1888–1915) proved in 1913–4 that for each series a surprisingly simple relation holds between ν, the frequency, and the position of the element in the series of atomic weights. The Nth element, starting from hydrogen as one and allowing for the gaps still left in the periodic table, was found to have both its frequencies proportional, with a different constant, to N^2.

At the same period, this quantity N ("atomic number") was coming to have another, concordant meaning. Soddy and Fajans, in tracing the "lines of descent" of the radioactive elements, began (about 1910) to put forward evidence of a new phenomenon—"isotopes," that is, elements of different atomic weight but identical chemical properties. Ionium (discovered by Boltwood, 1906) was found to be indistinguishable from thorium.

From the rules already elicited as to changes in charge and weight during radioactivity, Soddy predicted that the final term of the uranium series should have atomic weight 206, that of the thorium series 208; while both should occupy the same square in the periodic table, and so have the same "atomic number." This atomic number should be the one for lead (atomic weight 207·2). Richards (1913), Hönigschmid and others verified that this is so.

J. J. Thomson's positive ray apparatus began to be regarded as giving what we now call a "mass-spectrum" of the gases in the tube, with lines for definite, blurs for indefinite, masses. Examined in this way, the inert gas neon gave results (from 1912) suggesting that it might consist of two kinds of atoms of weights 20 and 22, of which the ordinary gas is a mixture (20·2). After the War (1919) Aston (1877–) began to perfect the mass-spectrograph with a view to examining the homogeneity of all the elements systematically. His essential improvement, like that of Fraunhofer a century earlier, lay in focusing the rays. A new micro-balance and a number of other refined devices had also to be perfected.

The result was that isotopy was found to be widespread. But the capital point was that all the isotopes, unlike the elements, have integral atomic weights.* This was one of the several points which favoured unitary theories of the atom. To understand these, we must shortly retrace our steps.

When the idea of atoms as made up of electrons and of positive charges came to the fore, various "models" were proposed. J. J. Thomson's, a development of one by Kelvin (1902), postulated a large positive body closely surrounded by electrons in successive rings, typically with eight in a ring (1903–4).† The more charges, the heavier the atom. The greater the atomic weight, the more rings of eight they formed. The outer ring would generally be incomplete. If most chemical properties, including valency, were due to it, the periodicity of Mendeléef received a general explanation. The idea could be followed into some detail on the chemical side, but Lenard (1903) showed that swift electrons could pass through thousands of atoms, suggesting that for physical purposes the large nucleus should be replaced by a small one and that most of the atom should be considered empty. Rutherford's atom, which superseded both of these, had been investigated theoretically by Nagaoka in 1904. Here electrons circulated as "planets" round a small positive "sun." It was Rutherford who, in 1911, convincingly strengthened the evidence for this model by following, with the Wilson method, the tracks of α rays in a gas. Any deflection which these suffer must be due to atoms: electrons would be too light. But it was found that deflections are far

* Apart from refinements to be noted later.

† Thomson gave grounds for supposing this to be called for by stability.

too rare to be caused whenever an α particle approaches within the distance which the kinetic theory would assign as the radius of an atom. It has therefore been concluded that the kinetic theory "detects" the planetary electron orbit, the α particle, the positive "nucleus"; and that that of the former is of the order 10^{-8} cm., that of the latter of the order 10^{-13} cm.

In Rutherford's atom, the behaviour and number of the planetary electrons, to which the chemical and simpler optical properties are due, depends only on the *net* positive charge on the nucleus. If this were made up of N unit positive bodies ("protons") with *n* electrons among them, all atoms with the same value of (N—*n*), that is, the same "atomic number," would be isotopes; while their atomic weights would be practically given by N only.

Thus, at about the time of the War, the picture of the atom began to grow definite in the minds of physicists. We shall see in the next chapter the fundamental theoretical problem with which the picture was faced. Here we proceed with the experimental side.

For, even on this side, there was no stability. Atomic physics, the domain of mathematical order, is certainly also the anarchic subject of present-day science. No sooner does it set up the nucleus as unassailable than it breaks down. No sooner does it confine radioactivity to a small region of the periodic table than it induces it artificially in other regions. No sooner does it reduce matter to electrons and protons than it flings several more ultimate particles into the cauldron. Indeed, no sooner does it erect elaborate mathematics on the constancy of the velocity of light, the absence of "ether drift," and the recession of the nebulae, than it alleges evidence against these doctrines. Yet "anarchy" is the wrong word. At no time has it been impossible to keep one's bearings, though not for a generation has it been possible to keep long on any one tack.

We may resume the story in 1919, when Aston's isotopes began to come and Rutherford obtained the first evidence of experimental disintegration of the atom. Some of the advances since then have sprung from increased virtuosity with existing technique, some from new devices, but especially from the relentless pressure towards more and more energetic particles with which to assail the nucleus. Electron tube amplifiers, electroscopes at the point of instability, Wilson cloud chambers, more and more efficient vacuum pumps—these have been factors.* But so have quite other advances, such as those in wave-mechanics (see next chapter) which have enabled us to predict new types of isotope or of ultimate particle.

In 1919 Rutherford noticed that new long-range particles were formed when fast α particles from radioactive sources are passed through nitrogen. The real difficulties in such investigations—such as making sure that no foreign influence can be causing the observed

* Thus, the Geiger-Müller automatic positron or electron counter consists of a tube containing low-pressure air and two electrodes between which a critical potential difference is maintained. The ionisation due to the entry of a particle causes a current, which is amplified by thyratrons to actuate a recording device.

effects—cannot be followed in an account like the present; but these particles were found to have the mass and charge of hydrogen nuclei or protons.

Between 1921–4, Rutherford and Chadwick disintegrated all the elements from boron to potassium, except carbon, oxygen, lithium, and beryllium. A new type of "chemical" equation thus became necessary; and in the nitrogen case the question arose, did the helium atoms of the α rays enter into the "reaction" or merely provoke it. This could be decided by finding whether oxygen or carbon was the second product. In 1925, Blackett used cloud-track methods to show that the former alternative obtained.

The matter had immediate relation to quantum and relativity mathematics, because, if exact atomic weights were used, a slight loss of mass, such as would agree with the relativity equation of mass and energy (see Chap. XXII), was found to have occurred. Further, the penetration of the nucleus by the particles was found to be a case of getting "through" rather than "over" the potential barrier (next chapter).

Theory had another dramatic application about this time. In 1924, Mecke observed that the band-spectrum of molecular hydrogen has a marked alternation of intensity in those lines due to rotational levels. In 1927 Heisenberg, and in 1928 Hund, showed that on the new wave mechanics hydrogen molecules might exist in two forms, according as the two nuclei were rotating in the same or in opposite directions; and that if ordinary hydrogen were a mixture of these, the spectrum would be as observed. The ambition immediately grew to separate the two, and in 1929 Bonhoeffer and Harteck achieved this. They used charcoal to accelerate the attainment of equilibrium at very low temperatures, when the antisymmetrical form is almost entirely converted into the other.

This was sensational enough, but a greater sensation remained. The discovery in 1929 of isotopes of oxygen, 17 and 18 in mass, upset the agreement with theory which had always been presumed for the mass of *hydrogen*. Now in 1920 Harkins, to complete certain empirical relations among the then-known isotopes, had suggested a hydrogen isotope 2; and this suggestion was now (1931) revived by Birge and Menzel. In the same year, Urey, Brickwedde, and Murphy in U.S.A., by fractional distillation of hydrogen near its triple point, prepared a gas which had the predicted spectrum. In 1932 Urey and Washburn reported that water which has been much electrolysed is heavier than ordinary water, and in 1933 Lewis and Macdonald thus prepared nearly pure "heavy water." The symbol D is used for the "2" isotope; and D_2O, HDO, and many other compounds, have since been the subject of active research. D_2 exists, like H_2, in symmetrical and antisymmetrical forms. In 1934 "tritium," with mass three, was discovered.

This had reactions on artificial disintegration. In 1932 Cockcroft and Walton obtained, by improved methods which cannot be detailed here, very high velocity protons, which were found to be very much more "destructive" than α particles. By their use, lithium was split into two heliums.* In 1933 Lewis, Lawrence, and Livingston found that the corresponding particles from D, deuterons, were even more effective than protons. Work done with these has once more made necessary (1935) slight modifications of the exact atomic masses ; since it has revealed values mutually incompatible if the relativity mass-energy equivalence is to be retained.

We must now turn back to examine advances as to even more fundamental entities. The two levels of size, proton and electron, could not fail to suggest, from symmetry, positive electrons and negative protons ; while as early as 1915 Harkins had suggested *neutral* particles. We could in fact conceive four more niches to be filled. How far this filling has actually occurred, we now consider.

It will be recalled that beryllium was one of the elements not disintegrated by Rutherford and Chadwick. This meant, of course, that it was still the subject of attack ; and in 1930 Bothe and Backer found that under fast particles it emitted radiation which was γ-like in that it was not bent by applied fields. Work by Bothe, and by Curie and Joliot, in 1931, and again by Webster in 1932, revealed that this radiation had a phenomenal range, greater than that of any known γ rays. It disintegrated hydrogen, helium, and carbon. In 1932, Chadwick showed that, in fact, no γ rays could have done this ; and definitely introduced a neutral particle, proton-like as to mass, to account for the absence of bending in fields.

The quantum theory had no room for any proton-electron combination except the hydrogen atom, so the " neutron " was taken as *sui generis*. In 1933 Heisenberg suggested that there are in fact no electrons in the nucleus, only protons and neutrons ; though the helium nucleus is perhaps another fundamental brick. Fermi in 1934 suggested viewing neutrons and protons as the same fundamental particle in two different quantum states. Another view is suggested by classifying the neutron with the atoms, as being another neutral entity. It then falls into the first place in the inert gas series. In view of the suggestion that helium, also, may be a unit for nuclear architecture, it thus appears that, whether at the nuclear, the atomic, or the molecular level, all the stable configurations are embodied in the inert gas series. At all events, another of the six niches is filled.

To understand the next niche, we must diverge into the question of cosmic rays, which we do not treat in detail in this book. As early as 1900, Elster and Geitel, and also C. T. R. Wilson, had realised that even the driest and cleanest air has a slight ionisation and conduction. Wilson suggested that this might be due to extra-terrestrial radiation.

* In this and most similar cases, other elements were successfully treated later on by other workers. We must confine ourselves to the first instances. A recent device to give very high velocity particles is the cyclotron. The particles are turned round and round in successive semi-circles by a powerful magnet. Between each semi-circle they are speeded up through a short distance by a static field which alternates just fast enough to have reversed when they reach the opposite end of their diameter. Owing to the simple laws of the magnetic bending, the increase of radius of the circles is exactly compensated by the increased velocity. Thus the frequency of the alternating field needs only to remain exactly constant.

Gockel in 1909 found by balloon ascents that, in fact, this ionisation of air does increase with height. Through this and other work it was widely held by about 1914 that there is a very penetrating radiation which comes from above the surface of the earth, from clouds or from outside the atmosphere. It has always been hard to get consistent results, but recent work suggests that the rays vary simply as might be expected from absorption by superincumbent matter. But they also vary with latitude. In 1930 Compton organised world-wide observations and obtained a reliable idea of this variation. Clay in 1932 showed it to be consistent with a radiation coming to the earth equally from all directions, but spread into a spectrum by the earth's magnetic field.

In 1929 Bothe and Kolhörster established that, whatever the "original" nature of the radiation, it consists when detected * of charged particles. The energy of these is, of course, very high, probably too high to be produced, for instance, by thunderstorms. The storm theory, too, could hardly account for the observed distribution; though (for instance) a galactic origin is not in this respect in much better case. E. A. Milne has shown that a cosmic radiation emerges naturally from his relativity theory, but we cannot here pursue speculations on origin, such as Lemaître's view of the whole universe as one huge super-radioactive atom. We must turn to the discoveries in other subjects which arose indirectly from the experimental investigation of cosmic rays.

In California in 1932 Anderson and Millikan, the better to analyse cosmic rays, were working with a Wilson chamber apparatus having an exceedingly powerful electromagnet (18,000 gauss). This apparatus had an unexpected effect: it revealed that there are particles of *positive* charge which have a range far beyond that of any proton and a curvature much suggestive of an electron. Now it will be seen in the next chapter that Dirac's relativistic wave mechanics gave, among other rather outré results, the suggestion of a new positive body. The suggestion was now eagerly taken up.

Work by Anderson in 1932, by Blackett and Occhialini in 1933, and by others later, showed that these "positrons" have, in fact, at least to about 10 per cent., the mass of the electron. But we must not let the niches carry us too far. On Dirac's theory the electron and positron are not simply opposites: positrons only exist very transiently in electron-positron pairs which arise from γ ray absorption. They swiftly reunite to give γ radiation once more. This latter transaction was verified in 1933 by Thibaud and Joliot. The whole train of events has been a remarkable triumph for Dirac's theory.

The other two niches are still *sub judice*. A neutral electron ("neutrino") would, as Pauli, and especially (1934) Fermi, have pointed out, make it possible to account for the continuous background of the β ray spectrum in radioactivity without abandoning, in detail, the conservation of energy. For they would share the energy with the electrons. But this possibility is not proof. The discovery of a pair

* I.e. having regard for secondary, tertiary . . . effects.

of radioactive substances (uranium Z and uranium X_2) the same in both atomic number and atomic mass could be explained (Gamow, 1934, 1935) if we had negative protons—heavy electrons—to build with. Very lately (1938) evidence is to hand of negative bodies, of mass intermediate between that of electrons and protons, present in cosmic rays, as suggested in 1937 by Neddermeyer and Anderson. We turn in conclusion to artificial radioactivity.

It had long been realised that we might, for symmetry, suppose that all elements, not only the radioactive ones, are in fact disintegrating, only immeasurably slowly. The concept that every atom, and indeed every group of such, ought to be supposed to have a length of life as well as its other attributes, is an interesting one naturally suggested by the kinetic point of view, and by such work as Paneth's on free radicals. But until the year 1934 it was also an academic one so far as spontaneous changes in atoms were concerned. In that year Curie and Joliot obtained a substance which, without external stimulus, emitted positrons at the usual exponentially decreasing rate. The substance was indeed *produced by* a strong stimulus—*a* ray bombardment—on aluminium. But it went on emitting positrons after the stimulus was withdrawn.

In this and similar cases the isolation of the radio-element has of course been the aim, and has sometimes proved possible. The simplest interpretation of the case of aluminium is that the bombardment produces a radioactive isotope of phosphorus which disintegrates into one of silicon. This one is stable. Boron gives a radioactive isotope of nitrogen ; magnesium, one of silicon. Empirical regularities, which may help in elucidating the type of architecture prevailing in the nucleus, have been detected among the mass-numbers and atomic numbers of unstable nuclei.

Fermi and others (1934, 1935) have reported that neutron bombardment of uranium gives a new element which, as it is produced without emission of rays or particles, must have a greater atomic mass than the parent element. If this is so, physicists have begun extending the atomic weight table upwards towards heavier and heavier atoms.

CHAPTER XXI

QUANTUM THEORY

IN this chapter we start from the Maxwell-Boltzmann statistical mechanics, and watch it developing into a theory of radiation. We then follow the application of Maxwell's electromagnetic theory to the idea of the electron, which fitted in so perfectly with the experimental discovery of the particulate nature of electricity as already recounted. The electron theory issued in an electrical approach to radiation which was expected to chime in with the thermal theory of the same subject. And indeed it did, to this extent, that both were found in the first decade of our century to be definitely erroneous. Our next subject, therefore, is the quantum theory brought in to correct them.

Certain other difficulties about the transmission of radiation—those which led to relativity theory—are deferred to the next chapter ; and with them, in part, go those modifications of quantum theory designed to unify it with relativity, as well as those modifications of this latter designed to reconcile it with recent astrophysics.

Though we have already followed out one major experimental line, we shall find progress in other lines conditioned by instruments and applications. Thermopiles made possible the refined radiation experiments of the late 19th century, as they had made possible the experiments on the laws of the electric current in the earlier decades. Fine gratings, and interferometers of great path-difference, made possible the observations which forced home the quantum theory and necessitated the special relativity theory. Wireless waves gave authority to Maxwell's theory.

We have mentioned in Chap. XIII the equipartition of energy among the degrees of freedom of an assembly of molecules. The immediate physical bearing of this was on the ratio of the two specific heats of gases. The theory gave a value 1·67 for monatomic, 1·4 for diatomic gases. Warburg in 1876 found that mercury vapour behaved on this hypothesis as monatomic. Up to then, no gas giving the value 1·67 had been known. In 1877 Boltzmann went on to show that the purely thermodynamic law of increasing entropy had its physical basis in the statistical improbability of the recurrence of any exception to equipartition large enough to be detected.* Maxwell, more simply, had already pointed out that a small enough being (" demon ") could

* Boltzmann implied that he had derived this view from Willard Gibbs.

MAP VI

BIRTHPLACES OF SCIENTISTS

Please consult Explanation on p. 10 before studying the Map

Workers in the exact sciences born in North England, North Ireland, and South Scotland, chiefly from about 1800

Number on Map	Name	Birthplace *
1	Airy	Alnwick
2	Andrews	Belfast
3	Balfour Stewart	Edinburgh
4	Bragg, W. H.	Wigton
5	Dalton	Cockermouth
6	Dewar	Kincardine-on-Forth
7	Eddington	Kendal
8	Fitzgerald	Dublin
9	Forbes	Edinburgh
10	Frankland	Lancaster
11	Graham	Glasgow
12	Hamilton	Dublin
13	Jevons	Liverpool
14	Joule	Salford
15	Kelvin	Belfast
16	Larmor	Magheragall
17	Leslie	Largo
18	MacCullagh	Strabane
19	Maxwell, Clerk	Edinburgh
20	Melvil	Glasgow
21	Peacock	Darlington
22	Ramsay	Glasgow
23	Rankine	Edinburgh
24	Reynolds, Osborne	Belfast
25	Stokes	Skreen
26	Stoney, J.	Oakley Park
27	Tait	Dalkeith
28	Thomson, J. J.	Manchester
29	Townsend	Galway
30	Tyndall	Co. Carlow
31	Wilson, C. T. R.	Glencorse

In addition, Boyle (17th century) was born in Ireland. Among recent English exact scientists not born in this area are Faraday, Rayleigh, and Dirac.

* In a few cases, work-places, or the nearest larger place, have been substituted.

MAP VI

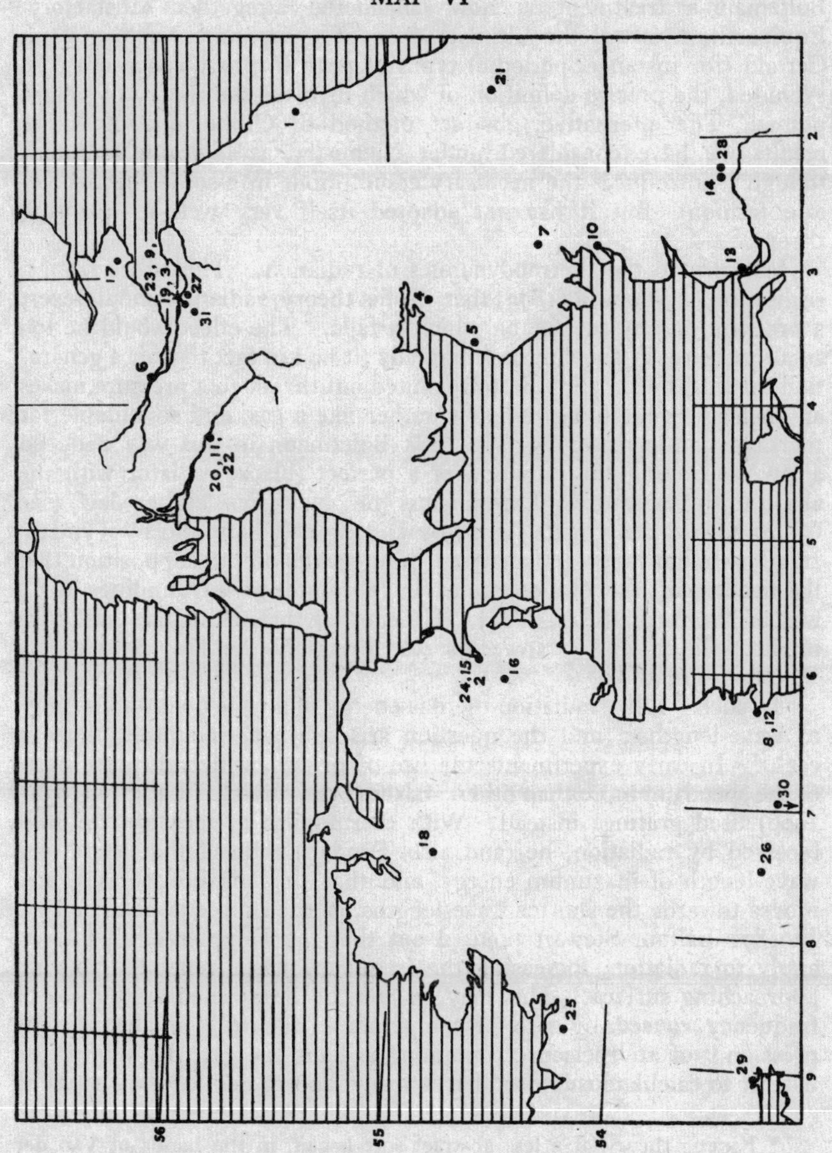

56
55
54
2
3
4
5
6
7
8
9
1
21
28
14
7
10
13
17
23, 9,
19, 3
27
31
4
5
6
20, 11
22
24, 15
2
16
8, 12
30
18
26
25
29

separate hot and cold molecules of a gas without expending energy. Thus the second law was, dynamically as opposed to thermodynamically, simply one of the laws of systems too fine-grained to be observed in detail.*

It should, however, be observed, that neither Maxwell's nor Boltzmann's treatment is now considered altogether satisfactory. Equipartition is *not* the ultimate state of every conceivable system. Certain (for instance, periodic) types of path in phase-space must be excluded, the precise definition of which involves the theory of sets of points. The alternative, abstract method of Gibbs, one of whose results we have considered under chemistry, is logically sounder; though it introduces the necessary assumptions in a somewhat wholesale fashion. But it has not adapted itself very well to quantum theory.

We turn to the thermodynamics of radiation. This arose from a realisation of Maxwell (1873) that, on his theory, radiation should exert a pressure on any surface on which it falls. The effect would be too small to be easily shown experimentally (it had in fact to wait a generation for this) but in 1875 Bartoli pointed out that such a pressure makes an enclosure containing radiation rather like a gas, and so suitable for thermodynamic reasoning. In 1884 Boltzmann in this way deduced a law connecting the radiation of a perfect (black) radiator with the absolute temperature. Experiments on Newton's earlier law (see Chap. X) had been made (1817) by Dulong and Petit and by Tyndall. In 1879 Stefan (1831–97) analysed these results on the supposition that the net loss of heat of a source at T_1 (absolute) to surroundings at T_2 was of the form $f(T_1)-f(T_2)$. He showed that, if so, the best $f(T)$ might well be kT^4. Now this was precisely the law deduced by Boltzmann in 1884.

In such "full" radiation there is energy of a wide continuous range of wave-lengths; and the question arises, what amount is there of each? In early experiments the use of prisms, which disperse parts of the spectrum more than others, falsified the results. Langley (1834–1906) used gratings instead. With thermopiles to measure the heat received by radiation, he (and also Tyndall) showed that there is a wave-length of maximum energy, and that this maximum rises, and moves towards the shorter wave-lengths, as the source is made hotter. In 1871 Balfour-Stewart pointed out that the Doppler effect † must apply to radiation, increasing the frequency when reflected from an approaching surface. This may, in fact, be interpreted as the rise of frequency caused by rise of temperature due to "adiabatic compression" of an enclosure containing radiant energy. In 1896 Wien used it to calculate this rise of frequency, and to deduce the shapes of

* Kinetic theory of a less abstract sort began, in the hands of Van der Waals (1837–1923), to show that the divergences from Boyle's law could be partly explained (1873) by assuming intermolecular forces and finite molecular size. More and more refined, yet never entirely successful, approximations to the observed laws have been made since.

† In 1845 the Austrian Doppler (1803–53) predicted a change of frequency with velocity of source for wave-motions—light and sound alike.

the curves of energy E against wave-length λ for all values of λ and of T. By pure thermodynamics he could get

$$E_\lambda = \frac{c^2}{\lambda^5} F\left(\frac{\lambda T}{c}\right), \; c = \text{velocity of light},$$

but he could not decide what function F must be. Making a minimum of further assumptions, however, he obtained an equation of the type

$$E_\lambda = \frac{k_1}{\lambda^5} e^{-\frac{k_2}{\lambda T}}$$

In 1899–1901 Lummer and Pringsheim showed that, experimentally, this formula fails. Paschen, Rubens, and Kurlbaum also reported divergences. These were of such a radical kind that Rayleigh (1842–1919) and Jeans (1877–) overhauled the whole question, while Planck (1858–) made a revolutionary hypothesis to account for them. This (1901) was nothing less than that energy cannot be equiparted or otherwise transformed as a continuous quantity, but only in definite parcels or " quanta," which were to be expressed in terms of a very small universal constant (" h ") with the dimensions of action. Since we cannot always divide a pile of parcels into any required number of groups of exactly equal size, divergences from equipartition were to be expected. In fact, Planck showed that this hypothesis would give the required result. But we leave this for the present.

Turning now to optical and electrical theory, we recall that Cauchy had given a theory of dispersion as due to the weighting of the ether by particles of matter. About 1840 Talbot discovered that dispersion becomes "anomalous " in the neighbourhood of a wave-length which the refracting substance does not transmit, n apparently ceasing to vary continuously with λ. In 1862 Leroux showed that Cauchy's theory cannot account for this. In 1869 Maxwell, and in 1872 Sellmeier independently, gave formulae based on the suggestion of Stokes in connection with the thermal side of this phenomenon : that both selective absorption and emission (line spectra) must be explained by the presence in atoms of oscillators of specific frequency like tuning forks.

The Maxwell-Sellmeier formula gave such very improved results that it had to be taken seriously. Definite views as to the causation of spectra began to be held. In the seventies Lockyer correlated the well-recognised division of natural spectra into band and line forms with the complex vibrations of a molecule and the simpler ones of single atoms. Kundt's discovery in 1880 that sodium *vapour* gave anomalous dispersion hinted that the oscillators in question could not be found in the interconnections of atoms and molecules so much as in their interiors. Thus came an impulse to subdivide the atom. About 1895 Lorentz took the step avoided by Maxwell, of making specific hypotheses as to the relation of charge to matter. He built up a mathematical theory of the behaviour of light and matter on the assumption of *atoms* of electricity. J. Stoney in 1874 and Helmholtz in 1881 had put forward the idea of such atoms to account for electrolysis. Stoney called them " electrons." In Lorentz's theory, the dispersion formula was only one detail. Radiation was the result of acceleration (or deceleration) of an electron ; magnetism was due to orbital circulation of electrons round positive charges. As such, light and magnetism should be connected. In a magnetic field, an emitter should have each normal spectral line split into three. When this effect was detected experimentally by Zeeman in 1896, this theory began to gain acceptance.

The sizes of the orbits gave the frequencies of the spectral lines typical of the substance : the circulating electrons were the vibrators and resonators of earlier theory. Unfortunately, this raised difficulties. For how, in fact, could any such schemes as those of Lorentz and Rutherford (last chapter) explain the line character of spectra ? How could planetary electrons behave as vibrators or resonators of fixed frequency ? In losing energy by radiation, a planetary electron orbit should shrink continuously ; and an assembly of them would give a band, not a fixed line, still less a set of fixed lines. Yet hundreds of spectra had been exactly analysed by the interferometer. The impasse was made worse by the fact that the experimental researches described in the last chapter were making electrons almost more real than atoms. It has been said by one writer that the whole state of spectral physics had become a scandal for science.

As the first decade of the twentieth century advanced it grew clear that the difficulty was closely related to that facing the thermodynamic theory of radiation. According to current mechanics, if the ether could respond to a continuously infinite range of wave-lengths, it must have an infinite number of degrees of freedom. Now matter has only a finite number of such, so that equipartition among degrees of freedom demands that the fraction of energy which radiating matter can retain in equilibrium with the ether is zero. Hence, as Jeans and Rayleigh pointed out, the only hope, if we are to avoid Planck's revolutionary idea, was that equilibrium would never be reached. They produced another formula by examining how long equilibrium would take to come : but, though different from Wien's, it was no nearer observation. It was, in fact, difficult to see not merely how planetary circulations could lead to line spectra, but how *any* Newtonian system could lead to either the lines or the energy-distribution. In 1912 the great Poincaré, as a last gift to science, gave a proof * that only Planck's hypothesis could account for the observed distribution of spectral energy.

Thus at last Planck's hypothesis, hard though it was to reconcile with Newtonian mechanics, had to be taken seriously. By then there were two other regions of physics, besides the distribution law, in which use had been made of it. Before coming to these, we may note that there was doubt among the early workers whether the parcelling concerned energy in matter, or in ether (Einstein " photons " 1905) or only (Planck) at the transformation between them. Vibratory energy always seemed to be in question ; for the parcels were of amount $h \times$ the frequency (ν) of the vibrations.

The first of the other uses mentioned had been started by Einstein

* A proof requiring further steps supplied by Fowler in 1921.

in 1905. In 1902 Lenard had shown that in the photo-electric effect (see last chapter), the velocity of emitted electrons had the extraordinary property of being independent of the intensity of the light causing it. Whether this was due to some fixed ejection mechanism within the atom or to some definite parcelling of the energy in the light, it was equally inconsistent with Newtonian mechanics and was equally a case for quanta. Lenard showed that the frequency of the light *did* affect the result. First, below a definite ν (ν_0 say) no electrons at all came off (another puzzle). Second, the velocity increased with ν. The quantum of such radiation being $h\nu$, Einstein suggested that the maximum energy of the electrons might be $h(\nu-\nu_0)$, where $h\nu_0$ represents the energy needed to get the electron outside the body of the emitter. In fact, the energy was to be a linear function of ν, with h as one constant. In 1914 this was shown by Millikan to be true to the facts. By this time, too, another consequence of the old equipartition law was being examined in the light of quanta.

This second use concerned specific heats. Dulong and Petit's constancy of the specific heat of atoms (see Chap. IX) should be exact, whereas it is only rough. Also the " constant " varies with temperature. In 1907 Einstein, assuming characteristic frequencies in the atoms, showed that Planck's quanta yielded a specific-heat-temperature curve much nearer to the type observed. Nernst, Lindemann, and especially (1912) Debye improved this. Debye argued from the frequencies of atomic vibration implied in the elastic constants of the materials used.

We now return to spectra. The interferometer had revealed not only innumerable lines for none of which classical theory could account, but, what was more baffling still, simple numerical relations among them (for instance, J. Stoney, 1871, Huggins, etc., 1880 onwards). In 1885 Balmer discovered that formula for a series of hydrogen lines which served as a model for similar ones found in other elements by Rydberg, Ritz, and many others. These series had certain common features. All crowded together towards the short wave-lengths. Ritz (1908) pointed out that, in any series, each frequency is the difference of two others in the series ; Schuster had shown that the formulae for such series always involves the difference of two terms of the type $\frac{1}{n^2}$, n integral. Rydberg had given $\nu=R\left(\frac{1}{a^2}-\frac{1}{(b+m)^2}\right)$, where R is the same constant in all cases.

In 1913 the Dane Bohr, working in Manchester, tried the effect of supposing that these terms represented electron orbits, and that radiation was given off not continuously but in Planckian quanta of $h\nu$, whenever an electron jumped between two of them, no orbits but those admitting of such definite differences being admissible. His simple theory has had complete success only in two simple cases : the hydrogen spectrum and the X ray spectra just then being discovered by Moseley. But even the first was enough to concentrate attention on the subject, especially when Franck and Hertz showed that Bohr's figures agreed with the critical potentials obtained in connection with the ionisation of gases.

Sommerfeld (1915) and others went on to show how some of the fine structure of spectra can be explained when we bring in the relativity doctrine (see next chapter) that mass varies with velocity. The elliptical electron orbits then cease to be quite closed.* But as soon as all the degrees of freedom began to be scrutinised with a view to quantising them, it began to be evident that this latter process can be carried out in a number of ways, not all equivalent, of which Bohr had made a fortunate but arbitrary choice. To limit this choice he brought in the "Correspondence Principle," according to which all other properties but frequency (for instance, intensity) of the lines were given by the classical theory, at any rate as the quantum numbers involved became large. This has the effect of limiting possibilities of quantisation.

The absence of any explanatory physical mechanism, however, kept Bohr's theory under constant criticism. Moreover, was it radiation as such which thus went in packets? This was practically a corpuscular theory. J. J. Thomson tried, like Maxwell in his early reasonings, to take Faraday's tubes of forces in a literal, particulate sense. Yet the standard wave-phenomena of optics did not seem to be quantised at all, for as early as 1909 G. I. Taylor had shown that fractions of a light quantum will give all the effects of interference. And yet in 1923 Compton recorded an experimental fact which seemed capable of most simple explanation as due to the collision of a light quantum with an electron. It consists in the appearance of a second line on the long-wave side of the original line when monochromatic X rays are subject to scattering.

The point of view, however, which made the exact location of quantisation a vital matter was rapidly falling out of date. It was the close, practical tussle with the hosts of spectroscopic results which set the peculiar tone of the coming period. Physical models fell into the background in comparison to equations which would work. Even when, in one of the two lines which we now trace, De Broglie and Schrödinger jumped off from a "physical" theory of waves, they were (as we shall see) hardly waves which commended themselves to the old-fashioned school. It is with the other line, that of Heisenberg and Dirac, even more abstract or practical according to one's view of it, that we shall start.

From about 1922, a number of workers had been struggling to apply Bohr's cumbrous dual process of quantisation plus correspondence to the fine-structure of spectral lines. They had thus begun to use, besides the quantum numbers giving the lines themselves, further "dimensions" to represent the fine structure of each. An empirical formula of Landé gave separate quantum numbers to the planetary electrons, the nucleus, and the whole atom. It covered a great many facts, but not all. In fact, the simplest two-electron case, that of neutral helium, defeated it. A more fundamental approach began to be desired by a group of brilliant young mathematical physicists burdened with a minimum of laboratory experience.

One desire of the mathematician, as old as Lagrange's generalised

* Later points of view have altered the interpretation of fine structure.

coordinates, is that of dispensing with needless knowledge of detail as to the problem concerned, and of interposing between data and solution not any mechanical model or scheme but only a formal, symbolic scheme of one's own. This chimed in with the tendency already mentioned; and in Heisenberg (1925) it reached its apotheosis. Instead of answering the demand for a physical model, he declined to use even what there was (orbits, etc.) and confined himself entirely to observable quantities,* such as frequencies and intensities of spectra. On the classical theory we should do this by a Fourier series to represent the wave-train. That is, we should use a singly infinite series of trigonometrical terms or (using old results) of terms like ae^{ib}, where each term represents a spectral line. But, on quantum theory, each line is a function of *two* variables, not one: of the nth and m^{th} energy levels between which an electron jumps. We thus get a *doubly* infinite array or matrix of these exponentials.† It will be remembered that if a and b are matrices, $ab \neq ba$ in general (multiplication non-commutative).

Heisenberg's preference for symbolic over physical models was soon justified by remarkable successes. But before pursuing this we note that Pauli had been urging (1924) that the three quantum numbers of the Landé formula should be redistributed, each planetary electron being endowed with *four* such numbers, no two in any one atom having all four the same ("Exclusion Principle"). This now worked in with the idea of the "spinning electron" introduced (1925) by Uhlenbeck and Goudsmit; for the spin could be taken as the fourth quantum number. In 1926 Heisenberg showed, with his new mechanics, that this cleared up certain difficulties with which the Zeeman effect had faced the cruder Bohr theory.

We have said that the mathematicians preferred their own sort of "models" such as matrices, to those of the physicist or mechanic. For, no less than the latter, the former have a suggestiveness of their own: ideas, as Hegel observed, have hands and feet. We have several times noticed how mathematics has taken a new lease of life from the discovery of a new model—algorithm (see Chap. IV), "formalism." But the suggestiveness of matrices did not mean that matrices are the only formalism which mathematics could suggest for this situation. On the contrary, almost simultaneously with Heisenberg, at least two other features of the classical formalism of mechanics were being seized on for generalisation or analogical reasoning. Dirac, the more closely related to Heisenberg, seized on the "Poisson Brackets" of classical mechanics as a suitable jumping-off point; and was thus able to avoid part of the complication of the matrix procedure and to work more simply with non-commutative algebra as such. Even this, however, was sufficiently awkward for the bulk of physicists trained in the branches of mathematics normally used by classical theory; and it was a great relief when the very different development of De Broglie enabled Schrödinger (1926) to bring the subject more nearly under the familiar differential equations of wave-motion.

* For earlier manifestations of this point of view, see next chapter.

† Infinite matrices had been the subject of a paper by the German mathematicians Courant and Hilbert in 1924.

The algorithm which De Broglie used (for he had one no less than the others) was the formal analogy which, nearly a hundred years before, Hamilton had detected between mechanics and optics (see Chap. XIII). The relativistic identification of mass and energy had a trend this way, while in 1923 Davisson and Kunsman had begun the experimental proof that electrons can show the wave-like property of diffraction. The proof was not completed until 1927 (G. P. Thomson in England, Davisson and Germer in U.S.A.), but by 1924 De Broglie had been led to associate a wave-motion of wave-number s with the motion of a particle of momentum p, where $p=hs$; like $E=h\nu$. Schrödinger's "wave-mechanics" viewed the velocity of matter as a group-velocity in certain waves underlying both matter and radiation. His system was rapidly shown to be equivalent to that of Heisenberg and Dirac.

The new wave-equation was more complex than the old (being like that for a non-uniform string) but still, it was a wave-equation, and it raised in the breasts of physicists the hope that a mechanical model might still be possible. So, while Heisenberg completed his philosophy of the subject by the "Uncertainty Principle" (1927), others hoped to identify De Broglie waves with variations in a continuous distribution of charge.

The "Uncertainty Principle" is the assertion that we can never know (for instance) *both* the position and the momentum of an electron exactly. Humanly speaking, of course, we can never know anything empirical exactly at all; but Heisenberg's was a further and precise assertion that the very physical means we take to get within p of the momentum ensure that we shall not get within q of the position, where

$$pq=\frac{h}{2\pi}.$$

It is sometimes said that modern physics is marked by the introduction of the observer, but this is inaccurate. It is his probes or instruments which are introduced, here and in the relativity theory: he himself recedes still further into the background. In one sense, therefore, Heisenberg's principle is very abstract; in another, it falls in with our stress, throughout this book, on instruments of observation. Unfortunately it is doubtful if it is accurate and complete, at least in the simpler interpretations round which most of its discussion has raged. It is perhaps safest to treat it, not as a profoundly new philosophy releasing man from determinism, but as another case of the old tendency of abstract science to negative modes of statement (as in the laws of thermodynamics). It is not so much a truth as an indication of the kind of truth which it is profitable to seek. This is true of most of the wider concepts of modern abstract science. What is remarkable is that, as long as they are viewed in a practical light and purely as working hypotheses, the diversity of their interpretations leads to no ambiguity of result in detail.

As regards the waves, the effort to view them as variations in a space-charge failed, and for "charge" there had to be substituted some such phrase as "probability of finding an electron." It was not a welcome change from the old-fashioned point of view, but it served to re-stress the fact that the whole rhythm of the new movement lay in avoiding such questions.

We may mention here that probability has an even larger part in quantum than in classical theory. But the subtle mathematical and philosophical problems which it brings in its train have therefore been carried over strongly into the new subject. Quite apart, however, from these, the statistics of quanta are different from those of classical theory. In fact more than one kind of statistics becomes possible. The quantum condition is equivalent to dividing up the phase-space into finite cells, which may or may not be occupied by points representative of the state of the system. It is clear that Pauli's principle reduces the number of possible states the enumeration and averaging of which is an essential feature in every investigation in statistical mechanics, but it may do so in more than one way, according (to use Kempe's point of view, see Chap. XIII) to the kinds of states which are reckoned different. The "classical" statistics reckoned every particle recognisably different, so that if we have two cells to fill with two particles we may do so in four ways (multiple filling being allowed), namely, *a*, *b*; *b*, *a*; *ab*, o; o, *ab*. The Bose-Einstein statistics (1924), which apply to neutral atoms and to photons, reckon particles as indistinguishable; so that $a, b=b, a$. The Fermi-Dirac statistics (1926), which apply to elementary charged particles, do not allow multiple filling; so that here only one alternative survives in the (absurd) case of two particles and two cells.* The three cases lead to approximately the same results when the number of cells is large compared to the number of particles. In the classical case, of course, this always obtains, for there is a "continuous infinity" of cells. At the other extreme is the case of "degeneracy," when (for any reason, such as low temperature and nearness to the origin in the phase-space) all the available cells are full and the transitions possible on the Maxwellian view are greatly reduced.

One result of quantum theory is that certain transitions past potential barriers, which on classical theory would be simply impossible, become matters of probability. As a result of this, Gamow and others (from 1928) have brought the emission of α particles for the first time within the ambit of the theory of the atom.

It will be seen that by 1927–8 the new theory, on which several lines had thus converged, was becoming a settled thing, from which hosts of highly significant detailed results, in chemistry as well as in physics, were deduced. We note a few of these in the concluding chapters. But much remained to be done in applying it not only to complicated atoms and molecules where the mathematical difficulties are extreme, but to simple cases which raise difficulties similar to those which the relativity theory was designed to meet.

As to the first, the well-known complexity of the three-body problem is enhanced by the quantum condition; and various approximate methods alone are available to work out even the case of the outer electronic structure. These methods are highly technical and will not be given here. Even with their aid the calculations involved in the

* See Lennard-Jones, *Proc. Phys. Soc.* 40, p. 328.

numerical integrations would be prohibitive but for the modern development of calculating machines (Chap. XIII), some of which use electromagnetic devices themselves. It is not entirely a coincidence that when mechanical models fly out of the window others, even earthier, come in at the door.

As to the second, the initial task was that of making the equations of wave-mechanics invariant under the Lorentz transformation of special relativity (see next chapter); for Schrödinger's equation does not possess this property. This was undertaken by Dirac (1928) as regards an electron in a field of force. By introducing new non-commuting variables he was able to reduce the quadratic Hamiltonian expressions to the needed linear form, and to develop an equation which reduces to that of Schrödinger if relativity is neglected. One feature of this theory is that electron spin * is an integral part of it, not one introduced arbitrarily from outside. This was a great advance. But there were peculiar results, also, for which room had to be found in nature: there were electrons with negative kinetic energy. Dirac was led to suggest an infinite distribution of these, filling all space and themselves unobservable. A hole in this distribution would show itself as a positive corpuscle. The theory thus had a further triumph when experimental evidence for the positron was forthcoming (see last chapter). In 1936, however, Dirac still considered that very high velocities could not be dealt with by existing theory with the conclusive neatness which had been achieved for other cases.

We leave to the next chapter Eddington's recent attempt to harmonise general relativity and quanta in a different sense.

* Or terms so interpretable.

CHAPTER XXII

RELATIVITY AND COSMOLOGY

WE have seen in Chap. V that Newton's generation raised the question of relativity of motion, and that the immediate success of Newton's views led man to acquiesce in an attitude which could not have borne much further examination. The aberration of light (1727) raised no difficulty on the corpuscular theory, while with the revival of the wave-theory about 1800, man seemed to be presented with the gift of an absolute reference system. Since the refractive index became the ratio of the velocities of light *in vacuo* and in matter, it ought to have been different for light coming to the moving earth from different directions, and so with different velocities relative to the glass. Arago tested the point (1818), but failed to find any change in the refractive index. Fresnel calculated the (partial) ether-drift necessary to explain Arago's absence of result, and in 1851 Fizeau tested Fresnel's formula by a sensitive interference method. A beam of light was split, and the two beams passed through two tubes of water connected to form one circuit. The beams were reunited and made to interfere. When the water was set in motion, a shift, concordant with Fresnel's formula, was observed in the interference fringes.

Yet even here, there began to peep out the peculiarly puzzling feature which ultimately pinned scientific attention upon this subject : the " drift " postulated varied with the refractive index, and so with wave-length. As we might now put it, *more than one ether* seemed from the current point of view to be needed, one, in fact, for each wave-length used, each drifting with its own velocity. At the time, however, scientific attention had not been pinned ; and for a generation the best minds were fully occupied in unifying electromagnetism and light.

In one vital respect, however, the course for later experiments was already set : interference methods were to be the source of information on the subject, and we may note here that about this time Nobert, an optician, was preparing diffraction gratings commercially. He did not disclose his methods, but in the sixties Rutherfurd and others in the United States began to develop much better ones, a very refined dividing engine being the main necessity. Rowland's gratings followed later.

In view of the puzzling feature in Fresnel's theory mentioned above, Michelson carried out in 1881 an experiment (repeated with Morley, 1887, and by others later) to see if he could detect a difference in the velocity of light in and at right angles to the direction of the earth's motion. This lay in getting interference between two light beams which had been, over part of their path, at right angles, and then seeing whether the interference bands moved when the whole apparatus was turned through a right angle. There was no change. This and other experiments (for instance, by Lodge) increased the atmosphere of

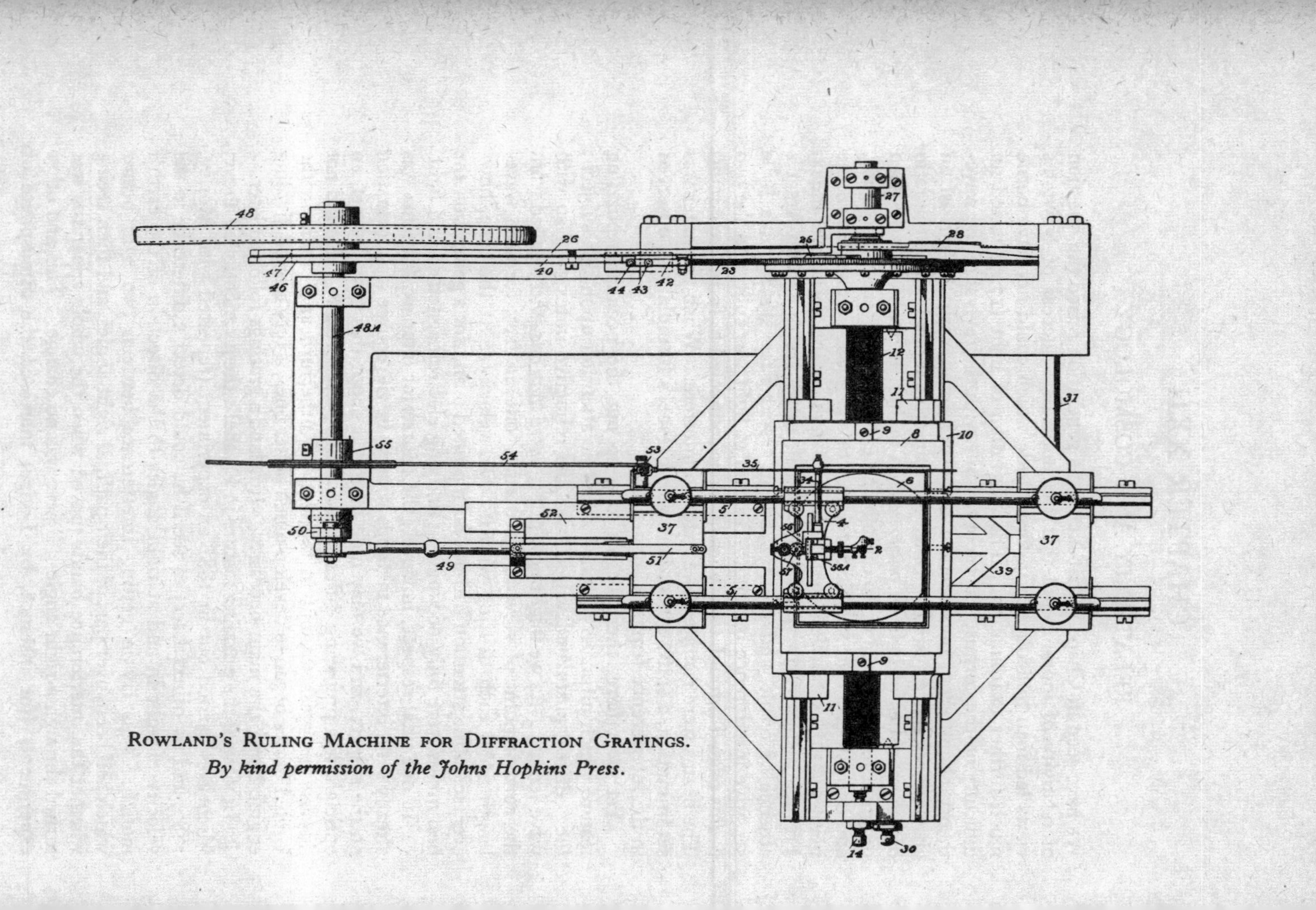

Rowland's Ruling Machine for Diffraction Gratings.
By kind permission of the Johns Hopkins Press.

perplexity, since they seemed to demand that each body should have its own private ether, a concept which, yet, was almost a contradiction in terms.

In 1893 Fitzgerald (1851–1901) pointed out that a contraction in length of all objects in the direction of motion would explain the result. If the velocity of the objects was v the contraction would reduce a unit to

$$\sqrt{1-\frac{v^2}{c^2}}$$

(c=velocity of light), which we shall call α.

Now by this time Maxwell's electromagnetic theory of light was gaining acceptance with those mathematically competent to handle it. Maxwell himself had not expressly considered the electrodynamics of moving bodies, nor "ether drift" nor its equivalent, but in 1881 J. J. Thomson had found the field of a charged particle in uniform motion; and in 1895 Lorentz took up the fact that Thomson's result was equivalent to replacing all unit lengths in the direction of motion by the fraction α. Independently of Fitzgerald he proposed this contraction as an explanation of the Michelson-Morley experiment. This he did in connection with the electron theory of matter.

Maxwell had left open the possibility of exhibiting charge and matter not as something different from field but as aspects or modifications of field. Larmor and J. J. Thomson were extending the related idea that tubes of force behave as if they have inertia. The suggestion was that *all* inertia (mass) is in this way electromagnetic, and would therefore vary with velocity like length. Calculations of the variation became possible; though different assumptions gave different results. Further, with the discovery of radium, Kaufmann and others (1901 and later) were able to test the theory by experiment on β particles. Before, however, these experiments became unequivocal, decisive developments had taken place on the theoretical side.

The theoretical situation was still complex and full of obstacles. Lorentz tried to deny drift; and so conflicted not only with Michelson and Morley's experiment but also with the whole orthodox feeling against a fixed ether, a preferred reference frame giving reality to absolute motion. Hertz had tried to treat the ether as an ordinary fluid with its own local motions in response to the motion of matter. This conflicted with Fizeau's experiment. In fact, the fundamental difficulty of the stationary or drifting ether had still not been faced.

The matter could be best put mathematically. If a law of nature is to obey the orthodox feeling mentioned above, its form must be unaltered by ("invariant under") the "Galilean" transformation $x_1=x-vt$ (the x axis being the direction of motion). Now Maxwell's equations are not unaltered; and Lorentz found that they are still altered if, to include the "contraction" which would account for the merely optical experiments, we put, instead $x_1=\frac{x-vt}{\alpha}$. In 1902, however, he proved that if we also put $t_1=\frac{t-\frac{vx}{c^2}}{\alpha}$ we get a transformation under which Maxwell's equations are invariant. This transformation was without adequate physical explanation. In fact, although by now there were discussions at the British Association and a general concentration of interest, a totally different point of view was needed to interpret it.

We may find a beginning of this point of view in Mach's analysis (1883) of the logical validity of the traditional concepts of mass, force, and the like. This was developed by Karl Pearson (1899) and others into an assertion, on the lines of the Locke-Hume philosophy, that only *observable* quantities are the proper subject of science, that unobservable abstractions in the background should be closely criticised, and that even such simple notions as that of measurement needed analysis. The even "simpler" notions at the bottom of geometry and analysis were, it will be remembered, being subjected just then to the first thorough scrutiny (Chap. XIII). In 1905 * there was read a paper by Poincaré (published 1906) in which he pointed out that the ordinary analysis of measurement as based on the equality of two superposed lengths must be replaced, if we accept the contraction hypothesis, by equality of time of traversing by light. Such perceptions had already, earlier in 1905, been developed by Einstein to a point where they cleared the difficulty up.

Einstein showed, in effect, that we had been careless in founding our measurements on rigid bodies, invariable clocks, and whatever lay conveniently at hand; that we had argued in a circle, and had created a confusion for ourselves. For instance, two observers in relative motion do not in general both see two events as simultaneous unless the events are also coincident in space. He showed that once this is admitted the Galilean transformation is not the one invariance under which secures that absolute velocity should be meaningless.

He made the assumption that we are incapable of detecting any change in the velocity of light due to the motion of its source or of its recipient. "Special Relativity"—the undetectability of *absolute* velocity—is then expressed mathematically by precisely invariance under Lorentz's x and t transformation given above. Thus subjectivity seemed to become, not the scientist's characteristic enemy, but a new ally in explanation. As with quantum theory later, however, it was not really the observer but his instruments, and especially the clock, which Einstein used as his new ally.

We have shown Einstein analysing our concepts and disclosing circular reasonings among them, but it should be noted that he by no means carried his analysis as far as it could have been carried, or has since been carried.† Nor was he concerned to do so. He was concerned not to disclose all the "axioms" of physics, but to hit on those

* See Windred, *Isis*, Vol. 20, pp. 192–219.

† In a series of works (1911, 3, 4, and 36) Robb has gone a stage or more further, denying meaning to (e.g.) distant simultaneity, where Einstein had only analysed it anew. Robb derives space from time, using as his basis the asymmetrical 2-term relation "before-and-after." He sets the matter out after the fashion of a formal Euclidean treatise, though the geometry at which he arrives is Minkowski's four-dimensional geometry, to which we next proceed.

which preserved relativity, and gave relativistic invariance to Maxwell's equations. And, for the physicist, this is the correct procedure. Laying down axioms has for him no value unless it clears up some difficulty or suggests new results. It is, in fact, far from clear that outside mathematics the problem of axioms is a determinate one.

With Einstein's work, the old substantial ether vanished from higher physics. In spite of the internal difficulties which had always dogged it, it was long mourned by the older school of physicists, who found the reasonings of Einstein perilous—and hard to follow.

The relativity theory carried a step further the question of the nature of mass. It related mass to energy : an addition of energy E to a body increases its mass by $\frac{E}{c^2}$, and a mass m is to be regarded as the same as a store of energy mc^2. This linkage of inertia and energy was important when it came to the formulation of the general theory of relativity.

In 1908 Minkowski pointed out that we could avoid the confusing relativity of our space and time measures by using a 4-dimensional space which includes all the needful 3-dimensional ones and so (in the "special" theory) is itself absolute. In this geometry time, the "fourth dimension," is not distinct, as formerly, from the three dimensions of space. Observers use time axes tilted at an angle varying with their relative velocity. Illustrating with only one dimension (x) of space, the line xx represents simultaneity on the old naïve ideas. Once we have realised our dependence on non-instantaneous signals, we

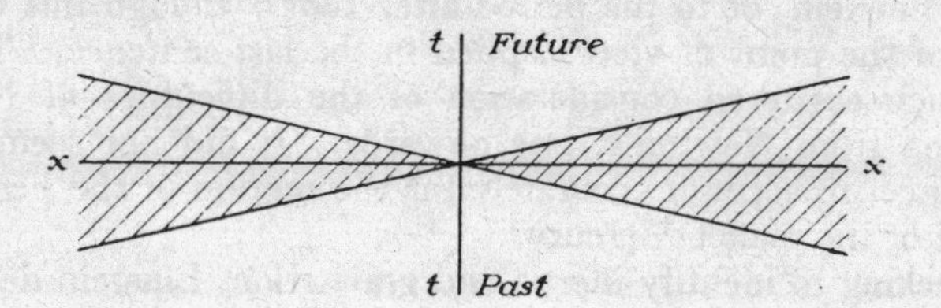

are forced to admit the existence of an *area* (that shaded) within which "before" and "after" have no absolute meaning. The angle of the sector is given by the maximum signal-velocity available, here that of light.

The geometry of this new space is not Euclidean. This is not the non-Euclidean space which Einstein brought in later, but Einstein was much helped in formulating this latter by the previous work of Minkowski. In Minkowski's space, Pythagoras' theorem $z^2=x^2+t^2$ must be replaced for the time direction by $z^2=x^2-t^2$ (t in suitable units) or by $z^2=x^2-c^2t^2$ (t in ordinary units).* Thus z can be zero for an infinity of values of x and t, and this geometry, unlike Euclid's, has no

* See Robb "Geometry of Time and Space," 1936, for the peculiar features of this geometry. The form $z^2=x^2+t^2$ can be restored by putting $t_1=ict$—by regarding time as "imaginary space." This recalls the fact that Hamilton (1833 and 5) had puzzled his contemporaries by the (mathematically irrelevant) view of algebra as the "science of pure time," which he connected with his number-couple view of i (see Chap. XII). Buée (1806) had also connected i with time.

unique shortest " separation " of two points (except zero). It has, in certain cases, a unique longest separation.

General Relativity

Einstein had shown that we can sum up a large class of facts simply by asserting that absolute velocity is indetectable. But after his work in 1905, there remained an important exception: though by optical and electrical methods we could never detect absolute velocity, it seemed that by using gravitation, as ordinarily understood, we could still do so. Einstein's attention was thus fastened on gravitation.

But there was another point. Absolute acceleration seemed to be on a different footing from absolute velocity. This, it seemed, *could* be detected—by means of the " inertial forces " which we feel, for example, when a tube-train slows down. Thus the " instinct " that all absolute motion must be undetectable remained unsatisfied. Now we have seen in the chapter on 17th-century physics that it was already realised by then that inertial forces have in common with gravitational forces, and with these alone among all the forces of nature, the curious property which may be expressed by saying that they are proportional to the " mass " of the body, or that they produce equal accelerations on all bodies in the same place. Einstein called attention in 1911 to this " equivalence " of inertia and of a local gravitational field. He sought a point of view from which the equivalence would become identity.

It must not be supposed that interest in these questions was confined to Einstein, or to the period after 1905; though this may fairly be said of the point of view implied in the last sentence. There had been much scattered consideration of the difficulties of Newtonian gravitation from Newton's time onwards. It did not seem possible, for instance, to account accurately for the motion of the perihelion of the orbit of the planet Mercury.

In seeking to identify inertia and gravitation, Einstein deserted the fruitful method of analysing kinematical and other concepts which had served him so well in the special theory. He took up, instead, Minkowski's purely geometrical view, and sought a modified Minkowski space to give him what he wanted. Now what he wanted was a geometry in which the path of a particle not " under no force " but " under no force except gravitation " would be the analogue of a straight line. The natural analogue of the straight line in more general metrical geometry is the " geodesic " or shape at which length goes through a stationary value, either maximum or minimum. In Euclidean space the latter, in Minkowski space (as we have seen) the former, is the determinate thing. Thus the geodesic replaced Newton's straight line, as the latter had replaced the circle of the ancients, in this method of approach.

This, however, does not sufficiently define the " geometrisation of physics " which Einstein undertook. For almost any law can be expressed by means of a suitable space-time and conventions of meaning. Nor was it merely that, since such a mode of expression has (as we have said elsewhere) a momentum of its own, it made Einstein prefer a slightly different law to that of Newton as demanding a simple space-time when Newton's demanded a very complex one. The real point was that an Einstein space-time was itself correlative to, and an expression of, the particular number and distribution of particles in it. It was not a curved space-time as independent as Newton's flat one of the particles in it. Its constants changed from place to place according to the distribution of particles in it. In fact the particles were only *aspects* of the space-time itself.

Thus Einstein's geometrisation of physics was in line with the " field " point of view taken up by Maxwell and Faraday ; though the intrinsic properties of the field were no longer to be those of an ether but those of space itself.* We have mentioned in Chap. XIII Riemann's prophetic work in this connection. In 1870 Clifford had suggested, much as in Einstein's general theory, that pieces of matter consist of departures from the Euclidean character of space ; but he had no special physical connection in mind where the idea would serve as a tool.

Einstein was much helped in his discovery of the best geometry to use by the prior existence of tensor calculus (see Chap. XIII), since clearly if natural laws can be expressed as tensor equations the relativity condition is satisfied. In particular, since he had already identified mass and energy, he had guidance in the fact that the energy-tensor must be concerned. It turned out (1915) that a Riemannian geometry would serve all these purposes ; and, making a further suitable assumption, he was able to obtain expressions for the quadratic differential form (in four variables) which replaced Pythagoras' theorem in our discussion in Chap. XIII.

When this law was tried on the perihelion of Mercury, it was at once successful. Two further tests, the gravitational bending of light and the (" apparent ") slowing up of electron vibrations in very massive bodies when viewed from another part of the field, have also strengthened it, especially (1919) the former.

The position left by its triumph, however, was far from simple. It was found that the distant bodies of the universe would produce a slight curvature of space-time in any neighbourhood. While nearby bodies would produce, locally, much greater ones, the awkward fact was that the greater number of the distant bodies would give them a predominant effect on the general curvature, and so " shape," of the universe. Relativity thus became bound up with astrophysics and cosmology, which, therefore, we treat briefly in the next section. We conclude this one with another, equally natural, development of the point of view of general relativity.

In fact, after the gravitational field had been reduced to geometry, it was natural to see if the electromagnetic field could be treated likewise

* It is not obvious, however, whether Einstein's relativistic equations determine the motions more or less fully than Newton's equations of motion. Einstein has lately shown (*Annals of Maths.*, Vol. 39, 1938, p. 65) that they are not more restrictive if matter is regarded as point-singularities of the field, while if charges are so regarded in Maxwell's theory, Maxwell's equations are less restrictive. But the need to introduce singularities shows that the organic connection of space and content is in any case still incomplete in both theories.

by finding a more general kind of non-Euclidean geometry which would yet be capable of representing the relativity theory also. In 1918 Weyl found one. It will be remembered that in Riemann's geometry a measuring rod, if it did not remain unchanged from place to place, could be *compared* from place to place; it changed according to a definite law. In 1918 Weyl abandoned this comparability, while retaining the postulate that the rod must be usable (for instance) in every direction at any one point. From Weyl's point of view, Einstein's work conflicted with the "relativity of size" which is perhaps an "instinct" as deeply-rooted as the relativity of motion. Weyl's new geometry had the properties of the electromagnetic field; while Eddington in 1921 showed that by abandoning the second postulate, also, we could get a space which exhibited the electric and magnetic fields even more naturally—as aspects of one tensor. Unfortunately, only charge-free space fell easily into line. We shall not pursue the subject, since in 1927–8 Einstein returned to a kind of Riemannian space (discovered by Levi-Civita in 1917) in which, in spite of the non-Euclidean nature of the geometry, "distant-parallelism" is possible. With the aid of this he has tried to formulate a "unified field theory." Of late, advantage has been taken of the fact, mentioned in Chap. XIII, that non-Euclidean geometries may be viewed as special cases of projective geometries, and "projective relativity" has made its appearance. These developments have, perhaps, hardly the urgency of the special and general theories, both of which aimed at the removal of definite inconsistencies.

Astronomy and Cosmology

What follows is not a history of astronomy since the 18th century but of certain themes in it, of which one is the process by which astronomy became a branch of physics as well as of mechanics. Advances in physics or in physical instruments have often been followed by advances in astronomy. Herschel's improvements of the telescope enabled him to see a double star in 1782, and to follow the mutual revolution of its components in correct Newtonian form in 1793. Laplace's nebular hypothesis of 1796, on the other hand, might be called an application of mechanics rather than of physics, although it contemplated changes of state. The increasing power of telescopes enabled astronomers to leave the well-understood planets and to probe the stellar system, of which, hitherto, they had only been able to grade the "magnitudes" or brightnesses. Only in 1832–3 (published 1839) was a finite stellar parallax (with the earth's orbit as base) discovered—by Henderson in South Africa. The star was a faint one; but from then on (e.g. Bessel and Struve, 1838) more and more stars were found to have observable distances, which could be estimated as upwards of 4000 times the diameter of our solar system. Then (1853 and 4) another side of physics began to make conquests in astronomy: Helmholtz and Kelvin began the series of theories by which men have tried to find an adequate source for the colossal emission of energy by the sun and the stars. Their suggestion was that of contraction under gravity.

With the advance of yet another physical subject, spectroscopy, a new classification of stars became possible. Fraunhofer had examined the spectra of celestial objects in 1817, Huggins resumed the subject in 1860. Between 1862–8, the Italian Jesuit Secchi classified a large number of stars by their spectra. The later immense researches (for instance, at Harvard) on the spectra of stars resulted in a classification under about nine types, which fall into a single sequence according to such marks as presence or absence of continuous background, of particular elements, of bands, etc. Comparisons of the intensities of different parts of a star's spectra have given indications of its effective

GRAPH III

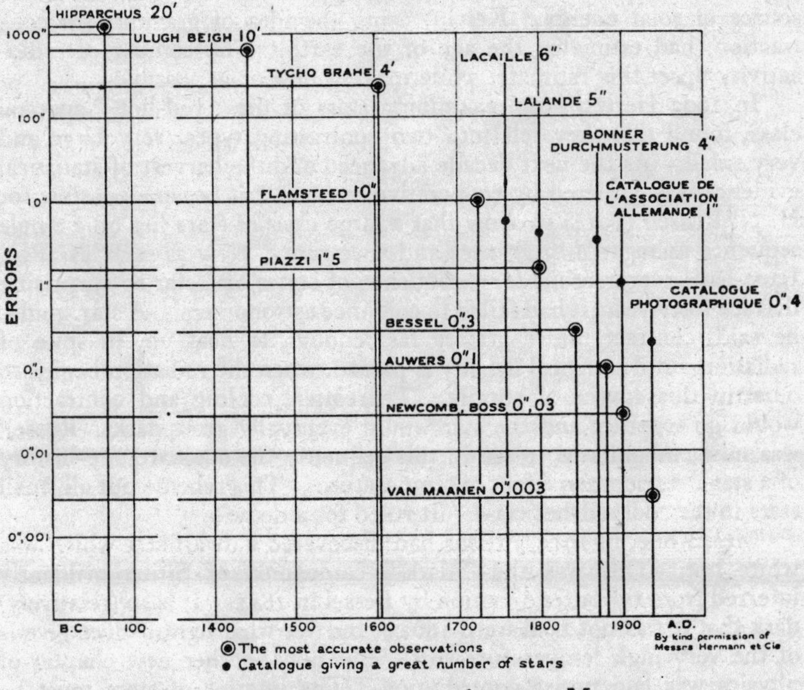

ACCURACY OF ASTRONOMICAL ANGULAR MEASURE.

From a graph prepared by H. Mineur.

surface-temperature, varying from about 1600° C. to about 23,000° C. for different main-sequence stars. Also through the spectroscope, Pickering in 1889 discovered the existence of many double stars of period too short, and mutual distance too small, to be separated by the telescope. Their spectral lines were split by the Doppler effect of their two components to extents which varied with time according to their positions in their orbits. From 1880 onwards the usefulness of all astronomical instruments began to be vastly increased by the use of photography to supplement the human eye.

The year 1900 was important. Certain work of Moulton and

Chamberlin, when combined, showed that Laplace's hypothesis could not account for the solar system. In part, this was a progress of the older celestial mechanics. Poincaré and others had discovered mathematically a new figure of equilibrium of a rotating liquid : an elongated ellipsoid which would be unstable and would split unequally into two smaller bodies. But in part it was also the result of applying a relatively new physical concept, Maxwell's radiation pressure, to astronomy. Chamberlin showed that in certain circumstances this latter would overbalance gravitation and thus prevent Laplace's condensation into planets. Later Jeans showed that this only holds for " small " entities like the solar system, and that something like Laplace's hypothesis might account for the birth of stellar aggregations out of nebulae. New physical conceptions were also being applied to the question of the source of solar energy. Kelvin, using the idea of gravitational contraction, had estimated the age of the earth. The coming of radioactivity upset this estimate, while making new ones possible.

In 1905 Hertzsprung, examining stars of the " red-hot " spectral class, found that they fell into two contrasting types, very large and very small. As the next decade advanced and the harvest of statistical evidence was gathered by cooperative research,* it became possible for H. N. Russell (1914) to show that a large class of stars lies on a single sequence as regards brightness and spectrum. Now as early as 1870, Lane had given reasons for a sequence of states of a star so surprising that for forty years it had failed to convince astronomers. A star would, he said, contract under gravity fast enough to heat up, in spite of radiation, until a critical density is passed, when the radiation begins to outstrip this source of supply. Thereafter, cooling and contraction would go together and the star would eventually grow dark. Russell was now emboldened to see in this sequence the standard life-history of a star : a rise, then a fall, of temperature. This scheme put all small stars in the cool red-hot class. It ruled for a decade.

Meanwhile, in 1914, Adams had discovered a dwarf star which was white hot. This was the " dark " companion of Sirius brilliantly inferred from the latter's motion by Bessel in 1844. It is so (relatively) dark that it was not seen until 1862 ; and yet what light it does give is of the very high temperature sort. But now another new chapter of physics was laid under contribution. The interior of stars must be much hotter than the surface temperature observed ; and it began to be evident (e.g. Jeans, 1917) that this would mean that the interior atoms would be ionised. In fact, they would be so far stripped of successive rings of electrons by the high-temperature radiation of the interior as to be reduced in size almost or quite to nuclear dimensions.† Thus (Eddington) a vastly further contraction, on gas-law lines, becomes possible, and with it densities altogether greater than any known on earth. This aroused doubt until Adams in 1925 found that the com-

* The great Mt. Wilson instrument came in 1908.

† X or γ rays would be the result of such temperatures. Nearly all these would, by collision in the outer layers, be reduced in frequency to those observed ; but a possible residue has been invoked as one theory of " cosmic rays."

panion of Sirius gave, on the evidence of the Einstein red-shift, just such densities.

In 1924 Eddington, on the basis of the new idea, calculated that if a star is gaseous its total radiation depends on its mass rather than on its diameter ; for if it contracts its decrease in available outlet-area just about balances its increased temperature, which, however, causes its light to grow whiter. Thus the case of the white dwarfs could be explained as soon (1926) as Fowler pointed out that, on quantum theory, such dense matter would be in a degenerate state. Jeans has more recently urged that, at the observed densities, the atomic nuclei must form a mass liquid rather than gaseous in properties, the radiation, like that of radioactive bodies, being independent of temperature. He has also urged that, if they were gaseous, the stars would be unstable.

Much of this new theory was made possible by new methods which, about 1914, began to be available for estimating the distances, and so the radiations, of stars too distant to give a parallax. Many of the stars of regularly varying brightness found in the sky have been found to be double stars which periodically eclipse each other, but this cannot be true of a certain ("Cepheid") type of varying star, because it would give a wrong brightness-time graph. The mechanism of Cepheids is a puzzle, but where their distances are known, their absolute brightnesses were found to be related in a precise way to their periods. This was discovered in 1912 by Miss Leavitt of Harvard, and was soon seized on (for instance, by Hertzsprung and Shapley, 1918) to give the distances of Cepheids beyond the parallax range. Another method of doing this ("spectroscopic parallaxes") was developed by Adams and Kohlschütter in 1914 and used (for instance) by Lindblad in 1922. Unlike terrestrial, artificially adjustable sources of light, stars cannot be identical in spectra (wave-lengths and intensities) without being closely similar in other respects. For every factor reacts on every other. We have just noted that the hottest star cannot radiate much unless it has a large area ; so that the brightest stars are always large. Now, as we have seen, stars fall into classes, in fact, into classes nearly enough identical for the above fact to be used to compare distances by comparing apparent brightnesses.

From these methods some estimate (for instance, Shapley, 1918) of the size and shape of our "galaxy" has become possible ; and it has been found to consist of a flattened region (the plane of which is that of the Milky Way), which includes our own solar system. It is of gigantic dimensions, but is small compared to the distances of the immensely numerous "spiral nebulae," some of which give signs of being over 10^8 light-years away. They are in slow rotatory motion, presumably from the tidal action of neighbouring bodies. Jeans has shown that, in view of this, a mass of gas of their size would spread into a lens shaped body which might ultimately break down at the edges into two spiral arms. It is uncertain if these arms can be explained entirely on ordinary dynamical lines ; but it has been shown that the condensation into the liquid state which is to be expected in them would yield stars of about the masses observed. These might then undergo Poincaré's fission. One of the surprises resulting from recent progress in determining the sizes of stars is that these bodies are rarely more than two or three times larger or smaller than our sun in mass, while varying enormously in radius.

Our increased numerical knowledge of the stars has enabled us to estimate the age of the stellar system. Some of the estimates amount practically to finding out how nearly it has so far approached the equipartition of energy. The approach is remarkably great, considering how slight are the opportunities for mutual influence ; and this is taken to show a great age. Before general relativity considerations came in to complicate matters, such methods led to figures of perhaps $5-10 \times 10^{12}$ years for the age of our galaxy. Such vast ages showed even radioactivity as inadequate to supply the radiation of energy needed ; but apart from such suppositions as super-radioactive elements of atomic numbers higher than those of the terrestrial series, the relativity equivalence of mass and energy came to the rescue. This could " explain," without unlikely diminutions of mass, the observed ages of the stars. It even furnishes a check on the estimate of these latter, by way of the observed fact that the older the binary star the nearer the masses of its components. It can be shown that the initially heavier constituent would transform the greatest mass per cent. into radiant energy.

In many of the arguments outlined above the proportion of fact to theory has been smaller than is usual in science, and the trains of deduction have been unusually long. Moreover, the estimates of several of the quantities, such as the age of the galaxy, depend on the view we take of such fundamental questions as the nature of space-time. To these questions we turn in conclusion. For we are now in a position to understand how cosmology and relativity have interacted.

From 1917 Einstein and De Sitter began to examine the kind of universes which general relativity suggested. They ignored the effects of local irregularities in the distribution of matter on space-time, assuming that the distribution was uniform through the universe, on an average, from a large enough point of view. Einstein's space-time embedded in a Euclidean 5-space would appear as a four-dimensional cylinder,* De Sitter's as a four-dimensional pseudo-sphere. It was later found that Einstein's model would not be stable. The weakness of De Sitter's was that, strictly speaking, it should not contain either matter or radiation. Both models, therefore, appeared as initial or other limiting cases of the set of possible universes.

In 1922 work by Slipher showed the existence of a red-ward shift in the light from the extra-galactic nebulae, apparently a Doppler effect due to their recession. By 1929 Hubble had found that the velocities of the nebulae away from us, thus disclosed, increase linearl with their distance, to a first approximation. This could be made to emerge quite naturally from de Sitter's universe, but had no intrinsic

* I.e. with a sphere instead of the circular section of the ordinary cylinder.

explanation upon Einstein's. By this time, however, their assumption that, though the shape of space-time might vary locally, its broad average did not change with time, had been abandoned. Friedmann had examined non-static space-time in 1922; Lemaître in 1927 applied the idea to reality. There emerged a model in which the recession of the nebulae was explained by the expansion of "spherical" or "hyperbolic" space-time, the distribution within the space-dimensions of this being still broadly homogeneous.

In all this, the essential idea of "geometrisation," the correlation of space-time shape and particle-distribution, was retained. From about 1933, E. A. Milne began examining the effect of dropping this and of returning, though with a difference, to the old familiar independence of the two, with Euclidean space. He pointed out (roughly speaking) that, collisions and gravity being neglected, *any* set of particles moving at random within a finite region of space * would eventually be found receding from one another according to the Hubble relation, the fastest having got farthest. But he did something else. In dropping Einstein's general-relativity process of geometrisation, he took up again Einstein's special-relativity analysis of kinematical concepts and concentration on observables. He derived space *from* these, so that its status was not the same as in pre-relativity days.

When both the "shape" of space-time, and the coordinate systems within any one shape, are at disposal, and there are physical, measurable quantities which have to be identified with geometrical features, the position becomes extremely complex. For instance, there is an ambiguity as to the meaning of length in ordinary expanding-universe theory. Again, consider the meaning of homogeneity in the assumption stated above. It relates to a universe many of our sense-data about which have taken thousands, even millions, of years to reach us. Is it to *be* homogeneous or to *seem* homogeneous? Has the contrast meaning—it arises over almost every question in this subject? General relativity leapt such hurdles with dizzying ease. Milne proposed a closer analysis and a new start.

He defined "equivalent observers" who, by a simple system of light signals, could calibrate their clocks into agreement. Starting like Robb, though diverging sharply, he derived space-measures from time-measures. He then easily constructed a system such that *every* member of it would, paradoxically, seem to itself to be surrounded on every side by an infinity of others receding according to the linear law—for the new definition of distance. The outermost would be travelling with the velocity of light. The whole sphere of them would be finite in the unsophisticated sense: but as the only available means of following them would be their light, the "universe" of measures of their distances would be infinite and the "outermost" would be inaccessible. In general, however, not all particles would be equivalent. Non-equivalents would appear as strongly concentrated round the equivalents—nebulae being the result. It ceases to be true that natural laws must be expressible as tensor equations, but they must be invariant

* Extraordinary types of distribution apart.

under the transformations leading from one equivalent observer to another. Like Einstein, Milne was not making an exhaustive analysis of axioms: he was fixing on axioms and identifications which worked, though his statements of them invited criticism.*

The momentum, however, of the geometrising point of view, coupled with the fact that no absolutely simple niche occurred in Milne's system for certain features of gravitation, has since tended to engulf the kinematic point of view in that of the general tensor analysis by which almost all this work has been done. Milne's theory then appears as a complement rather than a rival to general relativity.

Tensor analysis has had its apotheosis in Eddington's work of 1928–36. Dirac's relativistic wave-equation had a type of invariance the possibility of which had escaped the tensor experts; and Eddington, re-examining the point, found that he could obtain a "harmonisation" of relativity and quanta by using the quantity ψ (obeying the fundamental equation of wave-mechanics) in his tensor-analysis. He also showed the necessity of using *double* instead of single tensors, in order to take account of another sort of relation between the largest and smallest features of our world-picture. In what he has done there is something at least superficially reminiscent of ideas of Cusa and of Bruno mentioned in Chap. III. He showed, roughly speaking, that although quantum theory is sometimes actually *expounded* by saying that the action of the observing instruments affects the observed electron, yet it had not taken sufficient account of the fundamental relativity-perception that a measurement is not an absolute thing, but only relates one entity to another. The Uncertainty Principle, in fact, should be applied to the reference frame itself. Hence the double tensors. For, as Eddington observed, if the frame is made massive in an unsophisticated attempt to avoid this trouble, it bends space-time and needs correction of another sort.

One of Eddington's clues was the appearance of integer or other dimensionless relations among those fundamental constants of nature which modern physics has thrown up.† This suggested a deductive attack. To take one of his own examples, measuring length A against length B involves *four* points. Four is not the only number which measurement might be regarded as "involving"; but if it appeared in any special connection with measurement the fact would not be surprising. Eddington has in fact indirectly connected it (and so measurement itself) with the four dimensions of space-time. This looks like *a priorism*, but is not, since we do not know *a priori* that measurement is possible. Eddington similarly "deduced" the ratio of the masses of

* To philosophers, the scientist in a mood to analyse his own assumptions always appears to be trusting some the more, the more suddenly he finds it convenient to distrust others. But if the scientist's road to truth is a winding one, the philosopher has yet to find a straighter. It should, however, be mentioned, as a commentary on the vast fields of mathematics provoked by the linear recession, that its experimental discoverer, Hubble, does not admit that the red-shift is necessarily to be ascribed to the Doppler effect!

† The masses of the electron and proton, their charge, h, c, the gravitational and the "cosmical" constants: seven, reduced to four by eliminating the arbitrariness of units.

electrons and protons, apparently by showing that we could not treat them as both isolated and observable unless we treated them unconsciously as the roots of a certain quadratic equation. On similar principles he asserted a value for the number of particles in the universe: of " bricks " into which our present theories split it up.

There emerges the fact, not often realised, that progress in science may consist not only in knowing, but in finding out what it is that you already know. Like philosophy according to Wittgenstein, theoretical physics may end by making itself needless. It does not follow that it is needless yet. Both Eddington and Milne have been attacked for *a priorism*; but finding out what it is that you know has from the start been the best effort of relativity. If in the work of these men much space is taken up in erecting " imaginary " structures, their identification remains empirical. This should not obscure the fact that Eddington's scheme has not, any more than Milne's, met all the technical criticisms levelled against it.

CHAPTER XXIII

REAL MATERIALS. I

If we refer to the last mainly physico-chemical chapters, XIV and XV, we find them dominated by certain simple ideas : ions, surfaces, dynamic equilibrium, thermodynamics, the periodic law, valency as the simple geometry of certain lines in 3-space. The subsequent development of these ideas has usually been due to their taking up fresh elements from pure physics. Let us consider the points in roughly the order in which we shall treat them in the next two chapters.

X ray crystal researches completed the geometry of valency by introducing the idea of a space-lattice. The idea of valency, itself, drew from the periodic law and from the physics of the planetary electron a new subtlety which made it possible to approach such special points as the structure of metals, or of clays or proteins. These last two subjects were much clarified by a further physicalisation of old *surface* ideas. Turning to more kinetic questions, we find these, also, influenced by surface ideas. The ionic theory is corrected, like gas theory, by fresh drafts from the higher parts of physics. Physical, kinetic pictures of reaction, and of its promotion and inhibition, are developed.

These are but cases of our chapter's keynote, one keynote of recent science itself, namely, its growing capacity to deal with real, as opposed to " perfect " solids, liquids, and gases. We see its results applied to geophysics in the largest sense, with geology and meteorology as particular cases, and metallurgy, hydraulics, and aeronautics as industrial offshoots.

More Static

The process which we trace in this section is that of the development of a purely formal geometry of valency into an attack upon the whole problem of the architecture of the solid state. For until quite lately, and especially in pre-quantum days, these two subjects would by no means have been seen as essentially one. Questions such as that of shear strength were left for the future with a vague confidence in the adequacy of inverse-square attractions and repulsions, forces not instinctively connected with valency by chemists and physicists in general. It was probably, for instance, with no thought of such matters that, in 1892, Werner of Zurich began to publish his work.

This work concerned certain complicated metal-ammonia compounds which could not be accounted for by either of the two main constitutional and valency trends of the time. The older of these trends was an adaptation of the electrical dualism of Berzelius. It referred mainly to acids, bases, and salts. For these, certain simple valencies gave good results, merely by the old processes of regarding such compounds as made up of two parts, which might be atoms or radicals. The newer one had no electrical (" polar ") or dualistic implications. Its bonds were directed lines in space, and it was able brilliantly to account for the vast bulk of organic compounds, including their stereochemistry.

These two groups were each fringed with exceptions, but until Werner's work the exceptions were at least unrelated among themselves. What Werner did was to separate a further group, " coordination compounds," with a new type of regularity among them. Such were the platinum ammonium chlorides and similar compounds of cobalt and nickel. He found that the classification of these was easiest if he assumed that a " complex ion " was formed, in which a metal had six, or four (usually six) other atoms or molecules attached to it, whatever the ordinary valencies concerned. Thus a new type of valency had to be postulated. On the basis of space-formulae such as  (" coordination number " four) or 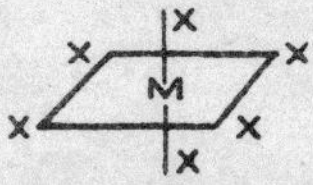 (coordination number six) he was able to predict and (1911) to find, stereoisomers, when the X's are different. This confirmed his ideas.

Meanwhile, in 1899, Abegg pointed out how often we can get a clearer picture of valency behaviour by supposing that each element has two valencies, " normal " and " contra," which always have a sum of eight. This idea bore fruit when a new generation arose which took it, and similar ideas, literally and physically instead of formally. One instance of such physicalisation was Debye's work of 1912 relating to the " polar " type of bond. He remarked that if molecules were held together by electrostatic attraction, they ought, when in solution, to orient themselves under a field and to show a measurable electric (" dipole ") moment analogous to Maxwell's and Faraday's " polarisation " in dielectrics. We shall see later that Debye turned his dipole moments into a valuable means of probing molecular structure.

Of like physical tendency was the work of two Americans, Lewis and Langmuir. For it was, as will be clear later, no accident that these men were also working on other physical subjects, such as thin films. We have said that the new valency theory, started by Lewis (and independently by Kossel) in 1916, was inspired by the new planetary atom theories. But in truth it was inspired quite as much by their chemical awkwardness. For how can we envisage a stable, " chemical," connection of spinning systems ?

All that could be done was to take up the idea of the planetary electrons' falling into " levels." As electron after electron is added, the successive levels get filled up. Each new electron gives us a new element, each full level a " period." Lewis' concern was with the outermost level of any particular atom ; and he supposed that chemical combination redisposed in definite ways the outer electrons of the combining atoms. This redisposition was always such as to make the outer levels, in *both* atoms, precisely similar to that of one of the inert gases, with eight (or in the earliest, two) electrons. These eight, Lewis disposed at the corners of a cube.

In 1919 and 1921 Langmuir, using Rydberg's empirical regularities (1914) in the numbers of electrons in successive shells,* built up on Lewis' lines a comprehensive scheme of atomic architecture to account for the periodic table. The rare earth elements were a difficulty here, for it was hard to account for their close similarity until the idea was brought in that an outer ring might in certain cases be completed before an inner one. The differences in the latter would then be masked by the outer rings and be of small chemical effect. Bohr in 1921 published a scheme based on quantum theory. Pauli's Exclusion Principle (1925) and later, wave-mechanics, have made it possible to bring the table into closer and closer relation with theory ; but we shall revert here to the valency question.

In its developed form the electron theory of valency regards non-polar (e.g. organic) compounds as due to the sharing of electrons (" homopolar " bonds. Langmuir's " covalency ") as in

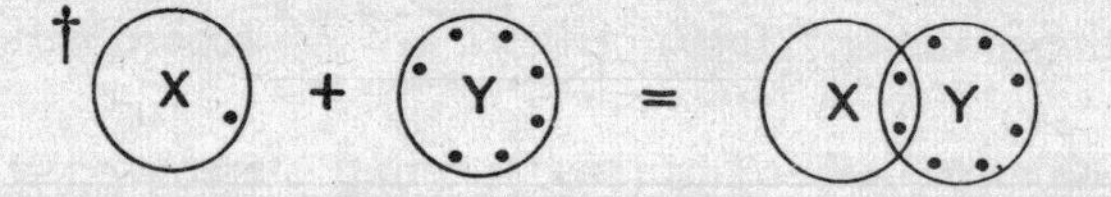

Polar compounds (" heteropolar " bonds. Langmuir's " electro-valency ") are due to actual electron transfer, each atom then having its outer ring separately complete but having an unbalanced electric charge.

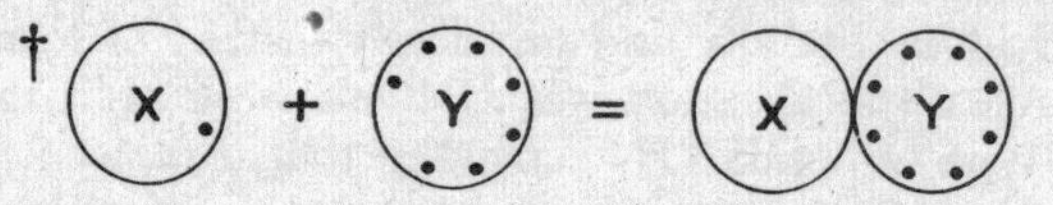

This type of combination, unlike the other, is purely electrostatic and so is not directional. It has been put on a quantum-mechanical basis by Born.

Finally, there is the " valency " obtaining in coordination compounds, or the " semi-polar double bond," which is in effect as if one of the shared electrons were of the polar, the other of the non-polar, type. It arises where *both* shared electrons come from one of the reactants. One, or even both, X and Y, may be capable of separate existence.

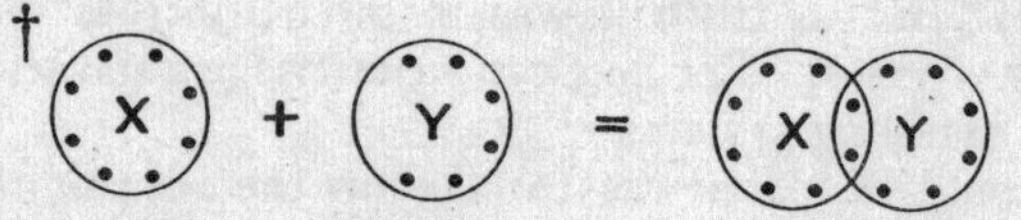

This theory has by no means closed the subject. For instance, coordination compounds, though usually near an inert gas configuration,

* Namely 2 (1^2, 2^2, 2^2, 3^2, 3^2 . . .).
† Only outermost level of electrons indicated.

do not always possess it exactly. Again, Lowry, retaining earlier views of Armstrong, persists that all reactions are between ions, sometimes formed very transiently from non-polar entities by means of catalysts. Moreover cases of abnormal valency continue to be unearthed.

There were some curious features about these Lewis-Langmuir ideas. One was that the bond always consisted of *two* electrons. In 1927 Heitler and London concluded from wave-mechanical calculations that this was a stable configuration if the two electrons were spinning in opposite directions (symmetrically) but not if they were spinning anti-symmetrically, that is, both in one direction. These writers and others have in fact achieved some success with the wave-mechanics of valency, especially of covalency. But we must turn to the position reached by about 1927 as to the " architecture of the solid (and other) states."

Supplementing valency in the narrower sense had come lattice ideas from the work of Laue and the Braggs, with the suggestion that " molecule " may be a meaningless abstraction. In common salt crystals, for instance, no one chlorine can be regarded as specially attached to any one sodium atom. In fact, unless we can draw distinctions among the bonds responsible for aggregations of matter, " molecule " can never have any strict meaning in the solid state. At the time mentioned, however, three general types of forces were in fact being broadly distinguished, apart from the metallic binding which we consider later.

First come the weak but long-range forces responsible for the Van der Waals corrections in the kinetic theory of fluids (see next section) and for the grosser types of adsorption. These are due to the mutual polarisation of molecules by induction.

Second come the mainly electrostatic forces responsible for ionic polar compounds, ionised both in solution and in the crystalline state (so that molecule *is* for them meaningless).

Third are the electrodynamic forces, of very short range, responsible for non-polar bonds. In crystals of compounds bound by these, there are still molecules, since an atom has this stronger binding with certain of its neighbours, with which it forms a group bound to neighbouring similar groups mainly by Van der Waals forces.

Considering a molecule by itself, it has, by comparison with an atom, not only quantum numbers belonging to the usual intra-atomic energy levels, but others arising from the relative rotations of the atoms and from their relative vibrations. Thus each line in a true spectrum comes to be replaced by a series each member of which is itself, on closer analysis, a series. The result observed is of course the familiar band spectrum.

The metallic state * is proving very difficult to handle. The classical idea of it as due to a " gas " of self-repellant electrons in the interstices of the ionised atoms was indeed simple, and was propounded (1898, 1900, Drude, J. J. Thomson, and others) very soon after the electron was discovered. It explained Wiedemann and Franz's old discovery of the constant ratio of conductivities for electricity and heat.

* As with colloids, the word " state " expresses the present point of view.

The number of free electrons appeared on optical grounds to be of the order of the number of atoms; but unfortunately these electrons seemed to have much too low a heat capacity.

Crystal lattice ideas brought a new definiteness into atomic pictures, and it became clear that *close-packing* is the characteristic of metal lattices: body- or face-centred cubes or close-packed hexagons, with 8, or the maximal 12, equidistant neighbours. Further, they suggested a reason for the abnormal valencies in intermetallic compounds. In normal compounds all the available valency electrons are used up. This cannot happen if the valency electrons are free. Hume-Rothery found that intermetallic compounds usually occur when the ratio of valency electrons to atoms is not an integer, but when, at the same time, this ratio is fairly simple $\left(\frac{3}{2}, \frac{21}{13}\right)$. This, however, left the conductivity problems, and the problems of cohesion and of strength, much where they were.

The difficulty as to heat-capacity was clearly one for quantisation, but the older form of the quantum theory could not deal with it. For the newer one, with its finite probability of escape through potential barriers, there was the difficulty of explaining why insulators do not conduct. Sommerfeld began a new era here in 1928. The Exclusion Principle, in the form of the Fermi-Dirac statistics, is invoked to show that in non-metallic types of lattice the array of possible electron-states is always full, while in metallic ones this is not the case. In the latter, indeed, there would be *no resistance* did not thermal agitation destroy the perfect periodicity of the lattice and scatter any beam of electrons passing through it. Even here, only electrons of nearly the maximum energy can find unoccupied states into which to "move." Supraconductivity, a subject not dealt with in this book, remains unexplained.

Metallic cohesion—the metallic bond—was pictured by Wigner and Seitz in 1933 as mainly the electrostatic attraction between the electron-cloud and the atoms; but the forces cannot be solely electrostatic, since even at the absolute zero quantum theory supposes the electrons to have high velocities. But this subject cannot be pursued here.

This work replaces the diagrammatic bonds of the chemical textbooks by the complex fields of wave-mechanics. But the diagrammatic bonds have not therefore ceased to be used. They suffice to represent a certain stage of information about a molecule, and even this stage has by no means been reached for many complex molecules. We shall briefly consider two of these latter, proteins and clays. But before doing so we give notes on certain experimental discoveries which have in common this, that they have become the basis of laboratory routines, especially in connection with this very subject of constitution.

An instance already mentioned is dipole moments. The non-polarity of CO_2, N_2O, CS_2 strongly suggests an A—B—A, linear, type of structure, while the polar nature of water forces us to put its bonds at an angle, although Debye (1929) did not succeed in giving the angle a definite value. The subject of dipoles is far from being completely understood. For instance, bromine and iodine are unexpectedly found to be polar.

One broad class of routine determinations is a development of the old ambition to find physical properties of compounds which are additive functions of the atomic weights of the atoms concerned. Such properties have obvious uses in constitutive work. An early suggestion was "molecular volume," $\frac{\text{molecular weight}}{\text{density}}\left(\frac{M}{d}\right)$ (Kopp, 1840 onwards). Less additive and more constitutive was "molecular refraction," $\frac{M}{d}\cdot(n-1)$ (1858, Gladstone and Dale) or $\frac{M}{d}\cdot\frac{n^2-1}{n^2+1}$ (1880, Lorenz-Lorentz), n being the refractive index.

In 1892 Ostwald showed that in dilute solution the absorption spectrum of a salt is the sum of that of its ions. In the organic realm it was early noticed that ring compounds show absorption spectra much more commonly than do open-chain ones; and since about 1906 such spectra have been largely used as evidence of rings. Ordinary *emission* spectra can only be obtained for compounds stable up to high temperatures. But, from 1923, McVicker, Marsh, and Stewart developed Tesla luminescence spectra, which can be photographed. Most such spectra are highly constitutive. Wave-mechanics has made their interpretation a fruitful, if immensely complex, field.

The physics of spectra have given us further guides to constitution. One is the Raman effect. The modern practice of photographing spectra enabled the Indian Raman to detect (1928) a faint radiation of lower wave-length scattered by liquids.* The magnitude of the drop in wave-length in this effect usually suggests that it concerns the vibrational, rather than the rotational or intra atomic, energy levels. It is thus often characteristic of a particular *bond*, a fact which makes its use in the present connection easy to see.

An important example of physical properties useful in determining structure lies in another region. Macleod in 1923 found that for non-associated substances the fourth power of the difference $(D-d)$ between the liquid- and the vapour-densities varies as the surface tension, γ. We may write $\frac{\gamma^{\frac{1}{4}}}{D-d}=c$, a constant.

Then, in 1924, Sugden found that Mc (M as before), which he called the "parachor," is additive for any one homologous series. For instance, adding $-CH_2$ usually causes a rise of 39 in the parachor. Unsaturation, ring-formation (Sugden, 1924, 1927), and the like, however, affect the result. The parachor furnished (1928) strong evidence for the truth of the idea of the semi-polar bond. It has become a subject in itself.

It should not, of course, be supposed that so complex a quantity is pure empirical. Relating surface tension to internal pressure, we see that the parachor is closely connected with the old molecular volume, but corrects it for the differing internal stresses in different substances.

There is another case. Debye and his co-workers managed in 1929 to get X ray diffraction photographs from *gases* by using them rarefied enough to make it possible to ignore inter-molecular effects.

* Since found in most gases and crystals.

But it has actually proved easier (1931 onwards) to use electrons scattered as waves. As with crystals, the method enables inter-atomic distances to be calculated. This often gives vital evidence as to constitution. The plane structure of benzene, the linear structure of CO_2, the tetrahedral nature of carbon, are confirmed. But the method is only in its infancy.

In concluding this section on methods, we deal more fully with a subject which, as concerned with surface effects, fitly precedes work on the structure of organic and silicate colloids. It had long been realised that surface tension is much affected by traces of impurities; but the first workers to develop an even approximately adequate technique for the cleaning of surfaces were Agnes Pockels (1891) and Rayleigh (1899). They found evidence that their methods had actually enabled them to produce uni-molecular films.

In the same year, 1912, as Debye's work on dipoles, a similar idea, that of oriented molecules, was applied by Hardy to films of substances spread (for example) on the surface of water. Among the substances examined were polar compounds such as fatty acids. From 1916 onwards, Langmuir, and also Harkins, began testing experimentally the idea that fatty acid films might actually consist of the long carbon chains of the fatty acids standing out at right angles to the surface, parallel and close-packed, attached by their acid groups.*

Langmuir, in the same year, also applied the unimolecular idea to gaseous films adsorbed on solids. These he was correlating with heterogeneous catalysis (see next section). In his view, the forces concerned here were chemical. We shall keep the gaseous and liquid cases running together.

In "Langmuir's trough," to return to the liquid case, the film was confined by paper strips. These were attached to a balance which measured the forces exerted. Air jets at the edges of the strips prevented the film from spreading beyond them. In 1921 Adam began his comprehensive work, using paraffined copper instead of paper. The result was to confirm the theory. For the minimum area occupied by a given molar quantity of unimolecular film was found to be independent, within limits, of the nature of the end-group or of the number of units in the chain. This conception of the behaviour of oily substances has had important applications in lubrication.

In 1921 Labrouste observed that changes of temperature sometimes produced marked effects on the films; and an "expanded state" was postulated, the nature of which was not at once clear. In 1925–6 Schofield and Rideal began viewing liquid films as two-dimensional gases, and produced a kinetic theory; while (also in 1925) Volmer advanced a theory of free mobile molecules for gas films also. In 1926 Adam and Jessop again improved their technique, using thin platinum strips instead of air-jets to confine the film. They were actually able, in the liquid case, to detect a transition from an "expanded" to a "gaseous" state. In the latter, a Boyle's law analogue is obeyed, and the molecules are evidently more or less flat on the surface.

* Polar substances tend to be soluble in ionising solvents, but in the case of fatty acids this solubility goes down as the chain lengthens.

In the gas-solid case, too, evidence has accumulated (1926–32) for mobility on solid surfaces, especially where only Van der Waals forces are concerned. Various observations suggest this, of which the simplest is the mere fact that drops often form when uniform films are laid down.

To account for the "expanded" state of fatty acids on liquids, Schofield and Rideal brought X ray evidence that the molecules become tilted, their (corrugated) chains intermeshing. A new measurable was found when, in 1931, Rideal and Schulman measured the changes in boundary potentials due to films. The result of all this work is that a complete series of two-dimensional states, with definite transition-points, has been postulated, analogous to the solid, liquid, gas, of three-dimensional states.

These new dimensional concepts apply not only to foreign films, but to surface regions of (say) crystals themselves. The transition values are further data as to lattice and molecular structure. Two-dimensional reactions are also envisaged, giving us new information as to (for instance) steric hindrance. Such approaches as have been made to an extension to one- and zero-dimensional states will be noticed under catalysis.

We left the proteins with Fischer's view of them as chains of amino-acid residues. The idea that rings might occur was by no means excluded, and evidence of specific rings had been brought forward by several workers before 1925, the year from which these notes may conveniently be dated. The first half of the twenties had seen one important point established, the stereochemical fact that all naturally-occurring amino-acids have the same configuration.* The second half saw several further advances.

Since 1923 Svedberg had been developing an appliance which was to give valuable data, the ultra-centrifuge. As with most devices, the invention of this was the invention of its last detail. In view of the enormous rotor speeds, Svedberg had expected vibration to be the worst trouble, but it turned out that local heating and convection currents were the hardest to eliminate. Water-cooling the bearings was not enough, and the crucial step was found to lie in covering the surface of the solution with a film of oil to prevent evaporation.

From about 1925 Svedberg began in this way to get consistent results for the molecular weights of proteins, using, in 1928, fields 100,000 times that of gravity. He found that some suggestive, though not invariable, rules prevail among these molecular weights, most of them being multiples of about 34,500, or at worst of 17,000; while a distinct class of exceptionally enormous molecular weights run in units of 140,000 or at worst of 70,000 (i.e. presumably $2 \times 34{,}500$). Endo-cellular pigments seem all to belong to the smaller, plasma pigments to the larger, group (with weights of millions). All vertebrate haemoglobins have 68,000.

A great requisite in protein research has been the choice of a suitable class of protein with which to begin. If plant and animal serum

* Not that they all rotate polarised light in the same direction. Space is lacking to explain the exact meaning of "configuration."

proteins are hydrolysed, the proportions of different amino-acids obtained may vary widely with time in any one individual, as well as varying from species to species. This variability does not seem to affect the high specificity of such sera when tested against each other. But it makes them very hard to examine conclusively by chemical methods. We now come to a method which centred the attack on less variable classes of protein.

From 1920, efforts had been made to secure X ray photographs of proteins, in the hope that they would reveal a crystal lattice. But these efforts were not at once successful. It was a non-protein, rubber, which, in 1925, pointed the way to success. Katz found that rubber, which gives no sign of lattice structure when unstretched, acquires a definite pattern when under tension, only to lose it again when the tension is taken off. The stress, it was supposed, oriented the normally chaotic chains of polymerised isoprene units of which rubber is made up.

Then, in 1928, Herzog and Jancke succeeded in making a complete crystal analysis of cellulose, a substance which, as we have seen, has, for the present purpose, much in common with some of the proteins. The X ray photographs revealed straight chains which (as also with starch) are made up of 6-membered rings like those of the simple sugars, joined by oxygen atoms. The molecular weights of celluloses seem (1936) to be considerably less definite than those of proteins.

Towards 1930, the attention of workers on proteins was concentrating on the stable fibre constituents. One of these, the fibroin of silk, gave the least unsatisfactory photographs with X rays. Its patterns suggested that it was made up of long straight polypeptide chains. The stage was set for Astbury's work, so often quoted as a case of " applied " research (in textiles) having results of theoretical importance.

The clue lay not only in the X ray analysis of wool and hair keratin achieved by Ewles and Speakman in 1930 (as well as by Astbury and Street) but in Astbury's correlation of this with a clue from the case of rubber, namely *contractility*. This property is widespread among natural fibre constituents, including those not of the stable class, like the myosin of muscle.* Silk and cellulosic fibres are in this respect *less* clearcut than wool and hair, showing elastic after-effects ; so that wool and hair proved to be the crux. Their patterns, like those of rubber, underwent a profound, but reversible, change on stretching. The change was from a state unlike, to one something like, that of silk. The suggestion (1932) was that their unstretched state was one in which the chains were puckered in some definite way. As usual, the unit was supposed to be a long chain, but adjacent chains, unlike those in rubber,† were held together " chemically " in a definite grid which survived puckering. Because keratin and fibroin (are supposed to) have only short side chains, this grid can be compact and can *exclude water*. This is a main point in the stability of the protective proteins, and is regarded

* Claimed by v. Muralt and Edsall in 1930 as rod-like in structure. Muscle " fibres " are not fibres in the same simple sense as hair or wool.

† Which are held to be joined only by cohesional, Van der Waals forces.

as partially explaining the difference between them and the highly metabolic types.

The constitution thus postulated for the fibre proteins must not be confused with (though it is no doubt related to) the spirality and " crimping " of wool, or with the wide prevalence of spirality among plants and other living things. A point of evolutionary interest is that avian and reptilian keratin (feathers, scales) is unlike that of the hair, nails, horns, etc., of mammals.

The suggestions of Wrinch and others (1934 onwards) are of interest as showing one trend of ideas on the more metabolic proteins; though very recent experimental work (1938) warns us that they *are* suggestions, not final theories. Perhaps the chief fact to be explained here is the " denaturation " which these bodies can undergo, sometimes reversibly, losing then some of their contrasts with the fibrous forms. Possibly, the change is attributable to the opening of rings into chains, but the interest of Wrinch's ideas lies in her suggestions of *lattices* of rings, with definite contractilities. A very few simple " bricks " suffice, C(OH), CHR, and N. Thus, in one class of case, there are laminar lattices. Owing to the property that all the amino-acids present have the same configuration, the R's will all be on one side of a lamina, the OH's on the other. Whether or not this particular theory proves true, the day of geometry seems to be dawning in this subject.

Under the " Feulgen " reaction for the nucleic acids, chromosomes * show a series of dark bands. These are attributed by Wrinch to the acid molecules strapped round the sheaves of linear protein units wherever these show mainly basic groups. A genetic significance is attributed to the pattern of these bands. The point serves to introduce a current idea on what is perhaps the central mystery of growth and reproduction, namely, how like produces like.

The suggestion is that the protein lattice may act as a " template " on which new atoms may be laid down. If these are the same as the old ones, further protein would be built up, until perhaps the structure becomes unstable and division ensues. But the new atoms need not be the same. An atomic grouping capable of fitting locally over that of the protein may be present in the environment. This would recall the action of enzymes and other physiologically potent substances; and in fact investigations of both enzymes and proteins have been inspired by this idea. For instance, artificial substrates of known constitution have been built up for proteolytic enzymes.

Laminar lattices, linked by hydroxyl groups, would bring proteins into analogy with another very important class of micro-crystalline colloids to which we have referred, the clays of the soil and the silicate minerals generally. For recent work, arising from X ray analyses by Pauling (especially 1930 and 1933), has disclosed layer lattices in many of these. As mentioned in Chap. XIX, base exchanges in the soil may be said to provide the first stage to the constant internal environment of land mammals. In both cases, free movement for water in the lattice is a point of great importance.

* Especially the giant ones in salivary glands investigated since 1934 by Painter and others.

CHAPTER XXIV

REAL MATERIALS. II

OUR knowledge of the states of matter proceeded in its early stages by the erection of " perfect " or " ideal " materials—solids, liquids, gases, crystals, solutions. Perfect gases obeyed Boyle's and Charles' laws, perfect liquids had no viscosity or shear strength, perfect solids obeyed Hooke's and other elastic laws. Towards the end of the 19th century came ideal solutions and electrolytes. The rôle of all these was like that of the straight line in applied geometry, or of the ideal cycle in thermodynamics ; they showed what deviations had still to be accounted for. But to account for them has not proved so easy, and it is one of the distinctive notes of very recent science to be able even to start on the process. The gas law corrections are perhaps the only exceptions to this recentness, and accordingly, in sketching the modern " science of real materials," we begin with them. We go on to the recent theory of electrolytes and of the kinetics and intimate structure of chemical reactions, to which they are related. We may perhaps view these latter as corrections of the " ideal " theory that reactions are complete and instantaneous. We then consider other aspects of real solids and fluids, and their rôle in the atmosphere and in the crust of the earth.

We saw in Chap. XV that by the sixties a fairly exact knowledge had been reached of the deviations from Boyle's law over what seemed, until lately, a very wide range of pressures. The kinetic theory of gases was then new, but by the seventies, it was being made the basis of a series of attempts to account for the deviations. The best known of these, that of Van der Waals, appeared in 1879. It took account of the two most obvious sources of error in the simple kinetic deduction : the finite size of the molecules and their mutual attractions. This gave

$$\left(p+\frac{a}{v^2}\right)(v-b)=\mathrm{RT}$$

a quartic equation not wholly incapable of bringing the matter into relation with another question, that of the continuity of state.

We shall not enter on the other attempts which followed (e.g. Clausius, 1880) because none have proved entirely satisfactory. Our point is that they laid down the lines on which the question of solutions and electrolytes was to be attacked. This is not wonderful, since, as we have seen in Chap. XV, the eighties were seeing the relevant ideal case set up by reference to ideal gases.

It must be remembered that in its early days the ionic dissociation theory of electrolytes had been dogged by the confused notion that the electrostatic attractions of the ions made dissociation impossible (by the same principle, gravity should make all gases a heap of molecules on the ground), and it was not until the dynamic concept was thoroughly established as accounting for this difficulty, that any real corrections which these attractions might call for could be considered.

Meanwhile, we may recall, the application of the Law of Mass Action to the balanced reaction of ions with unionised molecules had been considered by Arrhenius ; and in 1888 Ostwald had thence deduced his " dilution law " : $[B^{\cdot}][A'] = k[BA]$, or, if a is the degree of dissociation, c the concentration, $\frac{a^2c}{1-a}$ is a constant. This, it was found, failed for strong electrolytes ; and towards 1905 the idea gained ground that these latter must be regarded as *wholly* ionised, so that the action was not a balanced one and Mass Action did not apply. Much later again, as we have seen, this view of strong electrolytes was extended even to their crystalline state.

From about 1908 Lewis evolved the idea of " activity " : only a percentage of the ions could be supposed " active " if the ideal laws were to be kept. The concept issued in equations with coefficients which could be fitted to individual cases, and by 1921, at the hands of Lewis and Randall, activity reached the point of being recognised as depending only on ionic strength in any one solution. It had its root in chemical thermodynamics, which we shall continue to leave aside. We turn to theories which considered particular physical mechanisms to correct the " ideal " theory.

Now the chief difficulty, here and with other " real materials," is not to suggest new factors to be taken into account but to select ones which give results definite enough to be tested by experiment. At first, the matter was in the hands mainly of chemists. They suggested that complexes were formed, or the ions hydrated, hypotheses which issued in indefinite or false results. These workers did not usually abandon the Law of Mass Action, but doing so proved to be the vital feature in the theories which at last began to make definite headway. These were due to Milner, 1912–9, and especially to Debye and Hückel (1923, 1925, etc.). For these workers fixed on the mutual attraction as the best correction to start with, and this, of course, constitutes an abandonment of the Law of Mass Action. For this latter, like Kohlrausch's conductivity law (1876, see Chap. XV) involves that the effect of ions depends only on their number, not on their proximity. It was natural that strong electrolytes, where the ions are numerous and close-packed, should show deviations.

These theories succeeded in connecting the quantity $\sqrt{\text{concentration}}$ (which had appeared in Lewis' empirical work) with the inverse square law of electrostatic force between ions, and in this way very dilute strong electrolytes have now been largely mastered. We cannot describe the theories of Milner or of Debye in detail, but we may note that the secret of Debye's superior success lay in his image of each ion as surrounded by an atmosphere of others, in which those of opposite sign would predominate. When the ion moves, this atmosphere becomes calculably asymmetric, a finite " relaxation time " being needed for the same statistical preponderance of unlike ions to reassert itself in the new neighbourhood.

These workers did not follow Van der Waals in considering the finite size of the ions, and later attempts to do this have shown its great difficulty. As already noted, it is easy to suggest other corrections, such as the ionic association considered by Bjerrum in 1926. Another obvious one is the influence of the solvent molecules, which is without analogy in the simple gas case. Most of these have been the subject of bold, but not very successful, mathematical attack since 1920. A gas case which does present analogies to them, however, began to be considered at about the same time. This was the kinetics of *reactions* in gases, to which we proceed at once.

Reaction Kinetics

We have noted how the generation following the sixties, when the Law of Mass Action was enunciated, assimilated and applied the idea of dynamic equilibrium and of rate of reaction. In 1884, Van't Hoff reinterpreted the " order " * of a reaction in terms of the kinetic theory of matter, and deduced the law of dependence on temperature. Towards the end of the century, however, the complexity of the question of reaction-speeds became evident. Ionic reactions were indeed simple. They were virtually instantaneous—obviously, as we can now see, because they involve no rearrangement of oriented bonds, only of electrons. But reactions which took a measurable time only very partially followed the velocity-time curves to be expected from their order. Nor was order the only factor. Menschutkin in 1890 showed that the solvent may, quite apart from ionisation, influence speed enormously. But the greatest puzzle was that it seemed that only a tiny fraction of the molecules were at any given time in a condition to react. In fact, Arrhenius in 1889 gave a kinetic explanation of the temperature-velocity relation by assuming that a certain minimum of energy was needed to throw the molecules into an activated state. This he envisaged as reached by a balanced tautomeric, that is chemical, change.

Other obvious chemical possibilities of the type of complex-formation were much discussed in the next twenty years, as we have seen in connection with electrolysis ; but, as there, they offered too *many* resources to be easily tested.† It was the impact of two lines of *physical* thought which was largely responsible for the decisive change which may be dated from about 1915. One of these was Einstein's " quantum-yield " law in photochemistry (Chap. XVIII) which suggested to Lewis (1916–8) a now-abandoned theory that all reactions are photo-reactions. Some such idea was thought necessary (Perrin, 1919) to explain how unimolecular reactions (with velocity independent of pressure) can take place at all ; since without some such external

* I.e. the number of molecules on the left of the equation.

† Intermediate complexes, or the related transiently " free " radicals, have always been important. *Ionised* free radicals were early evidenced in positive ray experiments, while in 1900 Gomberg isolated $(C_6H_5)_3C$. In 1929 Paneth succeeded in securing evidence of the brief existence of the methyl radical in a certain reaction. The CH_3 groups soon combine in pairs, but only at the walls of the vessel (see later). In narrow tubes one hundredth of a second sees most of the methyls gone.

impulse as light, the isolated molecule seemed to have no cause for change. This photochemical episode had one important effect. It caused Lewis to develop (1918) a second physical idea which ran roughly thus.

A proportion of the collisions of molecules, calculable from Maxwell's distribution, would exceed any given level of severity, and it would be the exceptionally severe ones which would result in reaction. Thus Arrhenius' idea received a new and precise meaning, and " activation " became a function of *both* reactants instead of being envisaged as a *state* of one of them. From this point, success began to be achieved, especially as regards gaseous reactions.

For instance, several workers, but especially Hinshelwood (1926), showed that activation by collision could account for typical unimolecular reactions if a finite interval elapses while the energy acquired by collision distributes itself through the internal degrees of freedom (until, perhaps through resonance effects, enough accumulates in the one needful, to effect disruption). Such concepts are the key to recent work in this subject.

It is in complex molecules that internal degrees of freedom have the most obvious scope ; and in fact the method which converted unimolecular reactions from rarities into commonplaces was that of investigating complex compounds rather than simple ones. But these studies have shown that few reactions are simple. For instance, unimolecular reactions are " really " (that is, physically, as collisions) bimolecular. It has in fact become necessary to distinguish different criteria of " order." The primary experimental one lies in the relation of initial concentration to velocity.

Other contrasts among non-ionic reactions, besides order, had to be taken into account, and in particular two, gaseous reactions and those in solution, and homogeneous and heterogeneous. It was, for instance, natural to hope that the same kinetic theory would account for reactions in solution, but in 1924 Christiansen found that the velocity of reaction in solution might be only 10^{-5} of the kinetically calculated values. One obvious possibility was deactivation by collision with solvent molecules ; but then the slowing up did not always occur. In 1931 Hinshelwood and Moelwyn-Hughes, having found a number of reactions which they could perform both in gas and in solution, were able to compare various theories of the subject, such as that preliminary ionisation is necessary (1932).

The best view so far seems to be a union by Eyring (1935) and also by Evans and Polanyi, of the idea of a " complex " with that of " activation." Earlier work by Brønsted (1922) and Bjerrum (1924) had kept the idea of a complex from oblivion. The new workers supposed that a complex, even after collisions had both formed it and endowed it with exceptional energy, had still to wait while the energy got into the degree of freedom along which disruption could take place. But meanwhile there had been much development on the cross-division, that of heterogeneous and homogeneous.

The old assumption had been that, apart from exceptions such as catalysis and rare and obscure effects of the walls, all reactions are homogeneous—take place throughout the body of the mixture of reactants. But recent work has made the exceptions appear the rule, and has altered our view of homogeneous reactions themselves.

For both types, 1916 again makes a suitable start. In that year, as we have seen, Langmuir put forward his idea of an oriented unimolecular gaseous film at adsorbent surfaces. This had important results in the theory of heterogeneous reactions. The same year saw a new idea enter the homogeneous question also, that of chain reactions.* We shall begin with this.

Like others already noticed, this idea arose from the photochemical question then to the fore. An altogether abnormal quantum yield in certain exothermic reactions stood in need of explanation. The theory suggested was that of the transfer of the energy of reaction from a molecule of product to a neighbouring molecule of reactant. This would be thereby activated, would react, and do the same again, indefinitely. In 1918 Nernst considered that, with bimolecular reactions, a rather different type of chain mechanism might act :

$$Cl+H_2=HCl+H$$
$$H+Cl_2=HCl+Cl, \text{ and so on.}$$

It has, in fact, been observed that certain reactions, started by light, continue an appreciable time after the light is cut off. One type of effect particularly explicable by this theory is the sudden onset of explosiveness at a certain temperature in mixtures like hydrogen and oxygen. This is referred to branching chains. In 1924 Christiansen explained the action of certain negative catalysts by supposing that these can break the chains by taking part themselves—by taking up the activation energy. "Anti-knock" investigation for internal combustion engines is concerned with this fact. Other agents, such as neutral molecules and above all the walls, were also supposed to break the chains. Thus the effect of the walls, and so of the size and shape of the vessel, is a common feature of homogeneous and heterogeneous reactions, though the effects are in opposite directions. Recent work has shown the immense prevalence of chain reactions, especially among gases.

As to heterogeneous reactions, the hypothesis was that the reactants were adsorbed in oriented unimolecular layers on the walls, and that only then could they react. In some cases, then, the limiting factor would not be the rate of the reaction itself, but the rate at which the products could diffuse away.† The laws of diffusion were well known, however, and by varying the specific surface of the vessel much insight could be gained.

Films suggest an obvious explanation of both positive and negative catalysis, especially when the effect is proportional to the amount of the catalyst for small amounts and constant thereafter. For in these cases the catalysts clearly act (in part) by covering the walls with a layer of their own. We recur to this immediately.

* Less definitely put forward by Bodenstein in 1913.

† This does not apply only to wall reactions. A number of researches have been invalidated by ignoring it.

Different types of adsorption (Van der Waals, chemical, " activated ") have been distinguished, each with fairly characteristic behaviour. Pure effects scarcely ever occur, but analysis has been remarkably successful, in most cases, in excluding many of the factors which might be proposed. Among promoters of reaction for instance, mere proximity, due to adsorption, has been shown to have a rather narrow scope, activated chemical bonds a wide one. Unfortunately, the analysis of even one of the many types of reaction which have to be considered to gain a clear picture would be too long for this book. We must turn to the modifications brought in when the surface is actually catalytic.

Here the keynote of recent theories is the heterogeneity of the catalytic surface itself. This takes many forms, the presence of active centres being one, with active centres of promoters as a possible special case. In 1922, Vavon and Husson showed that a catalytic surface may be poisoned for one reaction, not for another. Several types of adsorbent surface of charcoal can be experimentally distinguished. Of enzymes, in particular, it has been shown that only a small proportion of the surface is active. We have recalled elsewhere the old knowledge of the value of scratching surfaces in promoting such actions as crystallisation. The *boundary* of two surface phases is one factor in catalysis but (Volmer, 1929) it is not a sufficient condition. In 1923 Norrish found that for certain reactions a polar surface was more effective than one made of a non-polar substance. In 1925 Taylor tried to get a definite physical picture by considering that the molecules at surface excrescences would be less strongly bound. In 1926, following Langmuir, he supposed that a catalyst could often bring about the atomic state of diatomic gases. Among further current ideas are, a strained condition of the chemical bonds at the boundary of incompatible lattices, or of lattice and fluid. This lowers the energy of activation of the reactants. Very refined methods have recently been developed for examining surface states and for following changes in them. Electron diffraction is perhaps the most promising.

All this justifies recent " dimensional " views in these subjects. Actions might be imaginatively classified as three-dimensional (homogeneous), two-dimensional (at walls, in films or in layer-lattices—base-exchange in clays), one-dimensional (some catalyses, chains, preferential adsorption on crystal edges), even nought-dimensional (collisions of chains with walls).

It should be remembered that not all catalysis is heterogeneous. Homogeneous catalysis by ions, for instance, has long been an important subject of research. We have mentioned in Chap. XV the catalytic effects of acids (for instance, in the inversion of cane-sugar). Ostwald suggested that it is the H (or OH) ions which catalyse in such cases. But from about 1907, it began to be clear that the other ionic component, and also the undissociated molecule, had an influence. As so often in these subjects, the position remained complex until an idea was brought in from higher physics. Lowry, and Brønsted (1923) suggested that the idea of acid and base be replaced, for this and other purposes, by that of proton donor and acceptor. This makes the *anions* of weak acids into "bases," and vice versa. In these new senses of the words, the catalysts in these cases are always acids or bases ; and the actions are referred to as " acid-base " catalyses.

Fluid Resistance

We have already considered the movement from ideal to real as regards gases and solutions in certain relations. A logical order in treating the remaining cases is difficult, the more so because the subjects cannot be separated from their applications, the imperative needs of which, in many cases, called them into being. For ideal fluids give little help to the meteorologist, or to the designer of ships, projectiles or aeroplanes. Ideal solids are of little use to the structural engineer, the road maker, or the geophysical prospector. Some unity is, however, lent to the subject by the nature of some of the troubles with these ideal cases, namely, the fact that both solids and gases behave in some respects as liquids. We are thus dealing with two *transition states*.

From early days the mariner and farmer evolved a proverbial weather lore, while Aristotle had a classification of winds in which Sir Napier Shaw finds suggestions of the polar front theory. He, and those who spoke in his name down to Galileo, had also a theory of fluids, some of the unfortunate features of which have been noted in the early chapters of this work. As early as 1506 Leonardo da Vinci was concerned with another application of the theory of fluids, namely, flight. In this and other connections he rejected the Aristotelian idea that motion " naturally " slows down, and set up a uniform motion slowed by a *resistance*, due to the air. Nevertheless, it was nearly two centuries more before Newton began the long series of attempts to determine this resistance. By his time the barometer had begun its long reign in meteorology ; while Halley, in 1686, had built on the already wide experience of sailors a map of the main winds of the world.

Newton considered fluids in both the obvious ways, as made up of particles and as continuous. It was the latter view which formed the basis of the ideal fluid set up in the 18th century by the systematisers of hydrodynamics, of whom the first was D. Bernoulli (1700–83). This ideal fluid was supposed to be devoid of viscosity. We cannot give a history of hydrodynamics as a whole, but it so happens that the section on which we concentrate, that of fluid resistance, was the very one which revealed the extraordinary peculiarity of this ideal fluid among other ideal materials. Most of these latter give inaccurate values, but they do give some values. Now D'Alembert was forced, in 1768, to conclude that the ideal fluid could by no means be induced to give a finite resistance at all. Newton's old particulate fluid could do so, though it did not give an accurate value. But this was not enough to bring it back. An impasse had been reached, and it is this impasse which forces us to devote so much of the present section to an ideal rather than a real material.

One obvious way out was that actual resistances were due to viscosity, but experimental work was against this. From 1761, English artillerists had been investigating resistance to projectiles. In 1777 D'Alembert, working with Bossut and Condorcet for the French government, published investigations of ships in canals. These gave at least one important result which was reinforced by work on *air*, with the torsion balance, by Coulomb (1784 and 1801). This was, that while viscosity may account for most of the resistance at small velocities, it cannot do so for fairly high ones. The impasse remained.

It remained, but it did not prevent development in other directions, some of it highly general and mathematical. As in electricity, the continental school considered, chiefly, forces and velocities of which the components were derivatives of a " potential " and were continuous from place to place. Stokes, however, made an important advance when, in 1847, he realised that the definition of an ideal fluid was not incompatible with the existence of discontinuities of velocity within it. One result of this was that it became possible for Helmholtz (1858) to consider vortex-motion, already used by Cauchy (1815) and others to give a picture of the luminiferous ether. For the existence of vortices of course implied discontinuities. Another result, as Rayleigh saw in 1876, is that we can have dead water behind an obstacle, and so in principle get a finite resistance. The subject, however, was still far from clear. For, as Helmholtz and Kelvin realised, the discontinuities would be unstable.

We may (now) see in this one of the earlier points at which the old " continuous " theories in physics were obliged to give way, whether to a single discontinuity or to the multitudes involved in Newton's atomic approach. No such interpretation would, however, have been seen at the time.

The next investigations of resistance did not directly attack the question of the ideal fluid, but were experimental and of a practical origin. From 1856, but especially from 1870, Froude had opened a new phase in experiments on ship behaviour, while Rankine had brought streamlines into use. Meanwhile, men like Penaud in France, Lilienthal in Germany, Langley in America were experimenting with gliders and aeroplanes. But failure with whole aeroplanes was soon forcing these men to analyse them, part by part, and especially to study the lift and drag of aerofoils. Langley introduced the whirling table, Phillips (1885) the wind tunnel. But the decisive advance in understanding was that of the English experimentalist, Lanchester (1868–), who, in 1894, first pointed out that if the downflow due to an aeroplane wing was not to be supposed permanent, there must somewhere be an upflow to cancel it out. He showed that what a travelling wing in fact induces is an accompanying air-circulation up behind it and over it.

His work, mainly qualitative, was ignored at home, but was taken up (especially from about 1906) by the continental mathematicians. Kutta and Joukowski exhibited Lanchester's circulation as a case of Helmholtz's vortices.

In all this, the invoking of viscosity to account for resistance had been avoided ; but for all that viscous flow had received much attention. In 1831, Poisson had given equations for it ; in 1839 Hagen, and in 1840–1 Poiseuille, had given its empirical laws. The flow, however, of real fluids is visibly far from simple, and in 1843 Stokes had noted that it may become rather suddenly unstable as conditions are varied. In 1883 Osborne Reynolds (1842–1912) defined these conditions. Laminar (streamline) passes into turbulent motion at a certain value of the "Reynolds number," $\frac{\text{viscosity}}{\text{velocity} \times \text{a length}}$, which is constant where geometrically similar systems are being considered.*

Reynolds' work did not make the mathematics of the subject any easier, and little progress was made until in 1904 Prandtl (1875–) found that most of the effects of a small viscosity are localised in a thin *boundary-layer* † or more generally in any place of high velocity-gradient. So here is yet another subject in which the idea of a surface film is vital.

Much of the practical importance of this boundary layer lies in the next point in the theory: that this layer may under certain conditions separate from the walls and "roll up" into the fluid as a vortex, thus playing the part of a Helmholtz discontinuity. A surface of separation made up of tiny parallel vortex filaments is indeed one natural way of envisaging a sharp change of velocity; and Blasius (1908) and Von Kármán (1911) contended that it is in fact the *only* way of getting a stable discontinuity. So that again non-ideal properties were invoked to account for resistance.‡

The names of Prandtl and Von Kármán were unknown outside Germany until 1921. They have since become identified with this arduous study of real fluids, a subject (still in its infancy) where mathematics has been distanced by costly experiment. One experimental technique, that of using photography to gain exact pictures of wakes and the like, dates back to about 1910. A fundamental difficulty still lies here, however, that of envisaging the nature of an eddy. An eddy is the unit of turbulent, dissipative, motion. It must not be confused with a vortex which, though it may pass into an eddy, is in this respect no different from an open streamline, and is in itself a form of "steady" motion. One point seems to be that a vortex is in essence 2-dimensional, an eddy 3-dimensional. Another, more practical, difficulty is that the boundary-layer theory offers no means of calculating the point where the layer ends, initial turbulence having an influence here.

In the above, no reference has been made to the distinction between compressible and other fluids. This is because it has been found that compressibility only becomes important, even in gases, for enormous velocities § or bulks. The serious theory of real compressible fluids has made relatively little progress and cannot be discussed here. Its applications lie in many spheres, in acoustics, in projectiles and explosives, in stellar and terrestrial atmospheres (to which last we return) and in turbine blades. At extreme velocities hollows may

* He applied this to lubrication.

† He made the assumption, about which there had already been dispute, that the liquid actually on the walls does not move.

‡ Levi-Civita (1873–) had previously shown that Rayleigh's wake would contain vortices, but had only invoked non-ideal properties to explain why the wake was finite, and did not extend backwards to infinity.

§ Contrary to a supposition of Leonardo's.

develop behind propellers, alternating with intense pressures which force the fluids into the pores of the blades. In the case of liquids, there is boiling into the spaces formed. The net result of this " cavitation " is chemical action, showing as intense corrosion. The point is mentioned to show the ramifications of the subject.

Meteorology

This subject is only mentioned because it is a branch of fluid theory and of general geophysics. The handmaiden first of the sailing ship, then of the aeroplane, it falls only partially within pure science as the line is commonly drawn.

The meteorologists of the early eighteenth century had reached a picture of the general circulation over the globe. Up to about 1850, there proceeded, on the theoretical side, the further analysis of this into a permanent earth-wide circulation and temporary, local, vortex-like cyclones. On the practical side came the refinement of instruments and of observational technique, and a faith in these as the only reliable means of advance.

Towards the sixties, two changes occurred. One was Helmholtz's progress in hydrodynamics, some aspects of which have already been mentioned. This stimulated a long generation of *theories* of the general circulation, which, however, had little success. The other was the electric telegraph, making possible the ideal of a " réseau mondial " of stations, collaborating in daily weather charts and tracing the genesis and the dissolution of cyclones. National networks, the first step to the worldwide one, began in France in 1863. The other large countries followed in the next decade or so. Even now, however, the réseau mondial is still gravely imperfect, especially as to three enormous regions, the oceans, the poles, and the upper air. As early as 1749 Wilson had raised thermometers by means of kites, and in the succeeding century and a half the technique of vertical exploration was developed. But even with modern commercial flight as a stimulus, not nearly enough has been done to give a general picture of the upper air.

The intractability of the air as a mathematical subject springs not only from the need of considering non-ideal materials but from the need of considering thermodynamics. Air is always losing or receiving heat, from other masses, from the sun or the earth, from its own volume- and pressure-changes as it flows. The fatal point is that these thermal changes may cause cloud-formation, which interferes in most complex ways with radiation. These difficulties have not prevented floods of theoretical work, but the history of this work from Halley onwards, as analysed by Sir Napier Shaw, illustrates their nature. It has usually consisted either of thermodynamics only (convection of warm air) or of hydrodynamics only. The difficulty is to combine the two.

From about 1900, following on the failure of comprehensive schemes of the general circulation, more limited attacks have been made, such as Rayleigh's use (1910–11) of methods of similitude. In 1902 Bigelow in North America proposed to regard the north temperate cyclone belt as a Helmholtz surface of discontinuity, with attendant trains of vortices. Rather less than twenty years later Bjerknes in Scandinavia went into more detail and suggested that cyclones develop from waves of instability in the surface, or "polar front," of discontinuity. The mathematical side of his theory is of the haziest ; but by combining this new idea with the older thermal picture of a cyclone, he has been able to build up a new "unit" of atmospheric structure into which typical observations fit much more naturally than into any previous suggestion. A more recent line has been to regard surface conditions as pertaining to a boundary layer in the Prandtl sense. The difficulty here is that Prandtl's work cannot easily be extended to non-isothermal conditions.

While some of the most fundamental questions about cyclones and about the general circulation are still unanswered, interesting suggestions have arisen as to more detailed matters. Thus in 1901 Bénard showed that a thermally unstable liquid (that is, one hottest below) may start circulating by breaking up into close-packed (therefore roughly hexagonal) vortex rings, and in 1925 Low and Brunt suggested that these, sheared by the wind, account for certain cloud-forms. It has recently become possible to demonstrate this phenomenon in gases on the laboratory scale.

Real Solids and Geophysics

We have so far been occupied with gases treated as liquids (that is, as incompressible) gas-liquid transitions (vapours) and liquids treated as partially solid (as viscous). The latter transition has another extreme, solids treated as partially liquid, as capable of flow. The subject of real solids is even bulkier, and has even wider applications in engineering, than that of real liquids. The microphotography and heat-treating of metals, their testing, the design of beams and of structures built up of them, are the most obvious of these subjects, but there are many others. One theory, which was started early but has only lately got even within sight of practice, is that of earth pressure, and of the behaviour of powders and aggregates. Coulomb was an early experimenter here. Macadam (1757–1836) focused attention on similar matters. Until his day there was a confused idea that a road can be said to *bridge* (say) boggy ground. He made it clear that roads *float* ; so that a solid, well-drained substratum is the only foundation for them. But the subject of real solids is so vast, and so largely technological, that we shall comment here only on two points, one concerned with "real" crystal lattices, and one with modern geophysics.

The lattice theories described in the last chapter are in the nature of ideal theories ; and, like all these latter, they fail where questions of hysteresis, after-working, and the like, have to be dealt with. When pure single crystals are prepared, highly geometric glide-figures are produced under shear. The plane of glide is usually, but not always, that most closely packed with atoms. Weakening by thermal agitation has been invoked to explain some features of this phenomenon (Becker, 1928), and also of after-working. Prandtl (1928) has correlated hysteresis with the possession by the particles of solids of more than one equilibrium position ; while Bridgman's work * on the effects of

* Over about twenty years.

very high pressures suggests that a similar phenomenon, the interlocking of molecules, is largely concerned in the viscosity of liquids.

In 1925 Smekal began work on real crystals, which, he alleged, possess a larger-scale periodicity than that of the lattice. It is indeed far from clear that ideal lattices would, theoretically, be stable ; * and actual materials are shot through and through with microscopic cracks or with other surfaces in exceptional states. Molecules of foreign material may " key " portions of lattices in position.

The publication by Beilby in 1921 of twenty years of research on polish and similar questions revealed that here was yet another region for " two-dimensional states of matter." Beilby claimed that polishing was superficial melting, with destruction of the crystal lattice and flowing of the material over cracks and peaks to give a smooth surface. The point was controverted, X ray analysis showing that polished surfaces sometimes possess a lattice structure. But recent work has revealed an intermediate fused condition with subsequent re-formation of the lattice.

The fact that the solid and liquid states are by no means mutually exclusive has important applications in geophysics ; as may be illustrated by work arising out of Pratt's discovery of isostasy (1859, name given in 1889).† Pratt, collating gravity observations, formed the impression that the extra weight of mountains is compensated by the less density of material at some or all levels below them. The stresses deep in the earth are easily seen to be gigantic compared to the test-bench strengths of solid materials ; and in the following half century the idea was formed that, for long-term purposes, the earth must behave as a liquid (of non-uniform density) with a thin solid crust. In fact, there began to be suggestions, made definite by F. B. Taylor (1910), especially by Wegener (1912), that parts of the solid crust—the continents—could move relatively to each other on the liquid substratum. This very slow " continental drift " of the lighter acidic rocks on the heavier basic ones was held to explain such obvious features as the resemblance of the eastern American and the western African-European coast-lines, the two land masses having once been one. This doubtful theory began to be actively discussed in the early twenties and reached a nadir of popularity in the late ones. It has obvious bearings on palaeontology. The discussion, with more favour, at least from geologists, still goes on ; but we cannot pursue it here. Our object is rather to show what modifications " real " materials have brought, at the hands of Jeffreys (1924–9) and other workers, into questions like that of isostasy.

Two senses of the word " liquid " are distinguished : use is made of the fact that solids commonly retain some springiness under short-term forces at a temperature much higher than that at which they begin to flow indefinitely under almost any long-term force, however small.

* A weakness of the same fundamental order as that noticed in connection with ideal fluids.

† Precursors were Babbage, Herschel, and Airy (1855).

For the first hundred kilometres or so, strength in the latter sense probably increases, owing to pressure, as we descend in the earth ; and Jeffreys has shown that exact isostasy only occurs when the stress is over a wide enough area to reach its maximum below the region where the "strength" is appreciable. Given a *wide* enough region, the height (that is, mass) to be compensated can be small : it still produces its effect. Thus the theory of real materials enables us to begin a second stage of approximation. Astonishingly far-reaching conclusions can thus be reached, such as, that the moon's birth could not have been responsible for the asymmetric distribution of the continents.

We are still far from that ideal science which would treat as one subject astrophysics, geophysics, and meteorology ; but a number of interesting questions can now be discussed, if not exactly answered, with some confidence. Such are, questions of the age of the earth, of the moon's separation, of the ocean. Early estimates of the age of the earth were completely upset by the discovery of radioactivity. Joly (1923–5), taking radioactivity into account, pictured heat accumulating in the earth until, at vast intervals, the strains set up were released by the crinkling of the crust into mountain chains, with lesser elevations of great stretches of land, and consequent ice-ages. We have already mentioned the question of geological and historical climates.

CHAPTER XXV

CONCLUDING DISCUSSION

We have ended our survey with real materials, rather than with the abstractions of relativity, to illustrate how close science now is to its applications. The sciences are the antennae of the great insect of industry. Yet for this very reason, among others, they are also closer to professional philosophy than they have been for a long time. For the relation of the growth of our knowledge to its material backgrounds, geographical or industrial, is giving rise to much concern, and any view of this relation amounts to a philosophy. In summing up here, so far as it is permissible to sum up a story essentially unfinished, we shall try to emphasise the variety of viewpoint necessary for any understanding of this question. Instead of returning over the story as a whole we shall twist it this way and that, running our eyes along several of its facets and leaving their resynthesis to another place. We will briefly consider the material aspects first. It is a subject which needs books to itself. Those of Lancelot Hogben have already been mentioned.

Consider, for example, the path of science over the map of the world. Commercial prosperity and especially industrial needs, with the rise of the middle class, obviously form one key to this. Science revived in the commercial republics of Italy, then spread to the more northerly trade region of the Hanse. It was for long plainly governed by the needs of navigation. It migrated in the 17th century to commercial England and Holland. It became notably German and physico-chemical with the rise of the chemical, electrical, and other German industries in the 19th century.

But the case of the Hanse, past its prime by the time of Copernicus, should give us pause. Wealth is a desirable but not a sufficient factor. The movement of the learned world lags behind that of wealth. Science and learning are far from being the same, but the North long looked to Italy for both. Thither went Copernicus, Vesalius, and Harvey. Further, though Holland and England in the 17th century, Geneva and Basle in the 18th, were no doubt relatively tolerant in the interests of trade, religious persecution elsewhere was needed to make this tolerance the powerful attraction that it was for science. To understand the map of science (for instance, Maps I–VI) both points must be kept in mind.

For some branches of science, regard must also be had for another "ecological" factor. Natural history, as opposed to abstract science, took its rise in regions of exceptionally varied or otherwise remarkable flora, fauna, or rocks. Such were the Alpine mountain system, where the natural forms appropriate to very different climates are seen in close contrast; or the islands with their long-isolated life, which gave the 19th-century evolutionists their clue; or Scandinavia, where the operation of cold and other ecological limiting factors is very evident. Baltic science in general has also profited, in accord with the above remarks, by being off the main track of military and religious aggression.

Another point to remember in assessing the influence of trade and industry is that the industrial community has always needed an external stimulus to the use of science. This was supplied at first by the princely courts and capitals, with their military or religious rather than commercial origins, and especially by that of France, which made Paris for so long the centre of the scientific world. Even now, the state usually takes the initiative before the private industrial community. Finally, if science follows trade, it also leads it. At the very start of our period came one of the greatest examples of this, when the compass and navigational astronomy helped the outer and western lands, Portugal, England, and Holland, to seize commercial dominance from the inner and eastern ones, Italy and Germany. Science, which is exceedingly hard to destroy, but easy to localise in an extreme degree, uses what stepping-stones it can.

Science and trade have, however, a certain similarity which is not always seen. Trade means exchange, communication. Science is the communicable part of the true, and is apt to forget how much truth there is of which the evidence perishes with passing events or is otherwise incommunicable. Both trade and science wish to avoid closing up avenues of further profit. Trade does not mean an isolated bargain. Science does not mean an isolated, finished truth. It means always finding a new puzzle to replace the old, sometimes to the destruction of workers who have too far specialised their technique. Finally, trade agrees on currency and on standard weights and measures. Science lives by these and by further symbols and instruments as to which all observers can agree.

Instruments are the offspring of the wedlock of a science and an industry, and are usually the parents of further ones. Abstract science is not the knowledge of nature but of the artefacts of industry from nature. Geometry, for instance, is only likely to progress rapidly when industry has made most lines around us straight or circular. Our last two chapters showed how much of their neatness physics and chemistry lose when they press into parts of reality filtered only through those rough artefacts of life, the human senses. We can make a scale or chain in the process of greater and greater filtration or artificiality, from the crudest environment through weapons, tools, and scientific instruments to symbols, the units or currency of the mind itself. To the research scientist in full practice his instruments are almost as familiar and flexible in use as the words of his mother tongue or the coin of the realm ; and his ideal is that they should become ever more so. Up to now the refining process has never run short of material to put in crude at the other end of the chain.

This chain is a great help for our immediate purpose. For no one, approaching the matter for the first time, can fail to be surprised at the large percentage of scientific history to which a very few devices give the key. Easily followed lines lead from the crude clock via the pendulum to the all-pervasive idea of periodic motion. Others, with glass as their theme, lead through telescopes and simple microscopes (17th century) to achromatic microscopes and interferometers (19th) and again through glass tubes and containers to the physics, then the chemistry (17th and 18th) of gases. Others again, as is well known, have the pump as centre: the circulation of the blood and air-pumps (17th), steam-engines (18th), and so thermodynamics and vacuum pumps (19th), with (glass) vacuum tubes and all that they have meant. The balance and the electromagnet are other cases in point. By a legitimate extension of meaning, we may follow the " device " of artificial crossing in horticulture and agriculture into the results, partly due to it, in the " sexual " classification of Linnaeus and in evolutionary and genetical theory. And we have only exhibited part of what hangs from all these. In large measure technology hangs from them too.

Yet they tell, plainly, only part of the story. Though we may say, for instance, that glass has been of dominant importance in almost every scientific discovery made outside pure mathematics, yet it has usually furnished only opportunity, not discovery itself. Again, the collection of items above—lenses, tubes, pumps, balances, coils—is a disconnected, miscellaneous one. We cannot see why one should lead to the next. To follow the continuity of history we must supplement this series of pegs with another, also taken from the scale which we set up, that of the *ideas* which lie at the back of them.

In passing, we recall the middle term between devices and ideas, symbolism, some forms of which, of course, long pre-dated science. It may be claimed that, like glass, the symbol x has had a hand in most scientific discoveries since navigational mathematics enabled descriptive scientists to tap the new continents and oceans. Besides that, symbolism is no small part of modern chemistry and genetics, and it has its part, as nomenclature, in the descriptive sciences themselves.

Turning now to ideas, and to the earthiest first, we recall our earlier stress on the idea of the physical—how chemistry and physiology grew from this stem, and how central, too, physiology has always been. We saw this even in the medical interest which led to botanical classification in the 16th century, and in the further physiological discoveries, in sex and elsewhere, which have swayed systematics ever since. We saw it in the iatromechanics and iatrochemistry of the 17th century, in Galvani as one founder of electricity in the 18th, in the vast chemical importance of the carbon and nitrogen compounds in the 19th, in the great name of Helmholtz, in the origins of the doctrine of energy and in its later connection with the discovery of vitamins, in the birth of physical chemistry and its historical connection with the living cell.

Not disconnected in origin from this slightly " barbarous " idea of the physical was another which, it has been suggested, was derived from the barbarians and not from the Greeks. This was infection, one form of those " influences," " effluvia," and the like, which in various ways crystallised into the modern ideas of gases, of ether, of field-physics, of general relativity. The infection-fermentation idea, in one hazy form or another, may be traced from quarantine in the Middle Ages, through Ambroise Paré in the 16th century, to inoculation and vaccination in the 18th and so to Pasteur in the 19th, with stimulus to Priestley and Black on the way.

The ideas of the last paragraph all stress continuity, of which they appear as variously vulgarised forms. But this is only half the truth, and, taking a wider range, we can observe the alternation of continuity and discreteness among the key ideas in the history of science. The 17th century is dominated by continuity: Ray and others have no difficulty in imagining the transmutation of species, the long upheaval of strata. Hooke and Huygens imagine a continuous ether. Newton comes out with fluxions and fluents. In Leibniz the reaction to the 18th-century idea of caste, of form, or array—discreteness, rigidity—is already visible. Formulation, classification, discrete atoms and their combination, sweep on to triumph respectively in mathematics with Laplace and Lagrange, in Linnaeus and in Cuvier's comparative anatomy, in the chemistry of Lavoisier and Dalton. By Cuvier's time another reaction is preparing, and the central 19th century is dominated by continuity: Cauchy's continuity in function-theory, the ether for light and for Faraday, infection leading up to Pasteur, Darwin's continuity of species. And yet already the seeds of a further reaction are sown, for by Darwin's time mathematicians are beginning to realise that continuity, as such, is not a suitable "brick," that they must build with discrete units. Cell-theory is advancing apace, while microscopists, also, are replacing "influences" in disease, by discrete entities, bacteria. By 1900, a dramatic change is at hand: Cantor's mathematical outlook is about to triumph; physics and physical chemistry, already using particulate electricity, are about to swing over to quantum theory; biology is about to swing over to Mendel.

We must not, of course, be led by such crude bird's-eye views to imagine any regular pulsation intrinsic to science. On the contrary, even the rough simultaneity which does seem to be observable between different sciences suggests the operation of external, social or economic, as well as of intrinsic, factors. But only these latter can be mooted here.

In general, possession of the mind by continuity is natural when great new syntheses are waiting to be made, its possession by the discrete when it becomes necessary to examine the last great synthesis to see what it has really accomplished and why insuperable difficulties confront it. Continuity is one of the most abstract and difficult of ideas. One or two similar ones, such as cause and effect and epigenesis, have always teased the human mind, acting always as an irritant and spurring common sense, in the end, to turn away from them and to venture, rather than be hypnotised by insoluble problems, on the long road of science. On this road, as hinted elsewhere, the very impenetrability of these problems fits them for the rôle of pivot or bearing. Like ultimates in general, they can take several equivalent forms. A portmanteau form, enshrining most of them, is that of space-time; and with a brief glance at the development of ideas on this subject during our period, we may fitly end.

We suggested in the first chapter that elbow-room rather than order came, with the Explorations, to be the dominant idea as to space. We suggested that, with its tacit implication of relativity, this led to Copernicanism in the 16th century. Further tacit implications led in the 17th to the first great advances in mathematics and physics, analytical geometry, calculus, mechanics, planetary astronomy. But the idea of space as order did not die ; we have adduced it as giving mathematics a new weapon, the perception of (two-dimensional) symbolic form—arrays of quantities, paper calculation itself. This went to men's heads in the 18th century, when formulation became the ideal in mathematics, and the neatly ordered arrays of Linnaean classification in biology and geology.

Yet another idea of space had been enshrined in the perspective taken up by the artists in the 15th century. In the main, this lay dormant in science until it blossomed in the 19th century into perhaps the most artistic range of mathematics, projective geometry. By the 19th century, the idea of time, too, which we can only notice in passing, had taken on a new life and meaning—the long avenues of history and of evolution. In a different connection, it experienced other changes from its association with entropy in thermodynamics and with space in relativity theory.

But meanwhile white man was being forced to realise that, on this planet, elbow-room is not unlimited ; and, as if reflecting this, the idea of order rather than of room began to come to the front again, with the added point that there might be types of order more general than any sort of space. For even the abstract spaces of Frechet recognise the limitation that there must be a meaning for the " vicinity " of points. In their different ways, the most refined forms of mathematical logic and of general relativity and quantum theory all recognise that the deeper levels of reality are most conveniently represented by non-spatial order. Thus we seem to return, with a weight of experience, to that early beauty and clarity of the Greeks which survived into the religious ages. At the same time we may claim not to have lost the other feature of scientific method, its aspect as brilliant imagining, impartial testing, which gives it its value as a human discipline.

BIBLIOGRAPHICAL NOTE

(*See also footnotes to text*)

THE literature of the history of science is now very large, and increases rapidly ; as witness the classified bibliographies appearing regularly in *Isis*, its principal periodical. This note is only introductory. The main subjects are treated, not their subdivisions.

The various subjects and periods have been very unequally studied. Mediaeval and ancient times demand, and are receiving, the devotion of many lives. On the other hand, no works on the history of physics, geology nor, until very lately, of chemistry, come to memory as do those of, say, CANTOR, CAJORI, SMITH, and LORIA in mathematics, or of NEUBERGER, GARRISON, CASTIGLIONI, SUDHOFF, SINGER and others in biology and medicine.

The intending specialist, or the reader needing recondite information on a specific point, may be well advised if he refers at once to the above bibliographies in *Isis* or to the two small works by G. SARTON, the greatest living authority on the subject : *The study of the history of science*, Harvard U. P., 1936, 75 pp., which, however, does not list books on particular subjects and periods ; and *The study of the history of mathematics*, Harvard U. P., 1936, 113 pp., which goes more deeply into individual aspects.

1. Even the general reader should bear in mind that the periodical literature is now as indispensable here as in any other study, scientific or historical. Three general periodicals mainly in English are, *Isis*, Editor, G. SARTON, Harvard, Cambridge, Mass., U.S.A., which covers all subjects and periods, but especially Arabic and other mediaeval studies ; *Osiris*, from the same source, which now (1936 onwards) gives the longer and more technical papers ; and *Annals of Science, Quarterly Review of the History of Science since the Renaissance*, Taylor and Francis. As the title implies, the last intends to specialise in the modern period.

BOOKS, chiefly in English and of recent date, oriental science mainly ignored :—

2. The only good short general work on the whole field is, DAMPIER, Sir W., *A history of science, and its relations with philosophy and religion*, C.U.P., 1930, 536 pp. A large fraction of this is philosophical. There is little mathematics, and for the modern period the scale of treatment is much smaller than in the present book.

Several larger works aim at covering the whole field, but have not yet done so. The fullest is SARTON, G., *Introduction to the history of science*. Vol. 1, *From Homer to Omar Khayyam*, 1927, 839 pp. ; Vol. 2 in two parts, *From Rabbi ben Ezra to Roger Bacon*, 1931, 1251 pp. ; Vol. 3, *Science and learning in the fourteenth century*. (In preparation.) Carnegie Inst. of Washington. This encyclopaedic work is hardly suited for continuous reading. BRUNET, P., and MIELI, A., *Histoire des sciences. Antiquité*, Payot, Paris, 1935, 1224 pp., is more suitable for this purpose if the extracts are passed over ; while the large work by Prof. WOLF of University College, London, has now reached its second volume.

3. Since these, at present, amount to treatments of periods before that of the present book, a few more of these latter may be conveniently given here. One is REY, A., *La science dans l'antiquité*. Vol. 1, *La science orientale avant les Grecs*, 1930 ; Vol. 2, *La jeunesse de la science grecque*, 1933, La Renaissance du Livre, Paris. Another is the very small HEIBERG, J. L., *Mathematics and physical science in classical antiquity*, O.U.P., 1922, 110 pp. These are for the ancient period. THORNDIKE, LYNN, *A history of magic and experimental science during the first thirteen centuries of our era*, 1923, 2 Vols., and *In the fourteenth and fifteenth centuries*, 1934, 2 Vols., both Macmillan, deals with the periods stated. Works on early periods of individual subjects are given under these latter below. In using the following sections the reader should remember the biographies in Section *II*. General attention may also be called to the *Science Museum Handbooks*.

4. *Mathematics :*

(*a*) For the whole period we have LORIA, G., *Storia delle matematiche*, Torino. Vol. 1, *Antichita. Medio evo, rinascimento*, 1929, 495 pp. ; Vol. 2, *I secoli XVI e XVII*, 1931, 593 pp. ; Vol. 3, *Dall'alba del secola XVIII al tramonto del secolo*, 1933, 605 pp., which is on a readable scale ; also CAJORI, F., *A history of mathematics*, Macmillan, 2nd edn., 1919, and BALL, W. W. R., *A short account of the history of mathematics*, Macmillan, 5th edn., 1912, which are going out of date. SMITH, D. E., *History of mathematics*, 2 Vols., Boston, Ginn, 1925, is mainly of value under (*b*). BELL, E. T., *Men of mathematics*, Gollancz, 1937, 653 pp., is a series of popular biographies, while ARCHIBALD, R. C., *Outline of the history of mathematics*, The Math. Assoc. of America, Oberlin, 1934, 58 pp., is very short. On the whole, the more valuable works show the subject falling into two halves, before and after 1800 ; demanding respectively mainly historical and mainly mathematical qualifications in reader and writer.

(*b*) For the whole period before 1800 we have CANTOR, M., *Vorlesungen über Geschichte der Mathematik*. Vol. 1, *From the beginnings up to* A.D. *1200* ; Vol. 2, *1200–1668* ; Vol. 3, *1668–1758* ; Vol. 4, *1759–1799*, Teubner, Leipzig, 1900–1908, which is too large for continuous reading. Altogether shorter and simpler is SULLIVAN, J. W. N., *The history of mathematics in Europe*, O.U.P., 1925, 109 pp.

(*c*) For ancient mathematics there is HEATH, T. L., *History of Greek mathematics*, 2 Vols., O.U.P., 1921, 1060 pp., shortened in his *Manual of Greek mathematics*, O.U.P., 1931, 552 pp.

(*d*) For the period after 1800 we have KLEIN, F., *Vorlesungen über die Entwicklung der Mathematik im 19. Jahrhundert*, Springer. Vol. 1, 1926, 400 pp. ; Vol. 2, 1927, 208 pp., by a brilliant mathematician. PRASAD, G., *Some great mathematicians of the nineteenth century : their lives and their works*, 2 Vols., 1933–4, Benares, is biographical in arrangement. Poor printing, but has much matter not elsewhere collected in English.

5. *Astronomy :—*

For ancient and mediaeval astronomy, the general works on those periods may be consulted, also HEATH, T. L., *History of Greek astronomy to Aristarchus*, O.U.P., 1913, 434 pp., and DREYER, J. L. E., *History of the planetary systems from Thales to Kepler*, C.U.P., 1906, 432 pp., who has also written *Tycho Brahe*, Black, 1890, 405 pp. For very early times MENON, C. P. S., *Early astronomy and cosmology*, 1932, Allen & Unwin, 190 pp., presents an interesting idea. For the last hundred years, WATERFIELD, R. L., *A hundred years of astronomy*, Duckworth, 1938, 526 pp., may be used.

6. *Physics :—*

BUCKLEY, H., *A short history of physics*, Methuen, 1927, 275 pp., and CAJORI, F., *A history of physics*, Macmillan, rev. edn., 1929, 424 pp., are treatments of moderate size. The most valuable contributions cover only particular aspects, and include WHITTAKER, E. T., *A history of the theories of aether and electricity*, Longmans Green, 1910, 475 pp., and MCKIE, D., and HEATHCOTE, N. H. de V., *Discovery of specific and latent heats*, Arnold, 1935, 155 pp., one of the works of the new English school of historians of science.

7. *Chemistry :—*

MEYER, E. von., *History of chemistry*, Macmillan, 1906, 720 pp., is going out of date. MOORE, F. J., *History of chemistry*, McGraw-Hill, 1918, 324 pp., is a compact account. Longer and more up to date is PARTINGTON, J. R., *A short history of chemistry*, Macmillan, 1937, 400 pp. Partington's magnificent *Origins and development of applied chemistry*, Longmans, 1935, 610 pp., should also be remembered, even by the pure chemist. With others he has founded *Ambix*, Taylor and Francis, for the study of alchemy and other early chemistry. Other works by the English school are : MCKIE, D., *Antoine Lavoisier, the Father of modern chemistry*, Gollancz, 1935, 303 pp. ; HOLMYARD, E. J., *The great chemists*, Methuen, 1928, 138 pp.

8. Geology, geography :—

ZITTEL, K. A. von., *History of geology and palaeontology*, Scott, 1901, 562 pp., and GEIKIE, Sir A., *Founders of geology*, Macmillan, 1906, 498 pp., are both rather out of date. BEAZLEY, C. R., *Dawn of modern geography*, 3 Vols., O.U.P., 1897–1906, and TAYLOR, E. G. R., *Tudor geography*, 1485–1583, 2 Vols., Methuen, 1930–4, may be mentioned.

9. Biology, with medicine :—

SINGER, C., *A short history of biology*, O.U.P., 1931, 607 pp., and *A short history of medicine*, O.U.P., 1928, 382 pp., are both on a readable scale. NORDENSKIOLD, E., *The history of biology*, Kegan Paul, 1929, 658 pp., contains a good deal of biography. SACHS, J. von., *History of botany* (Trans.) (1530–1860), Clar. Press, 1906, 584 pp. GREEN, J. R., *History of botany* (1860–1900), Clar. Press, 1909, 544 pp. are also useful. For the early periods, refer also to the general books in *2* and *3*. For medicine, see SINGER's book, supplemented by GARRISON, F. H., *Introduction to the history of medicine*, Saunders, Philadelphia, 4th edn. 1929, 996 pp., which is full on the modern period, but is too large to be read straight through. NEUBERGER, M., *The history of medicine*, Vol. 1, to middle ages, Frowde, 1910, 414 pp., deals with the early period, as its name implies. There are also CASTIGLIONI, A., *Histoire de la medécine*, 1931, Payot, Paris, 781 pp., and periodicals such as the *Bulletin of the Institute of the History of Medicine*, Johns Hopkins University, Baltimore, U.S.A.

Further biological works are, RADL, E., *The history of biological theories* (Trans.), O.U.P., 1930, 420 pp. RUSSELL, E. S., *Form and function. A contribution to the history of animal morphology*, Murray, 1916, 383 pp.

10. Selections of original documents in science :—

There are the *Klassiker der exakten Naturwissenschaften*, Akademische Verlagsges., Leipzig, and *Source-books in the history of science*, edited by WALCOTT, G. D., and published by McGraw-Hill. There have been issued so far, SHAPLEY, H., and HOWARTH, H. E., *A source-book in astronomy*, 1929, 412 pp., SMITH, D. E., *A source-book in mathematics*, 1929, 701 pp., MAGIE, D. E., *A source-book in physics*, 1935, 620 pp. The *Alembic Club Reprints*, Gurney & Jackson, in chemistry may also be recalled.

11. Biographies :—

These are numberless, though few are of high value as history. Collections include OSTWALD, *Grosser Männer* (e.g. *Van't Hoff*, *Abbé*, *Roscoe*, *Hofmeister*, *J. Müller*, *Fraunhofer*, *Arrhenius*, *Koch*). Many books already listed are in the nature of collective biographies. The *Encyclopaedia Britannica* and other national encyclopaedias are in many cases the only immediate references. A "Who's Who," for "exact scientists" only, is POGGENDORF, *Biographisch-literarisches Handwörterbuch zur Geschichte der exakten Wissenschaften*, Leipzig, 1863, 2 Vols., with later editions. Similar bare facts, but with further references, are given in SARTON, *Study of the history of mathematics*, p. 70 onwards, for that subject, and by GARRISON (see section *9* above), p. 885 onwards, for medicine and for much of biology.

12. Related subjects and background studies :—

The older workers tend to relate science to its backgrounds in philosophy and scholarship, the newer ones, to its backgrounds in economics and sociology, via technology, medicine, and agriculture. For recent studies on the relations of science and philosophy, see the periodical, *Philosophy of science*, Williams and Wilkins, Baltimore. Medicine has already been dealt with. Economic history carries us too far afield, but the names of ROSTOVTZEFF for ancient, of PIRENNE and BOISSONNADE for mediaeval, and of CUNNINGHAM, LIPSON, CLAPHAM for recent, periods may be recalled ; while there is a large periodical literature. There is no satisfactory history of agricultural technique ; histories of agriculture are usually primarily economic and social. Good short ones are, GRAS, N. S. B., *A history of agriculture in Europe and America*, Pitman, 1926, 472 pp. ; ERNLE, Lord, *English farming past and present*, Longmans, 1936, 575 pp.

As to technology, the *Science Museum Handbooks* may again be mentioned, with the *Newcomen Society Transactions* for papers. The yearly volumes of these latter contain large bibliographies of papers and books published elsewhere, agriculture being included. PARTINGTON's fine book on ancient chemistry has no real parallel for the modern world, though MIALL, S., *A history of the British chemical industry*, Benn, 1931, 273 pp., may be mentioned ; nor among English books on modern civil and mechanical engineering, though numberless works exist on this. Some of the best research is recorded in the biographies of *James Watt*, C.U.P., 1936, 207 pp., and of *Matthew Boulton*, C.U.P., 1937, 218 pp., by H. W. DICKINSON ; while C. MATSCHOSS has made the classic study of the steam engine. His *Männer der Technik* is also a very useful biographical dictionary.

The best general sources for early periods are the two books of FELDHAUS, F., *Die Technik der Vorzeit der geschichtlichen Zeit und der Naturvölker*, Berlin, 1914, 1419 pp., and *Die Technik der Antike und des Mittelalters*, Akademische Verlagsges., Potsdam, 1931, 448 pp., with NEUBERGER, A., *The technical arts and sciences of the ancients*, Methuen (Trans.), 1930, 550 pp.

13. The general relations of science, technology, and society have in the last decade become the subject of an increasing number of good books with bibliographies : CLARK, G. N., *Science and social welfare in the age of Newton*, Clar. Press, 1937, 167 pp. MUMFORD, L., *Technics and civilization*, Routledge, 1934, 495 pp. CROWTHER, J. G., *British scientists of the 19th century*, Kegan Paul, 1935, 344 pp., and *Famous American men of science*, Secker & Warburg, 1937, 414 pp. HOGBEN, L. (e.g.) *Science for the citizen*, Allen & Unwin, 1938, 1120 pp. For relations with more immediate cultural backgrounds, ORNSTEIN, M., *The rôle of scientific societies in the 17th century*, Chicago Univ. Press, 3rd edn., 1938. RASHDALL, H., *Universities of Europe in the middle ages*, O.U.P., New edn., 1936, 3 Vols., may be consulted.

SUBJECT INDEX

Roman numerals indicate chapters. Parentheses indicate chapters only partly relevant. For main subjects only runs of chapters are given. But see also Contents table.

NAME INDEX

Roman numerals standing alone indicate Chapters; when preceded by "P" they indicate Plates. "C 106", e.g., means Chart on page 106.

Only some authors are dated, not necessarily the most important. The Bibliography is not indexed. The Maps are indexed as to person, not as to most places.

Printed under the authority of HIS MAJESTY'S STATIONERY OFFICE
by PERCY LUND, HUMPHRIES & Co. LTD.
10/47 (24568) Wt. 4453/—/Ps.10901. 3,072. 3/48. P.L.H. & CO. LTD. G.943.

S.O. Code No. 29-1057*

SCIENCE SINCE 1500

The present facsimile reprint in no way discharges the responsibility which the author feels towards readers who have so kindly sent him criticisms and suggestions. A true revised edition is impossible at present, but a warning may perhaps be issued against the following pages among others: 144, 277, 290, 292.

ERRATA corrected in 2nd impression.

Page 37	*5th line*	*Read "become" for "became".*
„ 54	*17th from bottom*	*Read "Peregrinus" for "Pereginus".*
„ 130	*27th line*	*Read "Réaumur" for "Reaumur".*
„ 145	*27th line*	*Should read " . . . and go on to the . . . "*
„ 189	*17th from bottom*	*Read "case" for "ease".*
„ 224	*23rd from bottom*	*Read "is" for "are".*
„ 275	*23rd line*	*Should read " . . . neighbourhood of a wave-length . . . "*
„ 350		*[Herschel, J.] delete "142".*
„ 350		*[Herschel, W.] add "142" and "177".*

Plate XIII, opposite page 143b. The facsimile part should be inverted, top to bottom.

ERRATA to be corrected in both impressions.

Plate V is perhaps of Harriot, not Napier. See Robertson, J., The Library, 1941–2, **XXI**, *pp.* 168–76.

Page 17	*1st line*	*For "Ctesibus", read "Ctesibius".*
		1st Footnote. For "See Graph I", read "See Graph II, p. 126".
„ 47	*9th from bottom*	*For "great" read "last", and in the following line "$x^n + y^n = z^n$ is impossible in integers . . ."*
„ 51	*7th line*	*For "refugees both" read "a refugee".*
„ 52	*Above footnote*	*"As to (a) . . . ", omit "and Spinoza". Read "was" for "were".*
„ 90	*Thirteen lines from bottom. For "1899" read "1699".*	
„ 137	*2nd line*	*After "light" insert "by terrestrial means".*
„ 138	*4th from bottom*	*Two "l's" for Daniell.*
		In the formula read "r^2" for "r". Further revision is needed.
„ 143	*3rd para., 5th line from bottom. For "caloric" read "imperfect".*	
„ 146	*2nd para., 1st line.*	*For "completed" read "carried on".*
	3rd–4th line	*For "nearly all" read "sometimes".*
„ 148	*4th para., 6th line*	*For "predicted value" read "existence".*
„ 162	*19th line*	*For "worms" read "tunicates".*
„ 166	*4th para., 2nd line*	*For "then" read "than".*
„ 178	*17th and 18th from bottom. Reverse x and x_1.*	
„ 187	*4th para., 3rd line*	*Omit "a Jew convert".*
„ 199	*10th line*	*For "oxidation produces" read "is the source of".*
„ 211		*In the equation replace "$H_{2n}A$" by "$H_n A$", and, two lines below, for "C" read "BA".*
„ 216	*6th from bottom*	*Insert "but not generally true" after "fungi."*
„ 221	*5th from bottom*	*Omit "and also . . . sterile".*
„ 230 *and Map V*		*No. 13 is incorrect.*
„ 244	*19th line*	*For "1773" read "1873".*
		2nd Footnote. For "differing somewhat from" read "somewhat higher than".
„ 247	*2nd para.*	*Omit last sentence.*
„ 252	*3rd para., last line but one. For "dispersion" read "scattering".*	
„ 256	*19th from bottom*	*For "teaming" read "teeming".*
„ 266	*21st line*	*Insert "it" after "breaks".*
„ 269	*2nd para.*	*Omit sentence beginning "E. A. Milne".*
„ 276	*16th line*	*Omit "by the interferometer".*
„ 280	*10th from bottom*	*For "laws" read "second law".*
„ 292	*Footnote*	*For "1908" read "1918".*
„ 303	*19th line*	*For "have" read "has".*
	6th from bottom	*For "pure" read "purely".*
„ 347	*bottom*	*Insert "Daniell . . . 138".*
„ 349		*Insert "Grove . . . 138".*
„ 350		*The pages "87–152" under "Halley" refer to "Haller".*
„ 352		*For "Lewis, G. N. . . . 309–10" read "309–11"*
„ 357		*For "Williamson(1816–65)" read "(1824–1904)"*

THE SCIENCE MUSEUM
January 1946

LONDON: HIS MAJESTY'S STATIONERY OFFICE

24568 P.L.H. & CO. LTD. 3,072. 3/48.